# THERMODYNAMICS OF
# NATURAL SYSTEMS

# THERMODYNAMICS OF
# NATURAL SYSTEMS

## G.M. ANDERSON
University of Toronto

## JOHN WILEY & SONS, INC.

New York  Chichester  Brisbane  Toronto  Singapore

ACQUISITIONS EDITOR              C.J. Rogers/Eric Stano

MARKETING MANAGER               Catherine Faduska

PRODUCTION EDITOR               Deborah Herbert

MANUFACTURING MANAGER           Dorothy Sinclair

This book was prepared using LATEX and output using the Lucida ® Bright and Lucida New Math font sets, with illustrations created by the author, and printed and bound by Courier Stoughton. The cover was printed by New England Book Components, Inc.

® Lucida is a registered trademark of Bigelow and Holmes Inc.

*Library of Congress Cataloging in Publication Data:*

Anderson, G. M. (Gregor Munro), 1932-
        Thermodynamics of natural systems / by G.M. Anderson.
            p.      cm.
        Includes bibliographical references (p.   -   ) and index.
        ISBN 0-471-10943-6 (pbk.  :  alk.  paper)
        1. Geochemistry.  2. Thermodynamics.  I. Title.
        QE515.5.T46A53   1996
        541.3'69--dc20

                                                                95-23040
                                                                CIP

L.C. Call No.          Dewey Classification NO.          L.C. Card No.
ISBN 0-471-10943-6 (pbk)

Printed in the United States of America

10  9  8  7  6  5  4  3  2

FOR KHODJASTEH

# PREFACE

This book has evolved out of many years of teaching chemical thermodynamics in a geological context in two successive terms — an introductory course for undergraduate students having some chemistry and calculus but no thermodynamics as a background, and a second course for students who had finished the first one, with more geological applications of thermodynamics. This experience has instilled in me certain strong opinions about the best way to approach the subject for these two groups of students, i.e., those having no background in thermodynamics and little science training of any type, and those who have some acquaintance with the subject and perhaps some other science courses.

For those with little or no background in thermodynamics or in science, I believe it is best to skim lightly past the abstractions of the First and Second Laws, so as to begin the applications as soon as possible, and to do plenty of illustrative examples and problems. These students are not ready for discussions on the distinction between exact and inexact differentials, the implications of the Carnot cycle, or the necessity for reversible processes. They need to be shown how to get data from tables and to use them to solve realistic problems. The emphasis must be on thermodynamics as a useful tool.

For students who have absorbed this approach and wish to learn more, some of the more difficult aspects of the subject must be addressed. To do this I have found that it is useful to discuss (briefly) the role of models in science, and to make the point that in using thermodynamics we are invariably constructing simplified models of real systems, and that our thermodynamic parameters refer to these models, not to the real systems. This makes discussions of reversible processes, absurd standard states, and many other things more easily assimilated, because it can be seen that they are not intended to be part of real life. Nevertheless, even for the second type of student, it is best to concentrate on problem solving and examples of useful applications.

There is, of course, a third type of student, the graduate student or postdoctoral fellow, for whom a more fundamental or mathematically based approach may be appropriate. This book is not for them. These and other more advanced readers are referred to *Thermodynamics in Geochemistry — The Equilibrium Model*, an earlier book by David Crerar and me.

This book is intended to serve as a textbook for the two undergraduate courses mentioned above or for any similar introductory course in departments of forestry, soil science, biochemistry, or indeed any subject area in which chemical thermodynamics is applied to natural systems. It could also serve as a supplementary text for more broadly based courses that include some introductory thermodynamics. To satisfy a certain sense of logical presentation, I begin with a brief discussion of equilibrium thermodynamics itself, what it is and how it is used. As mentioned above, this might best be omitted from an introductory course. A discussion of some background and more advanced material is presented in Appendix C, for those with more curiosity.

G.M. Anderson

# Contents

# 1

# WHAT IS THERMODYNAMICS?

## 1.1 DEFINITION AND EMPHASIS

Thermodynamics is the branch of science that deals with energy levels and transfers of energy between systems and between different states of matter. Because these subjects arise in virtually every other branch of science, thermodynamics is one of the cornerstones of scientific training. Various scientific specialties place varying degrees of emphasis on the subject areas covered by thermodynamics—a text on thermodynamics for physicists can look quite different from one for chemists, or one for mechanical engineers. For chemists, biologists, geologists, and environmental scientists of various types, the thermodynamics of chemical reactions is of course a central concern, and that is the emphasis to be found in this book. Let us start by considering a few simple reactions and the questions that arise in doing this.

## 1.2 WHAT IS THE PROBLEM?

### 1.2.1 Some Simple Chemical Reactions

A chemical reaction involves the rearrangement of atoms from one structure or configuration to another, normally accompanied by an energy change. Let's consider some simple examples.

- Take an ice cube from the freezer of your refrigerator and place it in a cup on the counter. After a few minutes, the ice begins to melt, and it soon is completely changed to water. When the water has warmed up to room temperature, no further change can be observed, even if you watch for hours. If you put the water back in the freezer, it changes back to ice within a few minutes, and again there is no further change. Evidently, this substance ($H_2O$) has at least two different forms, and it will change spontaneously from one to the other depending on its surroundings.

- Take an egg from the refrigerator and fry it on the stove, then cool to room temperature. Again, all change seems now to have stopped—the reaction is complete. However, putting the fried egg back in the refrigerator will not change it back into a raw egg. This change seems not to be reversible. What is different in this case?

- Put a teaspoonful of salt into a cup of water. The salt, which is made up of a great many tiny fragments of the mineral halite (NaCl), quickly disappears into the water. It is still there, of course, in some dissolved form, because the water now tastes salty, but why did it dissolve? And is there any way to reverse this reaction?

Eventually, of course we run out of experiments that can be performed in the kitchen. Consider two more reactions:

- On a museum shelf, you see a beautiful clear diamond and a piece of black graphite side by side. You know that these two specimens have exactly the same chemical composition (pure carbon, C), and that experiments at very high pressures and temperatures have succeeded in changing graphite into diamond. But how is it that these two different forms of carbon can exist side by side for years, while the two different forms of $H_2O$ cannot?

- When a stick of dynamite explodes, a spectacular chemical reaction takes place. The solid material of the dynamite changes very rapidly into a mixture of gases, plus some leftover solids, and the sudden expansion of the gases gives the dynamite its destructive power. The reaction would seem to be nonreversible, but the fact that energy is obviously released may furnish a clue to understanding our other examples, where energy changes were not obvious.

These reactions illustrate many of the problems addressed by chemical thermodynamics. You may have used ice in your drinks for years without realizing that there was a problem, but it is actually a profound and very difficult one. It can be stated this way: What controls the changes (reactions) that we observe taking place in substances? Why do they occur? And why can some reactions go in the forward and backward directions (i.e., ice→water or water→ice) while others can only go in one direction (i.e., raw egg→fried egg)? Scientists puzzled over these questions during most of the nineteenth century before the answers became clear. Having the answers is important; they furnish the ability to control the power of chemical reactions for human uses, and thus form one of the cornerstones of modern science.

## 1.3  A MECHANICAL ANALOGY

Wondering why things happen the way they do goes back much further than the last century and includes many things other than chemical reactions. Some

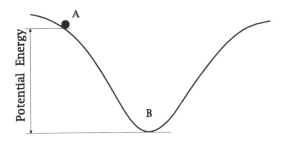

Figure 1.1: A mechanical analogy for a chemical system—a ball on a slope. The ball will spontaneously roll into the valley.

Figure 1.2: The ball has rolled into a valley, but there is a deeper valley.

of these things are much simpler than chemical reactions, and we might look to these for analogies, or hints, as to how to explain what is happening.

A simple mechanical analogy would be a ball rolling in a valley, as in Figure 1.1. Balls have always been observed to roll down hills. In physical terms, this is "explained" by saying that mechanical systems have a tendency to change so as to reduce their *potential energy* to a minimum. In the case of the ball on the surface, the potential energy (for a ball of given mass) is determined by the height of the ball above the lowest valley, or some other reference plane. It follows that the ball will spontaneously roll downhill, losing potential energy as it goes, to the lowest point it can reach. Thus it will always come to rest (equilibrium) at the bottom of a valley. However, if there is more than one valley, it may get stuck in a valley that is not the lowest available, as shown in Figure 1.2. This is discussed more fully in Chapter 2.

It was discovered quite early that most chemical reactions are accompanied by an energy transfer either to or from the reacting substances. In other words, chemical reactions usually either liberate heat or absorb heat. This is most easily seen in the case of the exploding dynamite, or when you strike a match, but in fact the freezing water is also a heat-liberating reaction. It was quite natural, then, by analogy with mechanical systems, to think that various substances contained various quantities of some kind of energy, and that reactions would occur if substances could rearrange themselves (react) so as to *lower* their energy content. According to this view, ice would have less of this energy (per gram, or per mole) than has water in the freezer, so water changes spontaneously to ice, and the salt in dissolved form would have less of this energy than solid salt, so salt dissolves in water. In the case of the diamond and graphite, perhaps the story is basically the same, but carbon is somehow "stuck" in the diamond structure.

Of course, chemical systems are not mechanical systems, and analogies can be misleading. You would be making a possibly fatal mistake if you believed that the energy of a stick of dynamite could be measured by how far above the ground it was. Nevertheless, the analogy is useful. Perhaps chemical systems will react such as to lower (in fact, minimize) their *chemical* energy, although sometimes, like diamond, they may get stuck in a valley higher than another nearby valley. We will see that this is in fact the case. The analogy *is* useful. The problem lies in discovering just what kind of energy is being minimized. What is this *chemical* energy?

## 1.3.1   Chemical Energy

We mentioned above that an early idea was that it is the *heat* energy content of systems that is minimized in chemical systems, that is, reactions will occur if heat is liberated. This is another way of saying that the heat content of the *products* is less than the heat content of the *reactants* of a reaction, so that the reaction liberates heat (Figure 1.3).

This view of things was common in the nineteenth century, and a great deal of effort was expended in measuring the flow of heat in chemical reactions. However, we don't even have to leave our kitchen to realize that this cannot be entirely correct. The melting of ice is obviously a reaction in which heat is *absorbed*, not liberated, which is why it is useful in cooling drinks. Therefore, despite the appealing simplicity of the "heat content" argument for explaining why chemical reactions occur, it cannot be the whole story. Nevertheless, the idea that some kind of "chemical energy" is liberated in reactions, or that "chemical energy" is minimized in systems at rest (equilibrium) is a powerful one. Perhaps heat is not the only energy involved. What other factors might there be? Not too many, we hope!

## 1.3.2   Plus Something Else?

Another very important clue we must pay attention to is the fact that some chemical reactions are able to take place with no energy change at all. For example, when gases mix together at low pressures, virtually no heat energy is lib-

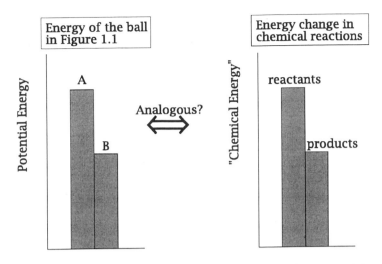

Figure 1.3: Mechanical processes always act so as to lower the potential energy content of the mechanical system. Perhaps, by analogy, chemical systems have some sort of "chemical energy" that is lowered during chemical reactions.

erated *or* absorbed. The situation is similar for a drop of ink spreading in a glass of water. These are spontaneous processes characterized by a *mixing* process, rather than by a re-organization of atomic structures like graphite→diamond, or raw egg→fried egg. Our "chemical energy" term will have to take account of observations like these.

At this point, we might become discouraged, and conclude that our idea that some sort of chemical energy is being reduced in all reactions must be wrong— there seem to be too many exceptions. It certainly was a puzzle during a large part of the last century. But we have the benefit of hindsight, and because we now know that this concept of decreasing chemical energy of some kind is in fact the correct answer, we will continue to pursue this line of thought.

## 1.4 MODELS IN SCIENCE

### 1.4.1 The Thermodynamic Model

The rest of this book will be concerned with the energy changes in chemical re-actions and the many useful concepts associated with this subject. We will be attempting to put chemical reactions into a framework of ideas that have some similarities to the ball-in-valley analogy. This is a difficult job, but it will be made easier by emphasizing one important fact at the outset. What we will be doing is making a *model* of chemical processes, rather than describing the pro-cesses themselves. This is the normal way to proceed in science. For example, we think of atoms as having nuclei composed of protons and neutrons, surrounded by electrons in successive shells or orbitals. But no one has ever seen atomic nuclei or electrons. What scientists have seen is a large number of spectral lines on photographic images, other lines on cloud-chamber photographs, and readings from countless other types of analytical and experimental instruments. From this huge pile of data, scientists using imagination, creativity, and mathematical skills at an awe-inspiring level, have constructed a model of atomic structure that accounts in a satisfying way for virtually all of the observations. At the other end of the scale, astronomers have accumulated data on luminosities, spectra, periods, and so on, and have constructed a model of our universe including galaxies, supernovas, black holes, neutron stars, and so on at unimaginable distances from us.

Some models are so well established that they will probably never be modified. A model of the solar system that allows us to put men on the moon and bring them back safely could be said to work pretty well. With models that work this well and that deal with such familiar objects, it can easily be forgotten that they *are* models. Just remember that the moon has never been weighed; we know its mass through a calculation process that relies on Newton's laws of motion, among other things. This calculation process is called *modeling* the mass of the moon—finding the mass that best fits our understanding of the laws governing planetary motion. We will be *modeling* chemical processes.

We won't insist on this point further here. The reason for even mentioning this rather philosophical point is that as we proceed, a number of procedures will arise that will seem rather bizarre, unless it is clear that we are not trying to describe what happens in real life but are constructing a model of what actually happens in real life. Because we wish to use mathematics, the model must have certain characteristics that distinguish it from everyday life.

## 1.4.2   Limitations of the Thermodynamic Model

This book outlines the essential elements of a first understanding of chemical thermodynamics, especially as applied to natural systems. However, it is useful at the start to have some idea of the scope of our objective—just how useful is this subject, and what are its limitations? It is at the same time very powerful and very limited. With the concepts described here, you can predict the equilibrium state for most chemical systems, and therefore the direction and amount of reaction that should occur, including the composition of all phases when reaction has stopped. The operative word here is "should." Our model consists of comparing equilibrium states, one with another, and determining which is more stable under the circumstances. We will not consider how fast the reaction will proceed, or how to tell if it will proceed at all. Many reactions that "should" occur do not occur, for various reasons. We will also say very little about what "actually" happens during these reactions—the specific interactions of ions and molecules that result in the new arrangements or structures that are more stable. In other words, our model will say virtually nothing about *why* one arrangement is more stable than another or has less "chemical energy," just that it does, and how to determine that it does.

These are serious limitations. Obviously, we will often need to know not only if a reaction *should* occur but *if* it occurs, and at what rate. A great deal of effort has also been directed toward understanding the structures of crystals and solutions, and of what happens during reactions, shedding much light on why things happen the way they do. However, these fields of study are not completely independent. The subject of this book is really a prerequisite for any more advanced understanding of chemical reactions, which is why every chemist, environmental scientist, biochemist, geochemist, soil scientist, and the like, must be familiar with it.

But in a sense, the limitations of our subject are also a source of its strength. The concepts and procedures described here are so firmly established partly because they are independent of our understanding of *why* they work. The laws of thermodynamics are distillations from our experience, not explanations, and that goes for all the deductions from these laws, such as are described in this book. As a scientist dealing with problems in the real world, you need to know the subject described here. You need to know other things as well, but this subject is so fundamental that virtually every scientist has it in some form in his toolkit.

## 1.5  SUMMARY

The fundamental problem addressed here is why things (specifically, chemical reactions) happen they way they do. Why does ice melt and water freeze? Why does graphite turn into diamond, or vice versa? Taking a cue from the study of simple mechanical systems, such as a ball rolling in a valley, we propose that these reactions happen if some kind of energy is being reduced, much as the ball rolls in order to reduce its potential energy. However, we quickly find that this cannot be the whole story—some reactions occur with *no* decrease in energy. We also note that whatever kind of energy is being reduced (we call it "chemical energy"), it is not simply heat energy.

For a given ball and valley (Figure 1.1), we need to know only one parameter to determine the potential energy of the ball (its height above the base level, or bottom of the valley). In our "chemical energy" analogy, we know that there must be *at least* one other parameter, to take care of those reactions that have no energy change. Determining the parameters of our "chemical energy" analogy is at the heart of chemical thermodynamics.

# 2

# DEFINING OUR TERMS

## 2.1  SOMETHING IS MISSING

We mentioned in Chapter 1 that an early idea for understanding chemical reactions held that spontaneous reactions would always be accompanied by the loss of energy, because the reactants were at a higher energy level than the products, and they wanted to go "downhill." This energy was usually thought to be in the form of heat, but this idea received a setback when it was found that some spontaneous reactions in fact absorb heat. Also, there are some reactions, such as the mixing of gases, where the energy change is virtually zero yet the processes proceed very strongly and are highly nonreversible. Obviously, something is missing. If the ball-in-valley analogy is right, that is, if reactions do proceed in the direction of decreasing chemical energy of some kind, something more than just heat is involved.

To learn more about chemical reactions, we have to become a bit more precise in our terminology and introduce some new concepts. In this chapter, we will define certain kinds of *systems*, because we need to be careful about what kinds of matter and energy transfers we are talking about; *equilibrium states*, the beginning and ending states for processes; *state variables*, the properties of systems that change during reactions; *processes*, the reactions themselves; and *phases*, the different types of matter within the systems. All these terms refer in fact to our models of natural systems, but they are also used to refer to things in real life. To be quite clear about thermodynamics, it is a good idea to keep the distinction in mind.

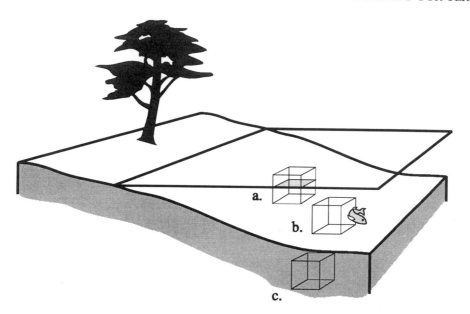

Figure 2.1: A seashore environment. The locations of three natural systems are shown.

## 2.2   SYSTEMS

### 2.2.1   Real Life Systems

In real life, a *system* is any part of the universe that we wish to consider. If we are conducting an experiment in a beaker, then the contents of the beaker is our system. For an astronomer calculating the properties of the planet Pluto, the solar system might be the system. In considering geochemical, biological, or environmental problems here on Earth, the choice of system is usually fairly obvious, and depends on the kind of problem in which you are interested.

Figure 2.1 shows a seashore environment with three possible choices of natural system. At **a**, we might be interested in the exchange of gases between the sea and the atmosphere (e.g., if the sea warms by one degree, how much $CO_2$ will be released to the atmosphere?). At **b**, we might be interested in the dissolved material in the sea itself (e.g., the reactions between dissolved $CO_2$ and carbonate and bicarbonate ions). And at **c**, we might be interested in reactions between the sediment and the water between the sediment particles (e.g., the reduction of dissolved sulfate by organic material in the sediment).

These are examples of *inorganic* systems. Thermodynamics can also be applied to organic systems, including living organisms. A single bacterium could be our system (Figure 2.2), or a dish full of bacteria, or a single organelle within a bacterium.

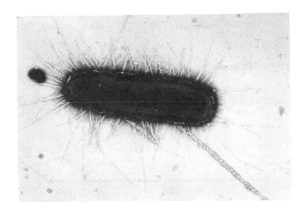

Figure 2.2: A common bacterium that lives in the human intestinal tract— *Escherichia coli*. It is about $2\,\mu$m in length. (Courtesy of Charles C. Brinton, Jr.)

The choice depends on your particular interests and is obviously very wide. However, they are all similar in one respect. Because natural systems exist in the real world, whatever system you choose is bounded by (in contact with) other parts of the world and may exchange energy and matter (liquids, solids, gases) with these other parts of the world. Systems of this type are said to be "open." All living organisms are thus open systems because they take in nutrients, and get rid of waste products. All three systems in Figure 2.1 are obviously open, because water can flow in and out of **a** and **b**, and even in **c**, compaction of the sediments squeezes water out, and diffusion allows solutes to move in and out.

## 2.2.2 Thermodynamic Systems

Our goal is to understand the energy changes in natural systems. We will do this by constructing models of these systems that contain the essential elements of the natural systems, but that are simpler and easier to handle. These models will not be material, but mathematical and conceptual. If we do it right, then the behavior of the model system will be very similar to (or will mimic) that of the real system. We will call this "understanding" the real system at the thermodynamic level.

Although most natural systems are open and are quite complex, our models of these systems can be much simpler and still be valuable. The kinds of thermodynamic or model systems that have been found to be useful in analyzing and understanding natural (real life) systems are as follows, and are illustrated in Figure 2.3. These thermodynamic systems are essentially defined by the types of walls they have. This is because we must be able to control (conceptually) the flow of matter and energy into and out of these systems.

- *Isolated systems* have walls or boundaries that are rigid (thus not permitting transfer of mechanical energy), perfectly insulating (thus preventing the flow of heat), and impermeable to matter. They therefore have a constant energy and mass content, since none can pass in or out. Perfectly insulating walls and the systems they enclose are called *adiabatic*. Isolated systems, of course, do not occur in nature, because there are no such impermeable and rigid boundaries. Nevertheless, this type of system has great significance because reactions that occur (or could occur) in isolated systems are ones that *cannot* liberate or absorb heat or any other kind of energy. Therefore, if we can figure out what causes *these* reactions to go, we may have an important clue to the overall puzzle.

- *Closed systems* have walls that allow transfer of energy into or out of the system but are impervious to matter. They therefore have a fixed mass and composition but variable energy levels.

- *Open systems* have walls that allow transfer of both energy and matter to and from the system. The system may be open to only one chemical species or to several.

As mentioned above, most natural systems are open. However, it is possible and convenient to model them as closed systems; that is, to consider a fixed composition, and simply ignore any possible changes in total composition. If what happens because of changes in composition is important, it can often be handled by considering two or more closed systems of different compositions. Thus we will be dealing mostly with closed systems in our efforts to understand chemical reactions. Basically this means that we will be concerned mostly with individual chemical reactions, rather than with whole complex systems. In other words, even though a bacterium is an open system, it can be treated (modeled) as a closed system while considering many individual reactions within it. The reactants may need to be ingested and the products eliminated by the organism, but the reaction itself can be modeled independently of these processes. This greatly simplifies the task of understanding the biochemical reactions. The same is true of most geochemical and environmental systems. Thermodynamic models of open systems are possible, but the closest we will come to looking at open systems is to consider the distribution of substances between two phases in a closed system. In this case, each separate phase is capable of changing composition and can be considered an open system, in other words, two open subsystems within an overall closed system, as in Figure 2.3b.

It is one of the paradoxes of thermodynamics that isolated systems, that have no counterpart in the real world, are possibly the most important of all in terms of our understanding of chemical reactions. You will have to wait until Chapter 4 to see why.

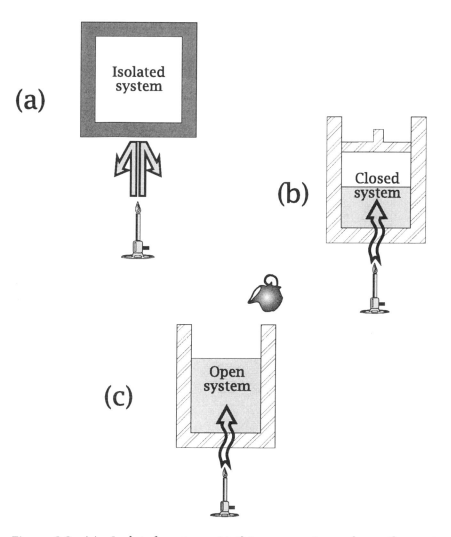

Figure 2.3: (a). Isolated system. Nothing can enter or leave the system (no energy, no matter). Whatever is inside the walls (which could be anything) will have a constant energy content and a constant composition. (b). Closed system. The closure is a piston to indicate that the pressure on the system is under our control. Energy can enter and leave the system, but matter cannot. The system here is shown as part liquid, part gas or vapor, but it could be anything. Both the liquid and the gas could also be considered as open systems, inside the closed system. Each may change composition, although the two together will have a constant composition. (c). Open system. Both matter and energy may enter and leave the system. The system may have a changing energy content and/or a changing composition. The pitcher shows one way of adding matter to the system.

## 2.3 EQUILIBRIUM

In studying chemical reactions, we obviously need to know when they start and when they have ended. To do this, we define the state of *equilibrium*, when no reactions at all are proceeding. This state has two attributes:

1. A system at equilibrium has none of its properties changing with time, no matter how long it is observed.

2. A system at equilibrium will return to that state after being disturbed, that is, after having one or more of its parameters slightly changed, then changed back to the original values.

This definition is framed so as to be "operational," that is, you can apply these criteria to real systems to determine whether they are at equilibrium. And in fact, many real systems do satisfy the definition. For example, a crystal of diamond sitting on a museum shelf obviously has exactly the same properties this year as last year (part 1 of the definition), and if we warm it slightly and then put it back on the shelf, it will gradually resume exactly the same temperature, dimensions, and so on that it had before we warmed it (part 2 of the definition). The same remarks hold for a crystal of graphite on the same shelf, so that the definition can apparently be satisfied for various forms of carbon. Many other natural systems just as obviously are not at equilibrium. Any system having temperature, pressure, or compositional gradients will tend to change so as to eliminate these gradients, and is not at equilibrium until that happens. A cup of hot coffee, for example, is not at equilibrium with the air around it until it cools down.

So if diamond and graphite are both at equilibrium, do we have two kinds of equilibrium? In our ball-in-valley analogy, the ball in any valley would fit our definition. What distinction do we make between the lowest valley and the others?

### 2.3.1 Stable and Metastable Equilibrium

Stable and metastable are the terms used to describe the system in its lowest equilibrium energy state and any other equilibrium energy state, respectively. Thus we say that diamond is a metastable form of carbon at Earth surface conditions. When we develop this subject further, we should be able to predict or calculate under what conditions it is the stable form of carbon.

In Figure 2.4, we see a ball on a surface having two valleys, one higher than the other. At (a), the ball is in an equilibrium position, that fulfills both parts of our definition — it will stay there forever, and will return there if disturbed, as long as the disturbance is not too great. However, it has not achieved the lowest possible potential energy state, and therefore (a) is a *metastable equilibrium* position. If the ball is pushed past position (b), it will roll down to the lowest available energy state at (d), a *stable equilibrium* state. During the fall, for

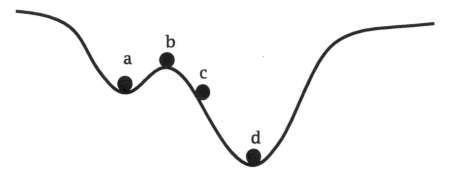

Figure 2.4: Four positions of a ball on a surface, to illustrate the concept of equi-librium. Position a — metastable equilibrium. Position b — unstable. Position c — unstable. Position d — stable equilibrium.

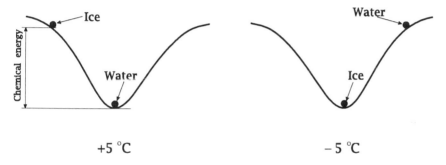

Figure 2.5: The mechanical analogy for $H_2O$ at $-5°C$ and $+5°C$ and atmospheric pressure. At $-5°C$, water is unstable and releases energy until it becomes ice at $-5°C$. At $+5°C$, ice is unstable and releases energy until it becomes water at $+5°C$. The problem is, what kind of energy is being minimized?

example, at position (c), the ball (system) is said to be *unstable*. In position (b), it is possible to imagine the ball balanced and unmoving, so that the first part of the definition would be fulfilled, and this is sometimes referred to as a third type of equilibrium, admittedly a trivial case, called *unstable equilibrium*. However, it does not survive the second part of the definition, so we are left with only two types of equilibrium, stable and metastable.

Of course, we find that the stable form of substances is different under different conditions. For example, the stable form of $H_2O$ is water at $+5°C$, and ice at $-5°C$ (Figure 2.5).

As suggested in Chapter 1, the analogy can be extended to include meta-stable substances, by supposing that there is more than one valley (Figure 2.6), and that substances that *should* change or react, but do not, are prevented from

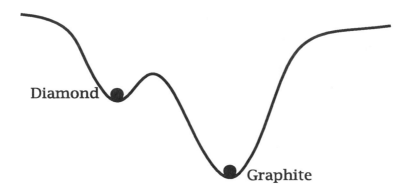

Figure 2.6: The mechanical analogy for carbon at Earth surface conditions. Graphite is the stable form of carbon because it has the lowest energy content of any form of carbon. Diamond has a higher energy content but is prevented from changing to graphite by an energy barrier.

reacting by an energy barrier; they are stuck in a valley above the lowest valley available. Again, as conditions change, the stable form may change. Thus deep in the Earth, at very high temperatures and pressures where diamond originates, it is the stable form of carbon and would occupy the lowest available valley.

## 2.4  STATE VARIABLES

Systems at equilibrium have measurable properties. A property of a system is any quantity that has a fixed and invariable value in a system at equilibrium, such as temperature, density, or refractive index. Every system has dozens of properties. If the system changes from one equilibrium state to another, the properties therefore have changes that depend only on the two states chosen, and not on the manner in which the system changed from one to the other. This dependence of properties on equilibrium states and not on processes is reflected in the alternative name for them, *state variables*. Several important state variables (which we consider in later chapters) are not measurable in the absolute sense in a particular equilibrium state, though they do have fixed, finite values in these states. However, their changes between equilibrium states are measurable.

Reference in the above definition to "equilibrium states" rather than "stable equilibrium states" is deliberate, since as long as metastable equilibrium states are truly unchanging they will have fixed values of the state variables. Thus both diamond and graphite have fixed properties. Metastable states are extremely common. For example, virtually all organic compounds are metastable in an oxidizing environment, such as the Earth's atmosphere. We may be grateful for those "activation energy barriers" that prevent metastable states from spontaneously changing to stable states; otherwise we would not be here to discuss the matter.

### 2.4.1   Total versus Molar Properties

Many physical properties, such as the volume and various energy terms, come in two forms—the total quantity in the system and the quantity per mole or per gram of substance considered. We use roman capital letters for "total" properties and the corresponding italic capitals for molar properties. For example, water has a volume per mole ($V$) of about 18.068 $cm^3$ $mol^{-1}$, so if we have 30 moles of water in a beaker, its volume (V) is 542.04 $cm^3$. This relationship for a pure substance such as $H_2O$ is $Z = Z/n_i$, where Z is any total property, $Z$ is the corresponding molar property, and $n_i$ is the number of moles of the substance. In our water example, above, $542.04/30 = 18.068$. In more complex systems where more than one substance is present, total and molar properties are related in the same way. A beaker containing, for example, a kilogram of water (55.51 moles $H_2O$) and 1 mole of NaCl occupies 1019.9 $cm^3$. The molar volume of the system is then $Z = Z/\sum_i n_i$, or $1019.9/(1 + 55.51) = 18.05$ $cm^3$ $mol^{-1}$.

These two types of state variables have been given names:

- *Extensive* variables are proportional to the quantity of matter being considered — for example, volume (V).

- *Intensive* variables are independent of quantity and include concentration, viscosity, and density, as well as all the *molar* properties, such as the molar volume, $V$.

Of course, many equations look much the same with total and molar properties because ratios are involved. That is, if $(\partial U/\partial S)_V = T$, then it is also true that $(\partial U/\partial S)_V = T$; or if $(\partial G/\partial P)_T = V$, then $(\partial G/\partial P)_T = V$, so that the distinction may seem to be unimportant. However, sometimes it *is* important. In general terms, we use the total form of our variables (roman type) in some theoretical discussions, and the molar form (italic type) in most calculations.

## 2.5   PHASES AND COMPONENTS

We must also have terms for the various types of matter to be found within our thermodynamic systems. A *phase* is defined as a homogeneous body of matter, having distinct boundaries with adjacent phases, and so is mechanically separable from the other phases. The shape, orientation, and position of the phase with respect to other phases are irrelevant, so that a single phase may occur in many places in a system. Thus the quartz in a granite is a single phase, regardless of how many grains of quartz there are. A salt solution is a single phase, as is a mixture of gases. There are only three very common types of phases—solid, liquid, and gas or vapor. A system having only a single phase is said to be *homogeneous*, and multiphase systems are *heterogeneous*.

The term generally used to describe the chemical composition of a system is *component*. The components of a system are defined by the smallest set of chemical formulae required to describe the composition of all the phases in the system. This simple definition is sometimes surprisingly difficult to use. To take a simple example, consider a solution of salt (NaCl) in water ($H_2O$), in equilibrium with water vapor. This might look like Figure 2.3b. There are two phases, liquid and vapor, and two components, NaCl and $H_2O$. A chemical analysis could report the amounts or concentrations of Na, Cl, H, and O in the system, but only two chemical formulae are needed to describe the compositions of both phases.

### 2.5.1  Real vs. Model Systems

Equilibrium, phases, and components are terms that appear to apply to real systems, not just to the model systems that we said thermodynamics applies to, and in general conversation, they do. But real phases, especially solids, are never perfectly homogeneous; they are only approximations to the ideal phases that the thermodynamic model uses. And real systems don't really have components, only our models of them do. Seawater, for example, has an incredibly complex composition, containing dozens of elements. But our thermodynamic models might model seawater as having two, three, or more components, depending on the application. As for equilibrium, real systems do often achieve equilibrium as we have defined it, but it is never a perfect equilibrium.

However, the fact that real phases are more or less homogeneous, and that real systems achieve an approximate equilibrium as opposed to the mathematically perfect model we will develop, is what makes thermodynamics useful. The model is perfect, but real life comes close enough in many respects so that the model is useful. In fact, the close similarity between reality and our models of reality, and the fact that we use the same terms to describe each, may lead to a certain degree of confusion as to that we are talking about. Usually no harm is done, and the distinction gets easier with practice.

## 2.6  PROCESSES

Finally, we get to something that looks more interesting. *Processes* are what we are usually interested in—changes in the real world. In geology, these might be igneous, diagenetic, or metamorphic processes. In biology, they might be cellular processes. In the environmental world, they might be potentially harmful processes near waste disposal sites—the possibilities are endless. However, most of the processes of interest to us have one thing in common—they are extremely complicated. The only hope we have of understanding them is to break complex processes down into their simpler component parts, and to construct simplified models of them. We have already begun to do this by defining several types of simple *systems* that we can use; we will now define a *process* in a way that will help us model real processes.

A thermodynamic *process* is what happens when a system changes from one equilibrium state to another. Thus any two equilibrium states of the system may be connected by an infinite number of different processes because only the initial and final states are fixed; anything at all could happen during the act of changing between them. A chemical reaction is one kind of process, but there are others. For example, simply warming or cooling a system is a process according to our definition.

In spite of there seeming to be an endless number of kinds of processes in the world, we find that in thermodynamic models there are only two.

- Processes that begin in a metastable equilibrium state and lead to a more stable state—irreversible processes.

- Processes that begin in a stable equilibrium state and proceed to a different stable equilibrium state, without ever leaving the state of equilibrium—reversible processes.

## 2.6.1 Irreversible Processes

We have defined a metastable state of a system as a state that has more than the minimum energy for the given conditions, but is for some reason prevented from releasing that energy and reacting or changing to the stable state of minimum energy. An irreversible process is one that occurs when whatever constraint is holding the system in its high energy state is removed, and the system slides down the energy gradient to a lower energy state.

The only example we have given thus far of a metastable system is the mineral diamond, that could lower its energy content by changing into graphite but does not, because energy is required to break the carbon-carbon bonds in diamond (which are very strong) before the atoms can rearrange themselves into the graphite structure. There are many other similar examples of metastable minerals. We have also mentioned that most organic compounds, such as all the ones in living organisms, are metastable. When the life processes maintaining their existence cease, they quickly react (decompose) to form more stable compounds.

But many other reactions or processes can be regarded in the same general way. If our system consists, for example, of a cup of coffee and a gram of sugar, then the two when separated constitute a metastable state of the system. The separation between them is the constraint preventing the reaction. When the sugar is added to the coffee the constraint is removed, and an irreversible reaction takes place—the sugar dissolves (Figure 2.7).

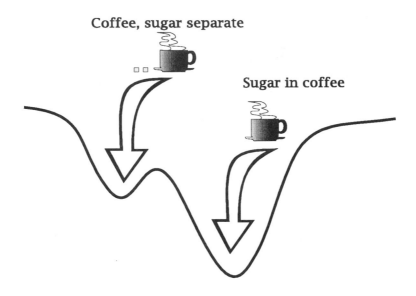

Figure 2.7: Sugar dissolves in coffee because the "chemical energy" of the dissolved state is less than that of the two coexisting separately.

This is analogous to most of the chemical reactions we will be considering—a combination of minerals, or minerals plus liquids or gases, reacts to form some different minerals under some given conditions. For example, the mineral corundum ($Al_2O_3$) is stable, considered by itself (i.e., there is no other form of $Al_2O_3$ that is more stable), but in the presence of water it reacts to form gibbsite ($Al_2O_3 \cdot 3H_2O$). The reaction is

$$Al_2O_3(s) + 3H_2O(l) = Al_2O_3 \cdot 3H_2O(s) \tag{2.1}$$

and the energy relationships are shown in Figure 2.8. We will use $(s)$, $(l)$, $(g)$, and $(aq)$ after our formulae to indicate whether they are in the solid, liquid, gas, or aqueous (dissolved in water) state.

Do not confuse the metastability of diamond at Earth surface conditions with the metastability of corundum or water. Diamond is metastable because the same carbon atoms would have a lower energy in the crystal structure of graphite. But corundum by itself is not metastable, and neither is water, at 25°C and atmospheric pressure. It is the *combination* of corundum and water that can be regarded as metastable, because their *combined* atoms would have a lower energy level in the form of gibbsite.

The essence of most irreversible reactions is that *energy is released* during the change (exactly what kind of energy we have not yet discussed). Therefore, unless energy is *added* to the system, the reaction cannot go in the reverse direction under the given conditions. In other words, the reaction or change

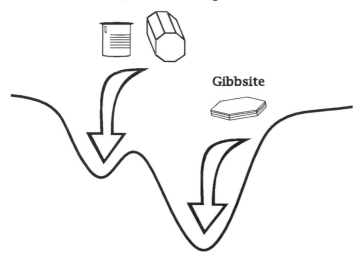

**Water, corundum separate**

**Gibbsite**

Figure 2.8: Water plus corundum can lower its energy content by reacting to form gibbsite.

is *spontaneous* in one direction only. The ball will never roll uphill of its own accord. This does not mean that the reaction can never go in the opposite way. It may very well go in the opposite way *under different circumstances.* Thus the corundum plus water reacts spontaneously to form gibbsite at low temperatures, but at high temperatures gibbsite spontaneously decomposes to form corundum and water. Similarly, we said that ice→water at 5°C, but water→ice at −5°C. Spontaneous or irreversible (these terms are synonymous) refers to a single set of conditions, such as a given temperature, pressure, and composition. If the conditions are changed, the reaction may become spontaneous in the other direction.

### Reactions Involving Organic Compounds

Reactions involving organic compounds, whether in living organisms or not, are no different in principle from any other kind of reaction, such as those between minerals. The only difference is that for organic compounds, the reaction usually proceeds from one metastable state to another metastable state of lower energy, rather than from a metastable state to a stable state. Consider for example the reaction

$$C_8H_{16}N_2O_3(aq) + H_2O(l) = C_6H_{13}NO_2(aq) + C_2H_5NO_2(aq) \qquad (2.2)$$

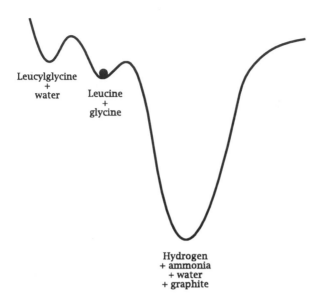

Leucylglycine
+
water

Leucine
+
glycine

Hydrogen
+ ammonia
+ water
+ graphite

Figure 2.9: Energy relationships between organic compounds. Most organic compounds have much higher energy contents than do combinations of simple inorganic compounds of the same overall composition.

which represents the breaking of a peptide bond between two amino acids, one of the more fundamental processes in biochemistry. The $(aq)$ here means that the compounds we are discussing are dissolved in water and, hence, the reaction takes place in water. If we use names rather than chemical formulae, this is

$$\text{leucylglycine} + \text{water} = \text{leucine} + \text{glycine} \tag{2.3}$$

This reaction occurs spontaneously, and the energy relations can be depicted exactly as for simpler compounds. The only difference is that rather than reacting to compounds in the lowest possible energy state, leucylglycine plus water reacts to form compounds in another metastable state (leucine plus glycine) of lower energy than the initial state, as shown in Figure 2.9. Virtually all organic compounds are metastable with respect to simple inorganic compounds and elements such as water, nitrogen, hydrogen, and graphite. Thus the reaction

$$C_6H_{13}NO_2(aq) + C_2H_5NO_2(aq) = 2\,H_2(g) + 2\,NH_3(g) + 4\,H_2O(l) + 8\,C_{graphite} \tag{2.4}$$

is also spontaneous, as shown in Figure 2.9.

Living organisms have developed mechanisms (involving enzymes) for overcoming the energy barriers separating products and reactants of reactions required for the life processes of the organisms. Obviously no enzymes have been developed to enable the breakdown of the organisms to the simple inorganic compounds of which they are composed, as this would be fatal.

## 2.6.2 Reversible Processes

We have already come across one meaning for the term "reversible process"—it could refer simply to a process that can go in either direction, depending on the circumstances, such as water freezing or ice melting. However, there is another, more important meaning that is usually meant when the term is used in thermodynamics.

A reversible process is one in which a system in a state of equilibrium changes to another state of equilibrium *without ever becoming out of equilibrium*. This type of process is not possible in the real world. For example, a crystal of diamond at 25°C is warmed to 50°C. What is so difficult about that? Although it is not difficult to warm a diamond to 50°C, it is impossible to do it without leaving the state of equilibrium. To change the temperature of the crystal, heat must be applied to it. This sets up a temperature gradient between the inside and the outside of the crystal, and heat travels into the crystal, raising its temperature. But while a temperature gradient exists in the crystal, it is not at equilibrium (a system at equilibrium can have no gradients in temperature, pressure, or composition). In a real heating process, the crystal of diamond is at equilibrium at 25°C, then it leaves the state of equilibrium for a time, and then it attains equilibrium later under its new conditions, 50°C. However, in a reversible heating process, the crystal is at all times at equilibrium with its environment, or at least never more than infinitesimally different in temperature from its environment, and changes from 25°C to 50°C in a continuous state of equilibrium.[1]

The reversible process as defined is impossible in the real world. However, it is quite simple in the thermodynamic model, because the temperature, volume, and all other properties of the diamond are just points on mathematical surfaces in the model, and there is nothing to prevent the point representing the temperature to move around on a surface representing the equilibrium values of various properties of the diamond.

Why in the world would we be interested in such a strange kind of impossible process? The reason is simple, but usually takes some time to fully grasp. We will state the reason here, but don't worry if it is not perfectly clear right now. The reason the reversible process (defined as a continuous succession of equilibrium states) is important in the thermodynamic model is that it is the only kind of process that our mathematical methods (differentiation, integration) can be applied to. Once our crystal of diamond leaves its state of equilibrium at 25°C, practically anything could happen to it, but as long as it settles back to equilibrium at 50°C, all of its state variables have changed by fixed amounts from their values at 25°C. We have equations to calculate these energy differences, but they refer to lines and surfaces in our model, and that

---

[1]We consider a different reversible process in more detail in §3.4.1.

means that they must refer to continuous equilibrium between the two states.

In other words, to *calculate* the energy difference between the two states, we must use a fictitious path (the reversible process) between the two states. The result is the real energy difference, no matter what actually happened to the system between the two states. The reversible process is another example of the difference between the real world and our models of the real world. Reversible processes are quite simple to carry out in our models, because the models are mathematical, not real.

### 2.6.3   Egg Reactions

We have not discussed all the examples we used in Chapter 1. To conclude our discussion of various common chemical reactions (§1.2.1), we should discuss the thermodynamics of frying eggs. At a simple level, we could say that the egg in the refrigerator represents a metastable state, and that frying it promotes a reaction to a more stable state, analogous to the leucylglycine + water→leucine + glycine reaction in Figure 2.9. Strictly speaking, however, we know that eggs in the refrigerator won't last indefinitely; they will eventually "go bad." This means that they are not in a truly metastable state in the refrigerator, but an unstable state. This helps to explain why a fried egg in the refrigerator does not change back into a raw egg; the raw egg occupies no "valley" for the egg components to roll into. Furthermore, the reaction involved in "going bad" is completely different from the fresh egg→fried egg reaction. Very complex unstable systems have quite a variety of reaction paths available to them, depending on the circumstances. In studying natural systems, such as eggs, it is often quite difficult to distinguish stable, metastable, and unstable states from each other without a considerable amount of work and ingenuity, but it can be done. When you get numbers from tables, as we will be doing, all this work has been done for you, although you have to realize that because of the difficulties involved, some of the data may not be accurate and may be revised at some future date.

Reactions in these complex systems are actually made up of a number of simpler reactions, and applying thermodynamics requires that the individual reactions be treated separately. The individual biochemical reactions in many organic systems still have not been figured out. Nevertheless, we are confident that any particular reaction, once defined, will follow the logic and the systematics described in this book.

### 2.6.4   Notation

#### Reaction Deltas

We have now set up the general framework within which thermodynamics is able to deal with processes. Any given process or chemical reaction within a chosen system will proceed from an initial equilibrium state (normally a meta-stable equilibrium state) to another equilibrium state more stable than the first one. During this process or reaction the system is out of equilibrium. The system has a number of properties or state variables, such as volume and energy content, that have fixed values in equilibrium states and that therefore have fixed amounts of change between equilibrium states. These changes are always written using a delta notation, where the delta refers to the property in the final state minus the property in the initial state. For example, if the system undergoes a process during which its volume (V) changes from $V_{initial}$ to $V_{final}$, we write

$$\Delta V = V_{final} - V_{initial}$$

If the process is a chemical reaction, a number of compounds may be involved. A generalized chemical reaction could be written as

$$a A + b B + \ldots = m M + n N + \ldots$$

An example is equation (2.1), where A is $Al_2O_3$, B is $H_2O$, and M is $Al_2O_3 \cdot 3H_2O$ (there is no N); $a$ and $m$ are 1 and $b$ is 3. The quantities A, B, M, and N are chemical formulae representing any compounds or elements we happen to be interested in, and each can be solid, liquid, gas, or a solute. One side of the reaction will be usually more stable than the other, and a reaction will tend to occur, unless there is an energy barrier preventing the reaction, or unless the compounds are all at equilibrium together. In this case, the volume change during the reaction is $\Delta_r V$ (we insert a subscript $r$ to indicate a chemical reaction) and is equal to the sum of the volumes of the reaction products (the final state) minus the sum of the volumes of the reactants (the initial state). Thus

$$\Delta_r V = m V_M + n V_N + \ldots - a V_A - b V_B - \ldots$$

where $V_M$ is the volume of a mole of compound M, and so on. For example, the change in volume for reaction (2.1) is

$$\Delta_r V = V_{Al_2O_3 \cdot 3H_2O} - V_{Al_2O_3} - 3 V_{H_2O}$$

Note that each volume must be multiplied by its corresponding stoichiometric coefficient in the reaction. Molar volumes are readily available for most pure substances.

### Example

The volume data in Appendix B are listed under $V°$, where superscript $°$ means standard state conditions, which we will discuss later. In the corundum-gibbsite reaction, then,

$$\Delta_r V = \Delta_r V° \quad = \quad V°_{Al_2O_3 \cdot 3H_2O} - V°_{Al_2O_3} - 3\,V°_{H_2O}$$
$$= \quad 63.912 - 25.575 - 3 \times 18.068$$
$$= \quad -15.867\,cm^3$$

There is therefore a net decrease in volume for the reaction *as written*. Note that although the $V°$ data are given in $cm^3\,mol^{-1}$, the result of the calculation is simply $cm^3$. If we said $-15.867\,cm^3\,mol^{-1}$, the question would arise, "per mole of what?" Because there are 3 moles of water in the reaction as written, the volume change per mole of water, for example, is $-15.867/3\,cm^3$.

Following this convention, the change in energy of the ball rolling down the hill in Figure 1.1 would be a negative quantity, as shown in Figure 1.3 (energy in state B minus energy in state A is negative). It follows, then, that the change in the "chemical energy" term we are looking for will always be a negative quantity in spontaneous reactions, as also shown in Figure 1.3 (energy of products minus energy of reactants).

### Chemical Equations

For the most part, when we write reactions such as (2.1) and (2.2), we use the = sign to indicate only that the reaction is "balanced", meaning that the same number and kinds of atoms appear on both sides, and that any electrical charges are also the same on both sides. If we want to emphasize that the reaction proceeds strongly or irreversibly we may use an arrow, as in A → B, and if we want to emphasize that the two sides are in equilibrium, we might use A ⇌ B. However, the = sign includes these possibilities, and all others. In Chapter 12, in discussing chemical kinetics, we define = and → a little differently (§12.2.3).

## 2.7  SUMMARY

If you look around the physical world today, you realize that there are an incredible number of chemical and physical *processes* going on all around you, and as you look into these in more and more detail, as science has done, you find more and more complexity at all levels, right down to the atomic and subatomic levels. How can we systematize and understand these processes in such a way as to be able to control some of them for our own purposes?

Thermodynamics is the net result of our attempts to do this. It is not a description of any real process but a rather abstract *model* that can be used for all real processes. Processes in the real world are incredibly complex, but our models of them are quite simple, containing a number of carefully defined concepts. *Processes* (reactions, changes) involve energy and/or mass changes, and these must enter or leave the place where the process is occurring; so thermodynamics begins by defining several types of *systems*, depending on how the energy and/or mass is transferred. Processes must be defined by beginning and ending states, so thermodynamics defines *equilibrium* states, some having more energy (*metastable equilibrium* states) than others (*stable equilibrium* states), and processes or reactions that are able to go from higher energy states to lower energy states (*irreversible processes*), just like a ball rolling down a hill. Of course, a state of lower energy (stable) under one set of conditions may be a state of higher energy (metastable) under other conditions (diamond is metastable at the Earth's surface, but stable deep in the mantle). Corundum and water are, by themselves, perfectly stable and unreactive, but together they have a higher energy state than does gibbsite.

The only thermodynamic difference between organic reactions (including those in living organisms) and inorganic reactions is that both the reactants and products of organic reactions are invariably metastable compounds; metastable, that is, with respect to simple inorganic compounds and elements. Inorganic reactions *may* involve metastable compounds, but more frequently they involve a metastable *assemblage* changing to a stable one (one having the lowest possible energy state).

Therefore, the determination of the energy states of substances and how they change under changing conditions is fundamental to understanding what processes are possible, and why they happen. The determination of the energy states of individual substances must be done by experiment and measurement, not by theoretical calculation, and the results are available in tables of data like those at the end of this book. Calculation of the change of these energy terms with changing conditions can be carried out only for hypothetical *reversible processes*, that are not possible in reality but are quite simple in the thermodynamic model.

As for the energy barriers that often prevent reactions from occurring, thermodynamics has nothing whatever to say about them. It pretends they do not exist. More exactly, thermodynamics simply deals with energy levels, energy differences. It does not concern itself with whether a system actually lowers its energy level or not.

The most important question now is what kind of energy is released during these reactions? If it is not heat energy, then what is it? We have called it "chemical energy," but this is just because we haven't said yet what it really is. This is the topic of the next two chapters.

# PROBLEMS

1. Calculate the volume change for the reaction

$$Al_2O_3(s) + 3\,H_2O(l) = 2\,Al(OH)_3(s)$$

$Al(OH)_3(s)$ is another way of writing the formula for gibbsite.

2. Calculate the volume change for the reaction

$$AlO_{1.5}(s) + 1.5\,H_2O(l) = Al(OH)_3(s)$$

$AlO_{1.5}(s)$ is another way of writing the formula for corundum.

3. Calculate $\Delta_r V°$ for reaction (2.2).

4. Why are the $V°$ values for all the gases the same in Appendix B? Calculate this $V°$ from data in Appendix A.

5. Calculate $\Delta_r V°$ for the reaction

$$HCl(aq) = H^+ + Cl^-$$

You should get zero. Why?

6. Calculate $\Delta_r V°$ for the reaction

$$NaCl(s) = Na^+ + Cl^-$$

Note that the standard volume of many ions is negative. How can any substance have a negative volume?

# 3

# THE FIRST LAW OF THERMODYNAMICS

## 3.1 TEMPERATURE AND PRESSURE SCALES

### 3.1.1 The Celsius Scale

One of the early triumphs of the study of thermodynamics was the demonstration that there is an absolute zero of temperature. However, there are several different temperature scales, for historical reasons. All you need to know about this is that the Kelvin scale (named after William Thompson, Lord Kelvin) has an absolute zero of 0 K and a temperature of 273.16 K at the triple point where water, ice, and water vapor are at equilibrium. The melting point of ice at one atmosphere pressure is 0.01 degrees less than this, at 273.15 K (Figure 3.1). The Celsius scale (named after Anders Celsius, a Swedish astronomer) has a temperature of 0°C at the ice point (273.15 K) and absolute zero at −273.15°C. This gives almost exactly 100°C between the freezing and boiling points of water at one atmosphere, so water boils at 100°C (373.15 K). Thus the numerical conversion between the two scales is

$$T \text{ K} = t°\text{C} + 273.15$$

Remember that *all* equations in thermodynamics use the absolute or Kelvin temperature scale, so that if you are given temperatures in °C, you must convert them to the Kelvin scale before using them. The "standard" temperature of 25°C for example is 298.15 K.

### 3.1.2 Pressure Scales

For many people, pressure is an easier concept to grasp than is temperature. Force is measured in newtons (N), where 1 newton will give a mass of 1 kg

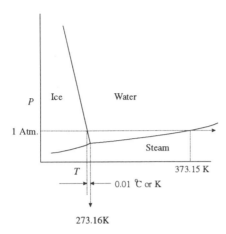

Figure 3.1: Schematic $P$-$T$ phase diagram for the system $H_2O$. The temperature of the triple point is defined as 273.16 K

an acceleration of 1 m sec$^{-2}$. Pressure is defined as force per unit area, and a pressure of 1 newton per square meter (1 N m$^{-2}$) is called 1 pascal (1 Pa). This is a very small pressure, and older, larger pressure units are still in use. The bar, for example, is 10$^5$ Pa and is almost equal to the standard atmosphere (1 atm = 1.01325 bar). Weather reports in many countries give the atmospheric pressure in kilopascals (kPa), and it is usually close to 101 kPa, or 1 atm, or 1 bar. These units are summarized in Appendix A.

The standard temperature and pressure chosen for reporting values of thermodynamic variables is now 25°C and 0.1 MPa. A pressure of 0.1 MPa is 100 kPa and 10$^5$ Pa, or 1 bar. It is convenient to use bars instead of pascals, because the bar is essentially the same as atmospheric pressure, and the notation is slightly simpler. We will use bars from now on in this text.

## 3.2  INTERNAL ENERGY

In everyday conversation we use words like heat, work, and energy quite frequently, and everyone has a sufficiently good idea of their meaning for our ideas to be communicated. Unfortunately, this type of understanding is not sufficient for the construction of a quantitative model of energy relationships like thermodynamics. To get quantitative about anything, or, in other words, to devise equations relating measurements of real quantities, you must first be quite sure what it is you are measuring. This is not too difficult if you are measuring the weight of potatoes and carrots; it is a more subtle problem when you are measuring heat, work, and energy. Historically, it took several decades of effort by many investigators in the nineteenth century to sort out the difficulties that you are expected to understand by reading this chapter!

## 3.2.1 Energy

Everyone knows what energy is, but it is an elusive topic if you are looking for a deep understanding. In fact, a Nobel Prize-winning physicist has affirmed that "It is important to realize that in physics today, we have no knowledge of what energy *is*." (Feynman et al., 1963, p. 4-2). Fortunately, in science it is not always necessary to have a deep understanding of what things *are*; we can leave that to the philosophers, and just deal with developing useful equations.

If you consult a dictionary as to the meaning of energy, you find that the scientific meaning is "the ability to do work, i.e., move a body."[1] In physics, work is not what you do from 9 to 5 every day, but the action of a force moving through a distance. So if you lift a book from the floor and put it on the table, you are performing work (the mass of the book times the distance from the floor to the table), and we say that we expended energy to lift the book. It has proved tremendously useful to take the view that the energy we expended has not disappeared, but has been transferred to the book. In other words, the book on the table has more energy (potential energy) than it had on the floor, and the increase is exactly equal to the work we did in lifting it. Thus we can use energy to do work, and we can do work to increase energy. Work and energy are thus very closely related concepts (note that they have the same dimensions in Appendix A).

If things were only that simple. However, we know that they are not, because the energy in a stick of dynamite on the table is not equal to the work expended in lifting it from the floor. Similarly, the energy in water is not the same as in ice, whether on the floor or the table. These complications are actually of two types. The first is that there are many ways of doing work, because there are many kinds of forces, and we are particularly concerned with those involved in chemical reactions. The second is that although work and energy are indeed closely related, doing work is not the only way of changing the energy of something, and changing the energy of something does not always produce work. For example, we could change the energy in our book by warming or cooling it. We have to consider both work (in all its forms) and heat to get a consistent picture of energy changes.

### Only One Kind of Work Considered

There are many different ways of doing work on a system, and many different ways of having a system do work, depending on what kinds of forces are available. For example, you might lift a heavy metallic object to the table with ropes and pulleys, or if you had a strong magnet, you might lift it by magnetic attraction. Thermodynamics can accommodate all kinds of forces and types of work, but because they are in principle all the same, and are treated in the same way, it is simpler to develop the subject by considering only one kind of work. This kind of work (pressure-volume work) will be introduced below. You will

---

[1] *The New Shorter Oxford Dictionary*, 1993 edition, p. 819.

notice that although we have been illustrating energy and work by using the ball-in-valley idea (Chapter 2) and the book-and-table idea (this chapter), which emphasizes *potential* energy, this particular kind of energy/work is actually irrelevant in thermodynamics, except as an analogy. We will define the energy content of systems of importance to us to be the same whether they are on the floor or the table.

## 3.2.2   Absolute Energy

In discussing energy, we always seem to be talking about *changes* in energy. The book has more energy on the table than on the floor, and presumably more energy on the roof than on the table. And we add energy by warming the book, too. But how much energy has the book *got* in any particular state—say, on the table at 25°C? What is the absolute energy content of the book? This was a difficult question until 1905, when Einstein postulated the essential equivalence of mass and energy in his famous equation

$$E_r = mc^2$$

where $E_r$ is the rest energy of a system, $m$ is the mass, and $c$ is the speed of light. Therefore, the energy contained in any macroscopic system is extremely large, and adding energy to a system (for instance by heating it) will in fact increase its mass. However, ordinary (i.e., non-nuclear) energy changes result in extremely small and unmeasurable changes in mass, so that relativity theory is not very useful to us, except in the sense that it gives energy an absolute kind of meaning, which is sometimes helpful in trying to visualize what energy *is*.

Thus in considering ordinary everyday kinds of changes and chemical reactions, we will continue to deal with energy *changes* only, never with how much energy is in any particular equilibrium state. This is entirely sufficient for our needs, but it does introduce some complications that would be avoided if we had a useful absolute energy scale.

## 3.2.3   The Internal Energy

All that is required to develop our model of energy relationships is that every equilibrium state of a system (such as our book on the table or the stick of dynamite on the table) have a fixed energy content, called the internal energy, U (or $U$, the molar internal energy) of the system. The numerical value of this energy content is not known, and not needed. It could be thought of as identical to the rest energy $E_r$, if that helps, or as some small subset of $E_r$; it doesn't really matter. All that matters is that when the system is at equilibrium, its energy content or energy level is constant. Formally, the relation between the total or rest energy and the internal energy used in thermodynamics is

$$E_r = U + constant$$

where the value of the constant is unknown (and unimportant). Since we do not use absolute values of U or $U$, we cannot use absolute values of any quantities having $U$ in their equations of definition.

Somewhat paradoxically, in spite of being possibly the most fundamental of thermodynamic quantities, Internal Energy or even changes in $U$ are little used in geochemical applications. It is never listed in tables of thermodynamic values, for example, and one rarely needs to calculate $\Delta U$. The reason for this will become apparent as we proceed. It has to do with the fact that we, the users of thermodynamics, have a great predilection for using temperature, pressure, and volume as our principle constraints or measured system parameters. This requires that we use $\Delta U$ in slightly modified forms, that is, $\Delta U$ modified by what are often relatively small correction factors (such as $P\Delta V$), and these modified forms are given different names and symbols. It is then quite possible to rarely think about $\Delta U$, since it seems only to arise in the development of the First Law. For a better understanding of the subject, however, it is best to realize that in most energy transfers in the real problems that we will be considering, $\Delta U$ is by far the largest term involved. Just because we do not usually calculate its value does not mean it is not important.

## 3.3 ENERGY TRANSFERS

In the discussions in the previous chapters, we proposed the idea that changes or reactions occur because systems can lower their energy by such changes. However, we mentioned that the most obvious kind of energy, heat energy, was not the right kind of energy. There is another very common kind—energy expended as *work*, as when dynamite is used to break rock. However, work energy is not the answer to our questions either, nor is the combination of heat and work. Nevertheless, they are extremely important, and together form the basis of the First Law.

- *Heat* ($q$) is the energy that flows across a system boundary in response to a temperature gradient.

- *Work* ($w$) is the energy that flows across a system boundary in response to a force moving through a distance (such as happens when a system changes volume).

Heat and work are therefore not separate entities as such but are forms of energy that are transferred in different ways. An enlightening analogy has been offered by Callen (1960). In Figure 3.2 we consider the water in a very deep pond (the amount of water is thus very great but finite and in principle could be exactly measured) to correspond to the internal energy U of a system.

Water may be added and subtracted from the pond either in the form of stream water (heat) or precipitation/evaporation (work). Both the inlet and

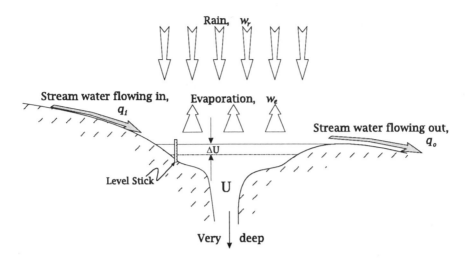

Figure 3.2: The pond analogy for the First Law.

outlet stream water can be monitored by flow gauges, and the precipitation measured by a rain gauge. Evaporation would be trickier to measure, but we may assume that we have a suitable measure for it. Now if the volume of stream inlet water over some period of time is $q_i$, the stream outlet water $q_o$, the rain $w_r$, and the evaporation $w_e$, then if there are no other ways of adding or subtracting water, clearly

$$\Delta U = (q_i - q_o) + (w_r - w_e)$$

where $\Delta U$ is the change in the amount of water in the pond, that could be monitored by a level indicator as shown. Thus

$$\Delta U = q + w$$

where

$$q = q_i - q_o$$

and

$$w = w_r - w_e$$

Once water has entered the pond, it loses its identity as stream or rain water. The pond does not contain any identifiable stream-water or rain-water, simply water. Similarly systems do not contain so much heat or work, just energy. Just as the water level in the pond can be raised *either* by stream water alone *or* by rain water alone, Joule showed in the last century that a temperature rise in a water bath of so many degrees can be caused *either* by heating (transferring energy due to a temperature difference) *or* by thrashing a paddle wheel about in it (transferring energy by force through distance, i.e., by deformation of the system boundary).

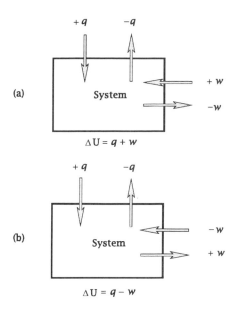

Figure 3.3: The two commonly used conventions for the sign of $q$ and $w$, leading to two formulations of the First Law.

Another implication or assumption in our pond analogy is that water is conserved, that is, it cannot simply disappear as if by magic. The same proposition regarding energy is known as the First Law of thermodynamics.

## 3.4  THE FIRST LAW OF THERMODYNAMICS

The First Law of thermodynamics is the law of conservation of energy. If U is the energy content of a system, and it may gain or lose energy only by the flow of heat ($q$) or work ($w$), then clearly, as in the pond analogy, $\Delta U$ must be the algebraic sum of $q$ and $w$. In order to express this algebraically, we must have some convention as to what direction of energy flow $+q$, $-q$, $+w$, and $-w$ refer to. In the pond analogy we assumed implicitly that addition of water to the pond was positive, whether as stream water or rain water. Thus heat added *to* a system is positive, and work done *on* a system is positive. This convention may be represented as in Figure 3.3a and is what we call the "scientific" convention—scientists like it because it is internally consistent. It results in the equation previously found,

$$\Delta U = q + w \qquad (3.1)$$

Another convention (Figure 3.3b) is to say that heat added to a system is positive, but that work done *by* a system is also positive, or that work done *on* a system is negative. This we call the "engineering" convention, because engineers prefer to think in terms of heat engines, and an engine doing work is something positive. This results in the relation

$$\Delta U = q - w$$

and also results in slightly simpler equations expressing pressure-volume work (the minus signs in equations (3.3), (3.4), (3.5), and so on would be missing). In this text we will use the scientific sign convention. Any additions of matter and energy to the system are positive in sign and all losses are negative.

Note that we have not "proved" the First Law. It is a principle that has been deduced from the way things work in our experience, but the fact that it has never been known to fail does not constitute a proof. Neither does the fact that the sun has never failed to rise in the east constitute a proof that it will rise in the east tomorrow, but I wouldn't bet against it.

### 3.4.1 Work

There are many configurations for the operation of moving a force through a distance, depending upon whether electric, magnetic, or gravity fields, surface tension, and so on are involved. As stated earlier, though, we may consider for now that these sources of work are not present, leaving only the most common sort of work in natural environments, pressure-volume work.

Pressure-volume work is always discussed using a piston-and-cylinder arrangement as shown in Figure 3.4. This may seem rather artificial or even useless to someone interested in processes that happen in nature or in the environment, but you have to realize that virtually *all* processes in *all* systems involve some change in volume, and therefore work is done against the pressure on the system, whatever that is (it is very often atmospheric pressure). We use a piston-cylinder arrangement for convenience—any system that changes volume could be used. Once we have found the appropriate equations for pressure-volume work, we can use them in our models of any system, whether or not they have pistons and cylinders.

Consider a thermostatted (constant temperature) piston-cylinder arrangement as shown in Figure 3.4. The cylinder is fitted with some devices that can hold the piston in position at various levels. When the piston is held stationary, the forces tending to move the piston are balanced (force pushing up equals force pushing down). If this were not the case, the piston would move. The two forces are acting on opposite sides of the same piston, having the same area, so the pressure of the gas, $P_{int}$, is exactly balanced by the external pressure, $P_{ext}$. The external pressure is provided partly by the stops that are holding it in place and partly by the weight of the piston itself, plus any weights on the piston. If the stops are removed, then all of a sudden $P_{ext}$ is reduced to that produced by the piston and weights only, $P_{int} \gg P_{ext}$, and the piston moves up until it encounters more stops — WHAP! — and all of a sudden $P_{int} = P_{ext}$ once more,

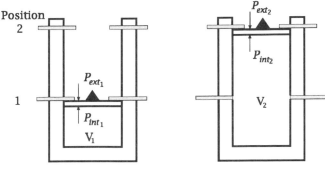

T constant

Figure 3.4: Irreversible expansion of a gas from external pressure $P_{ext_1}$ to $P_{ext_2}$. During expansion, external pressure is fixed by the weight of the piston plus the weights on the piston.

though at a different (lower) pressure (the experiment has been arranged such that the gas pressure is 10 pressure units at the upper stops, which is position 2, and 20 pressure units at the lower stops, position 1). Real gases tend to cool during expansion, so some heat will flow from the thermostat into the cylinder.

If the piston is well-lubricated and well-constructed, we can ignore friction effects, and the pressure-volume history of the change can be illustrated as in Figure 3.5. The external pressure during expansion is constant, since it is fixed by the mass of the piston. The work done during the expansion is[2]

$$w = \text{force} \times \text{distance}$$
$$= -(P_{ext} \times \mathcal{A}) \cdot \Delta \mathcal{L}$$

or

$$w = -P_{ext} \cdot \Delta V \tag{3.2}$$

where $\mathcal{A}$ is the area of the piston and $\Delta \mathcal{L}$ the distance it travels, which is seen to be the area under the path of expansion or expansion curve in Figure 3.5. If we repeat the process, but this time we place a larger weight on the piston, exactly the same thing will happen, but more work is done because a greater mass was lifted through the same volume.

---

[2]The work done will also include a term ($\frac{1}{2}mv^2$) for the work done in accelerating the piston. If we let the stops be part of the system, this kinetic energy is returned to the system at the upper stops, and can be neglected. Kivelson and Oppenheim (1966).

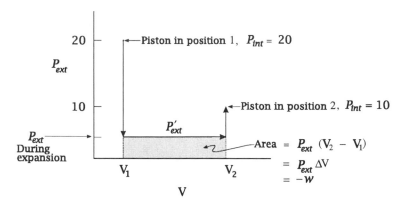

Figure 3.5: External pressure ($P_{ext}$) versus volume (V) plot for the irreversible expansion of the gas in Figure 3.4.

---

### Example

If the mass of the piston plus the weight on the piston give a $P_{ext}$ of 5 bars during the expansion, and $V_1 = 1000\,cm^3$ of ideal gas, how much work is done?

On decreasing $P$ (at constant $T$) by half, an ideal gas will expand to twice its volume ($PV = $ constant), so $V_2 = 2000\,cm^3$. Then

$$
\begin{aligned}
w &= -P_{ext}(V_2 - V_1) \\
  &= -5 \times (2000 - 1000) \\
  &= -5000\,bar\,cm^3
\end{aligned}
$$

To convert this to joules, Appendix A gives the conversion 1 bar = $0.10\,J\,cm^{-3}$, so

$$
w = -500\,J
$$

Note the minus sign, which indicates the system is doing work. If $V_2$ were less than $V_1$, $\Delta V$ would be negative and $w$ would be positive, meaning work is done on the system.

---

If another weight is added for the next expansion, we may have a total weight that is too great to allow the piston to reach the upper stops (position 2) and it will come to rest (equilibrium) somewhere in between. Then if the second weight is removed, the piston will proceed upward again as before, giving an expansion path as shown in Figure 3.6. If we use a lot of weights and remove them one at a time, letting the piston come to rest after each step, we will get a path such as shown in Figure 3.7.

Clearly we are approaching a limit of maximum work obtainable from the expansion of our gas, and clearly too, the maximum will be when we take an infinite number of infinitesimally small incremental steps from $V_1$ to $V_2$.

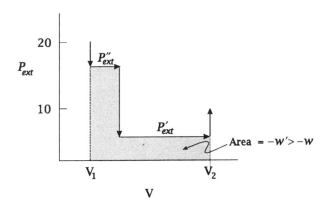

Figure 3.6: External pressure ($P_{ext}$) versus volume (V) for a two-stage expansion of gas. After an initial expansion at $P''_{ext}$, some weight was removed from the piston and the expansion continued at $P'_{ext}$.

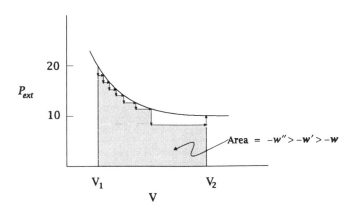

Figure 3.7: External pressure ($P_{ext}$) versus volume (V) for a multistage expansion of gas. After each constant $P_{ext}$ expansion, some weight was removed, allowing a further expansion.

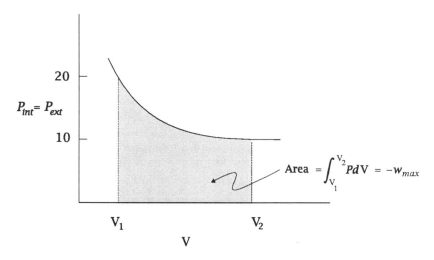

Figure 3.8: Pressure versus volume for the reversible expansion of a gas. The limiting case where an infinite number of constant $P_{ext}$ steps are taken gives the maximum area under the curve. During the expansion, internal pressure and external pressure are never more than infinitesimally different, or $P_{int} = P_{ext}$ at all times.

Since we have been letting the piston come to rest or equilibrium after every weight removal, in the limit we will have an infinite number or continuous succession of equilibrium states, giving us an example of a reversible process. In this particular case the name "reversible" is particularly appropriate since at any stage in the expansion the direction of movement can be reversed by changing the external pressure infinitesimally.

In the limit when infinitesimal increments of V are taken, the work of expansion is (see Figure 3.8)

$$w_{rev} = w_{max} = -\int_{V_1}^{V_2} P \, dV \qquad (3.3)$$

Here we need make no distinction between $P_{ext}$ and $P_{int}$ because they are never more than infinitesimally different in our continuous succession of equilibrium states. Again, note the negative sign required to comply with the scientific sign convention.

Since the end positions 1 and 2 of our expansion in every case consisted of our gas at stable equilibrium at a fixed $P$ and $T$, then according to the First Law there is a fixed energy difference $\Delta U$ between the two states. We have gone to some length to show that there is no fixed "difference in work," or work available from the change from one state to the other. Thus we are led to believe that the amount of heat flowing into our thermostatted cylinder must at all times, once equilibrium was established, have compensated for the variations in work performed, giving the same total $q + w$ in every case. We could verify this, of course, by making calorimetric measurements, but this is basically what Joule and many other workers have already done.

Our intent here is not so much to illustrate the constant energy change between states, but that this energy change, while accomplished by heat and work, can be made up of an infinite variety of combinations of heat and work. When the process is made reversible, we get the maximum work of expansion, and this will be given by equation (3.3), but even so, we are unable to calculate this amount of work (evaluate the integral) without more information (we need an equation of state for the gas, in order to know $P$ as a function of V). The integration of (3.3) at constant (external) pressure results in

$$w = -P_{ext}(V_2 - V_1)$$
$$= -P_{ext} \Delta V \qquad (3.4)$$

as in equation (3.2). The *internal* pressure necessarily varies during this expansion, as discussed above. Integration of (3.3) at "constant pressure" often means constant system pressure, or internal pressure, which can be done only if the heat flowing into the cylinder at all times exactly compensates for the expanding volume, that is, if the expansion is reversible. In this case we must write

$$w_{rev} = -P(V_2 - V_1)$$
$$= -P \Delta V \qquad (3.5)$$

where unsubscripted $P$ is understood to be the system or internal pressure.

A point worth emphasizing is that in any real or nonreversible expansion, as shown in our example, the work obtained is less than the maximum obtainable (from a reversible expansion). Thus in general, rewriting (3.3),

$$w \leq - \int_{V_1}^{V_2} P \, dV \qquad (3.6)$$

where the $<$ part of the $\leq$ sign refers to any irreversible change in V. This can also be expressed as

$$w \leq w_{max}$$

or

$$w \leq w_{rev}$$

For the opposite case of compressing the gas from 2 back to 1, the inverse series of steps can be employed. Thus, if at position 2 a heavy weight is placed on the piston, it will WHAP down to the stops at 1, describing a path such as in Figure 3.9. Obviously, much more work has had to be done in compressing the gas than we obtained, even in the reversible case, from expansion. However, by adding a larger number of smaller weights one at a time we can reduce the amount of work required for the compression, gradually approaching the stable

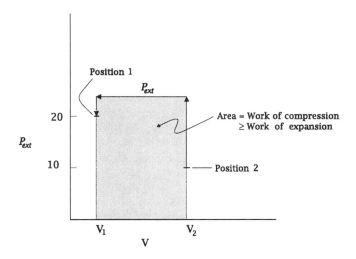

Figure 3.9: External pressure ($P_{ext}$) versus volume for the irreversible compression of gas at constant $P_{ext}$.

equilibrium curve from above, rather than from below as before. In the limit, of course, we find that for a reversible compression the work required is exactly the same as the work available from a reversible expansion.

### 3.4.2  Heat

It might be expected that since

$$\Delta U = q + w$$

and

$$w \le -\int_{V_1}^{V_2} P \, dV \tag{3.7}$$

perhaps there is a very similar story for the heat transfers in the gas expansion cases we have been considering. That is, perhaps

$$q \le -\int_{Z_1}^{Z_2} T \, dZ \tag{3.8}$$

where Z is some property of the gas. This is indeed the case (except for a sign change), but we must await the development of the Second Law, which will introduce us to entropy ($-Z$ in the above equation).

## 3.5  THE MODEL AGAIN

In this chapter we have discussed some very practical operations. There is nothing particularly theoretical about gases expanding in cylinders and performing
work. It happens countless times every day all over the world. Equations such
as (3.2) belong to the real world. However, the result of the limit-taking, when
the number of expansions or compressions in a single cycle is increased without limit, is a reversible process that belongs not to the real world but to the
thermodynamic model. This is another illustration of the point made in §2.4.2,
that energy differences between states can be calculated only for *reversible* processes.

The equation

$$w_{rev} = - \int_{V_1}^{V_2} P \, dV \tag{3.3}$$

is an extremely simple one, considered mathematically. If $P$ can be expressed
as an integrable function of V, then the integration is carried out and $w_{rev}$ is
determined for a given change from $V_1$ to $V_2$. This presents absolutely no
conceptual difficulties (beyond those in understanding calculus) if $P$ and V are
mathematical variables. However, if $P$ and V represent measured pressures and
volumes from a real system in the real world, then even if $P$ has been determined
as an integrable function of V for a number of individual measurements of $P$ and
V, the integration represents a variation of $P$ with V that is impossible to carry
out in the system. It is, however, simple to carry it out in the thermodynamic
model, that is essentially mathematical and in that $P$ as a function of V is simply
a line in $P$–V space. This line represents a reversible process, a perfectly simple
and understandable facet of the thermodynamic model.

### 3.5.1  Applicability of the Equations

Don't forget—this conclusion about the work done due to a change in volume
is not only applicable to piston-cylinder arrangements. Virtually all chemical
reactions involve some change in volume between reactants and products, and
the equations are applicable no matter what the physical form of the reactants
and products. In other words, when corundum and water react to form gibbsite
(Figure 2.7), the gibbsite occupies a different volume than does the sum of the
volumes of the water and the corundum; therefore, some work is done during
the reaction, and this work can be calculated using equations (3.3) and (3.4).
Even in reactions in living cells there will generally be a difference in volume
between products and reactants, and a constant pressure environment, and
so some work is done during each and every biochemical reaction. This work
energy may be relatively small compared to the heat evolved or absorbed during
the same reactions, but it must always be considered. In reactions at higher
pressures, it of course becomes even more important.

### 3.5.2  How Far Have We Got?

We have defined internal energy as some unspecified subset of the total energy in a system and considered the two common ways of changing this energy content. Along the way, we have noted that energy never disappears, and this is called the First Law of thermodynamics. How far have we got toward finding the "chemical energy," that always decreases in spontaneous changes?

Well, we've made the first vital step, but if you think about the previous chapters you'll realize that we cannot have the answer yet. Why? Because we noted that some reactions occur spontaneously with *no* energy change (ink spreading in water). Obviously, then, clarifying our thoughts about energy changes will not help in explaining processes that happen with no change in energy of any kind. Actually, we will note in the next chapter that the internal energy $U$ is in fact the energy we need to predict which way reactions will go under certain unusual conditions, but it is rarely used in this sense. We still have some way to go toward defining a useful "chemical energy."

## 3.6  SUMMARY

This chapter attempts to make precise our use of the terms *energy, heat*, and *work*. The line of thought we are pursuing has to do with systems that spontaneously decrease their energy content, and so we have started to get quite clear about what kinds of energy we mean. Relativity theory tells us that the *total* energy of all kinds contained in any system is given by multiplying the mass of the system by the square of the speed of light, but this approach is not very useful except in the study of nuclear processes. None of the chemical reactions we are interested in are of this type. However, apart from relativity theory there is no way of knowing the energy content of a system, so we have to be content with knowing *changes* in the energy content.

When we consider by what means the energy content of systems can change, we find that there are only two—we can heat/cool the system, or we can do work on the system/have the system do work. There are several ways of doing work on systems, depending on the forces we choose to consider (magnetic, electrostatic, surface tension, etc.), and so we start out by choosing the most common, pressure-volume work. The others are all handled in the same way and can be brought in when the situation calls for them.

Then by appealing to long experience with energy transfers, we propose the First Law of thermodynamics, the law of conservation of energy. Systems (that is, *any* system) can change their energy content by having energy subtracted or added in the two forms—heat and work. Any combination of the two can result in the same total energy change; there is no specific "difference in heat" or "difference in work" between two different states of the same system.

Defining what we mean by energy and energy transfers is, of course, important, but it does not by itself answer our questions about why reactions go one way and not the other.

## PROBLEMS

1. Calculate the work done (in joules) by the expansion in Figure 3.6 (of an ideal gas) if $V_1 = 1000\,cm^3$, $P'_{ext} = 5\,bar$, and $P''_{ext} = 18\,bar$. After the expansion at $P''_{ext}$, the piston is at equilibrium, and $P_{ext} = P_{int}$. However, during the expansion at $P'_{ext}$, the piston hits the upper stops, at that point the $P'_{ext}$ suddenly increases to 10 bars.

2. Calculate the work done if there are four expansions having $P_{ext}$ values of 18, 16, 14, and 10 bars.

3. Calculate the maximum amount of work available from the expansion of this gas.

4. Calculate the work done if the gas is compressed from $V_2$ to $V_1$ by placing a large weight on the piston equivalent to $P_{ext} = 22$ bars.

5. Why don't we use real gases instead of ideal gas in this type of question?

6. How much work is done when one mole of corundum combines with 3 moles of water to form one mole of gibbsite [reaction (2.1)], at atmospheric pressure?

# 4

# THE SECOND LAW OF THERMODYNAMICS

## 4.1  THE PROBLEM RESTATED

Having taken a couple of chapters to get our terminology settled and to get used to discussing energy changes in systems, we must now get back to our main problem—what determines whether processes will go or not go?

We have seen that the first great principle of energy transfers is that energy never disappears; it simply moves from place to place. It is the second principle or law that more directly addresses our main problem. It is observed that once the conditions of the beginning and ending states are decided upon, processes can proceed spontaneously in only one direction between these states and are never observed to proceed in the other direction unless they are "pushed" with an external energy source. Thus for beginning and ending conditions of $P = 1$ bar and $T = 5°C$, ice will melt, but water will never spontaneously change to ice. We are looking for a "chemical energy" term that will always decrease in such spontaneous reactions and will enable us to systematize and predict that way reactions will proceed under given conditions. This seems like a simple problem, but it is not.

Possibly the greatest single step forward in the history of the development of thermodynamics was the recognition and definition of a parameter, entropy, that enables such predictions and systematizations to be made. And yet, entropy still is not the energy term we have been looking for: the energy that always decreases in spontaneous reactions. In fact, it is not even an energy term. Nevertheless, it is the secret to an understanding of spontaneous reactions.

### 4.1.1   What's Ahead

In this chapter, we will try to explain why this is so. As with the First Law, there is no way of proving the Second Law. It is a principle that is distilled from our experience of how things happen. It can be stated in a lot of different ways, usually having something to do with the impossibility of perpetual motion or with the availability of energy, topics that seem to have little to do with the problem we have set for ourselves—that of finding an energy term that always decreases in spontaneous reactions. We will choose to state it in a way that emphasizes its role as a directionality parameter. This leads to the shortest possible path to the practical applications we wish to consider.

So here's the plan. We will define entropy as a parameter in our model systems, having certain properties. This definition is not, of course, pulled out of the blue, but is based on many years of work by many scientists. It should be accepted at first on faith, as simply a useful parameter, because it will not have any intuitive meaning as do our other terms such as energy, work, and so on. Then we will show how the "chemical energy" term we have been looking for is related to entropy. Finally, in Chapter 5, we will discuss how to measure the entropy, and what it means.

## 4.2   ENTROPY

One way to define entropy would be to simply say that the Z-term in equation (3.8) does indeed exist, where entropy is called S, and $Z = -S$. This provides a useful analogy between pressure-volume and temperature-entropy, and we will see these terms linked together in many equations. They represent work and heat energy in many processes we will be considering.

This way of defining entropy is also useful in explaining a somewhat puzzling feature of thermodynamics. In the next section, we will see that although entropy is a state variable of the kind we are looking for (one that can be used to tell which way reactions will go), it is unfortunately one that *increases* in spontaneous reactions, rather than decreasing, as we had supposed. This turns out to be simply because entropy was historically defined as $-Z$ in equation (3.8), rather than as Z. In other words, equation (3.8) is actually written

$$q \le \int_{S_1}^{S_2} T \, dS \qquad (4.1)$$

that is, without the minus sign. This is a sort of "historical accident." Because of the complete analogy between work-pressure-volume and heat-temperature-entropy (except for the minus sign), we can also write

$$q_{rev} = T \, \Delta S \qquad (4.2)$$

analogous to (3.5).

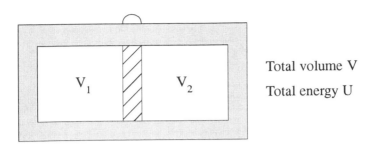

Total volume V

Total energy U

Figure 4.1: A composite isolated model system. The movable partition is impermeable to matter but conducts heat. The volume V of the system is the sum of the volumes of the two subsystems, that is, $V = V_1 + V_2$

However, a better way to define entropy is as follows. If there is indeed "something missing," that is, only one thing missing from the energy-decreasing analogy, it is something that causes reactions to "go," *even when no energy change whatsoever occurs*. Now, we have defined a type of system (the *isolated* system) that does not permit energy changes to occur in the system. Therefore, all we have to do is define a parameter with which we can predict reaction directions in this kind of system, combine it with the energy-decreasing idea, and we should have our answer. This is what we do with the following definition, paraphrased after Callen (1960). It can also serve as a statement of the Second Law of thermodynamics.

> There exists an extensive property of systems, entropy, which for isolated systems achieves a maximum when the system is at stable equilibrium. Entropy is a smoothly varying function of the other state variables and is an increasing function of the internal energy U.

We insert the postulate that entropy increases with U to ensure that the other directionality parameters to be derived decrease (have minima) rather than increase. This can be shown by considering the composite isolated (model) system in Figure 4.1.

The exterior wall is impermeable to energy and rigid, so the system is of constant U and V.[1] The piston is movable and can be locked in any position. It is impermeable but it conducts heat so that the two sides are at the same temperature. If there are equal amounts of the same gas in the two compartments, the equilibrium position of the piston when it is free to move is where $V_1 = V_2$. Also, according to our definition of S, the equilibrium position of the piston is one of maximum entropy for the system, and any other position has lower entropy. Then if we consider the same situation but with successively greater energy contents $U'$, $U''$, and $U'''$, we will have entropy-volume curves as

---

[1]The discussion and diagrams in this section use the total quantities U, S and V of the system in Figure 4.1. We could equally well use the molar properties $U$, $S$ and $V$.

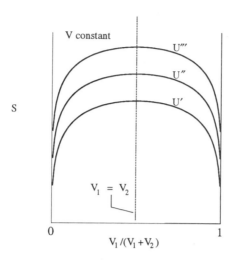

Figure 4.2: Entropy (S) versus volume fraction $V_1/(V_1 + V_2)$ for the system in Figure 4.1 at three different energy levels, where $U''' > U'' > U'$. Volume V is constant.

in Figure 4.2, where S is plotted against $V_1/(V_1 + V_2)$, which varies between 0 and 1. The maximum value of S is at $V_1/(V_1 + V_2) = 0.5$, where $V_1 = V_2$, and the curves for $U'$, $U''$, and $U'''$ are arranged with increasing entropies because we defined S to be an increasing function of U. In Figure 4.3, the curves of Figure 4.2 are drawn in three dimensions, and in Figure 4.4 the complete surface is shaded with a number of contours—the horizontal ones being contours of constant S and V, the vertical ones contours of constant U and V (recall that the whole diagram is for conditions of constant V). In Figure 4.5 two of these contours are abstracted to show more clearly that the two contours, which meet at a point, have a common tangent.

This tangent is located at the extremum in both curves, and so is in mathematical terms both $dU_{S,V} = 0$ and $dS_{U,V} = 0$. In other words, the condition that at equilibrium S is a maximum for given U, V *implies* the condition that U is a minimum for given S, V at a given equilibrium point.

Not only is S a parameter that always *increases* when a metastable state changes to a stable state at given values of U and V, but U is a parameter (a *state variable*) that always *decreases* when a metastable state changes to a stable state at given values of S and V. Thus as long as we consider only systems at constant U and V or constant S and V, both S and U are parameters of the type we have been looking for, but these kinds of systems are very rare. Still, we're getting closer.

Figure 4.6 shows two different states of an isolated system, one having more entropy than the other. A spontaneous process (mixing) occurs when the partition is removed, with an entropy increase. The importance of the isolated sys-

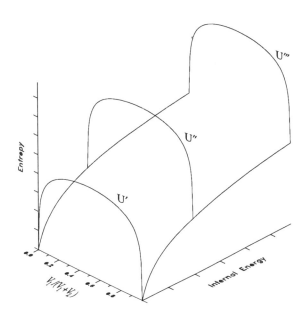

Figure 4.3: The U–S–V curves of Figure 4.2 in three dimensions.

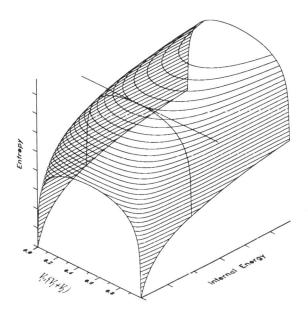

Figure 4.4: The U–S–V surface. This surface has been calculated for an ideal gas, but every system has such a surface. All points on the surface other than those at maximum S, minimum U represent metastable states of the system.

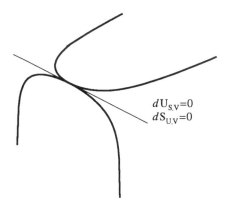

$$dU_{S,V}=0$$
$$dS_{U,V}=0$$

Figure 4.5: Constant S, V and constant U, V sections from the U-S-V surface shown in Figure 4.4, with their common tangent, which is simultaneously $dS_{U,V}$ and $dU_{S,V}$. The tangent point represents a position of stable equilibrium for the system.

tem is that it *prevents* energy changes from taking place in systems undergoing reactions, and it is therefore the clue to the missing factor we have mentioned several times (e.g., §1.3.1). We knew there was something missing because some spontaneous reactions (like melting ice) take place while absorbing heat energy, and some (like mixing gases) take place with little or no energy change at all. If reactions can take place with *no* energy change, and if entropy is the directionality parameter that predicts which way reactions will go when there is no energy change, then perhaps *combining* entropy with our other parameters such as heat and work will lead to more useful directionality parameters. This is exactly the case.

In other words, the reason we were a little off-base with suggesting that there is an analogy between a ball rolling in a valley and spontaneous chemical reactions is that some reactions can happen with no drop in energy at all. Chemical systems are more complex than simple mechanical systems, and analogies are dangerous. Entropy is a state variable that always increases in spontaneous reactions in which there is no energy change (those which take place in isolated systems), and is the missing factor. Combined with other state variables, we will have directionality parameters (our "chemical energy" term) for all kinds of systems.

Despite the apparent usefulness of entropy, we have not yet discussed what entropy actually *is*. We have no intuitive feeling for it such as we have for energy, pressure, and the other terms we have used so far. It is in fact quite difficult to develop such an intuitive understanding of entropy. We must for the moment just use it as a defined parameter in a mathematical sense, and let the understanding come later.

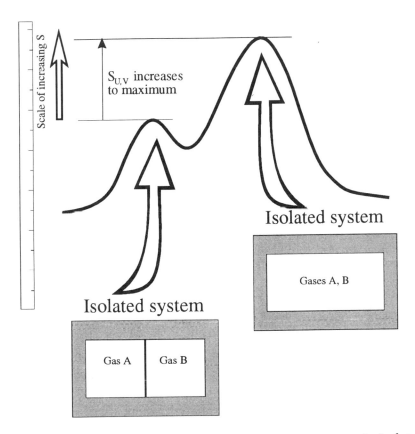

Scale of increasing S

$S_{U,V}$ increases to maximum

Isolated system

Gases A, B

Isolated system

Gas A | Gas B

Figure 4.6: Entropy increases in spontaneous processes in isolated systems.

## 4.3  THE FUNDAMENTAL EQUATION

The First and Second Laws can be combined into a single equation, which lies fairly close to the very heart of thermodynamics, called the Fundamental Equation.

In Chapter 3 we saw that at constant $P$ (and switching to molar variables, i.e., italic symbols),

$$w_{rev} = -P \, \Delta V \tag{3.5}$$

and, by analogy with these, we wrote

$$q_{rev} = T \, \Delta S \tag{4.2}$$

at constant $T$. Combining these equations (3.5) and (4.2) with the First Law

$$\Delta U = q + w$$

we have

$$\Delta U_{T,P} = T \, \Delta S - P \, \Delta V \tag{4.3}$$

This can also be written in differential notation as

$$\boxed{dU = T \, dS - P \, dV} \tag{4.4}$$

This is probably the single most important equation in thermodynamics, and for this reason it is called the *Fundamental Equation*. At this point we will simply outline some relationships of immediate use, leaving more detailed discussion of this equation for later chapters. From (4.4) we see that

$$(\partial U / \partial S)_V = T \tag{4.5}$$

and

$$(\partial U / \partial V)_S = -P \tag{4.6}$$

### 4.3.1  Geometrical Meaning of the Fundamental Equation

Many students can follow the development of the ideas so far presented up to equation (4.3), but changing to differentials (equation 4.4) seems to precipitate confusion. Why do we suddenly introduce differentials? Why must we bother with infinitesimal quantities?

This is a crucial point. We can answer in two ways. First, many students confuse differentials and derivatives, and infinitesimals spread their confusion over both. In fact, in changing from deltas to differentials ($\Delta S$ to $dS$), we have not changed anything. The differential of a variable such as $S$ is defined as an increment of *any* magnitude, not necessarily an infinitesimal increment. It is in defining the derivative and during integration that the differential increment is considered to decrease to infinitesimal size. These relations are explained further in Appendix C, where the meaning of equation (4.4) is presented in geometrical terms.

Secondly, and more significantly, we prefer to use differentials because they allow us to use the methods of the calculus. If we want to build a mathematical model of energy transfers, we have to use the language of mathematics. It's as simple as that.

## 4.4 ADDITIONAL DIRECTIONALITY PARAMETERS

In practical applications we rarely have occasion to deal with isolated systems, that is, those having constant U and V. It turns out, though, to be quite simple to develop additional thermodynamic potentials that give directionality information for systems having other types of constraints, using a mathematical procedure called the Legendre Transform. This relatively simple but powerful procedure provides a way to define new functions that have different dependent variables, but that have just as much information or usefulness as the old functions. You could think of it as rather like changing from rectangular to polar coordinates. The Legendre Transform is explained fully in Appendix C, but an understanding of the details is not really required.

We start with the fact that in the Fundamental Equation (4.4), we know $U$ as a function of $S$ and $V$. In general notation this is written

$$U = U(S, V)$$

### 4.4.1 Internal Energy and Volume

We rely on the discussion and diagrams of the preceding section to convince you that $U$ is a state variable exhibiting a minimum at equilibrium for systems of given $S$ and $V$. Calculus tells us that at the minimum or maximum of a function, the derivative and the differential of the function become zero, and thus

$$dU_{S,V} = 0$$

(See Appendix C for a review of this stuff.) Don't be unsure of equations like $dU_{S,V} = 0$. They refer to surfaces and lines in our thermodynamic model. In this case, we simply refer to the minimum in a curve such as those in Figures 4.4 and 4.5. It is a position of stable equilibrium for that (model) system, as any other point on the curve has a larger value of U. It follows too that, on changing from any position on the curve (a metastable state) toward the minimum (the stable state), U must decrease, and so $\Delta U_{S,V} < 0$, or $\Delta U_{S,V} < 0$.

### 4.4.2 Gibbs Energy

This is where the Legendre Transform becomes useful. Starting with the function $U = U(S, V)$, it provides a new function (we will call it $G$) that has the independent variables, not $S$ and $V$, but the *partial derivatives of U with respect to S and V*, that we know to be $T$ and $-P$ [equations (4.5), (4.6)].

The way it is done is simplicity itself (why it works is discussed in Appendix C). To get the new function, you subtract from the old function ($U$) the product of each of the independent variables ($S$ and $V$) times the partial derivative of the old function with respect to that variable. This definition in words is more difficult than the operation itself, which is simply

$$G = U - S\,(\partial U/\partial S)_V - V\,(\partial U/\partial V)_S$$

and, because $(\partial U/\partial S)_V = T$ and $(\partial U/\partial V)_S = -P$,

$$G = U - TS + PV \tag{4.7}$$

which is the definition of our new state variable, the Gibbs free energy. We are assured by the mathematicians that $G$ is a function with the same properties as $U$; that is, it is a state variable, and it achieves a minimum in systems at equilibrium. However, whereas $U$ reaches a minimum for system having a given $S$ and $V$ (Figure 4.5), $G$ reaches a minimum for systems having a given $T$ and $P$.

It is not easy to see from the definition that $G$ is in fact a function of $T$ and $P$. To show this, simply write the differential of $G$,

$$dG = dU - T\,dS - S\,dT + P\,dV + V\,dP$$

and then expand $dU$ by inserting equation (4.4),

$$\begin{aligned} dG &= T\,dS - P\,dV - T\,dS - S\,dT + P\,dV + V\,dP \\ &= -S\,dT + V\,dP \end{aligned} \tag{4.8}$$

From (4.8) we see that the independent variables for $G$ are $T$ and $P$. We can also derive a couple of very useful relationships by writing the general relationship between any state variable and its independent variables (see total differentials, Appendix C),

$$dG = (\partial G/\partial T)_P dT + (\partial G/\partial P)_T dP$$

Comparing this with (4.8) we have

$$(\partial G/\partial T)_P = -S \tag{4.9}$$

and

$$(\partial G/\partial P)_T = V \tag{4.10}$$

and

$$dG_{T,P} = 0 \tag{4.11}$$

Equation (4.11) is simply the condition for a minimum (or maximum) in $G$, that is, the tangent is horizontal (see Figure 4.7). It follows too, that $\Delta G_{T,P} < 0$ for spontaneous processes in systems having the same $T$ and $P$ before and after the process, because any spontaneous process must head toward this minimum from some higher point (some point of greater $G$ value).

### 4.4.3 The End of the Road

We pause here to note that, in case you hadn't noticed, we have arrived at the answer to the question posed in Chapters 1 and 2. The question was, what controls whether a reaction or a process will happen or not happen? Why does water freeze below 0°C and ice melt above 0°C? What is the "chemical energy" term that always decreases to a minimum, like the ball rolling down the hill? In answering this question, we first had to define fairly carefully some terminology such as system, equilibrium, and process. We then noted (Chapter 3) that systems have fixed energy contents (U, or $U$) at equilibrium, but this didn't help, because although this energy is conserved, it doesn't distinguish at all between directions processes take (bricks could cool themselves, and use this energy to fly, as far as $U$ is concerned).

The missing ingredient to understanding why reactions go one way and not the other is entropy. Entropy is defined as a state variable that always *increases* in spontaneous processes in isolated systems. But a parameter that is useful only in isolated systems is not of much practical use, so we pulled out a mathematical trick (the Legendre Transform) and converted entropy into another state variable, the Gibbs free energy, that always decreases in spontaneous process in systems at a given $T$ and $P$ (see Figure 4.7). This is the parameter we have been looking for. Figure 4.7 shows two different states of a system at the same temperature and pressure. A spontaneous process (the formation of gibbsite) occurs when the constraint keeping the reactants separated is removed (corundum and water are mixed together).

Similarly, in Figure 2.5, the "chemical energy" term is in fact the Gibbs free energy. At +5°C, 1 bar, $G_{ice} > G_{water}$, and at −5°C, 1 bar, $G_{water} > G_{ice}$. In Figure 2.6 $G_{diamond} > G_{graphite}$, assuming both have the same $P$ and $T$. It is not particularly important for you to understand *how* entropy concept gets transformed into the Gibbs free energy concept (the Legendre Transform). You need only know that mathematically, one implies the other, as shown in Figure 4.8. Looking at it in still another way, you see from Figure 4.5 that a single system can have $dU_{S,V} = 0$ and $dS_{U,V} = 0$ simultaneously. If we were not limited to three dimensions, we could show that the same system has $dG_{T,P} = 0$ at the same time. Each condition implies all the others.

In biochemistry, processes having a negative $\Delta_r G$ ($\Delta_r G < 0$) are termed *exergonic*, and those having a positive $\Delta_r G$ are termed *endergonic*. For some reason, these terms have not caught on in geochemistry.

The problem at the moment is that these new state variables $S$ and $G$ will have no "feeling of reality" for a reader new to the subject. That is, what *is* entropy or free energy, and how does one measure these things? Only by actually using these concepts will one become familiar with them. The next section is a first attempt at describing these variables in more familiar terms.

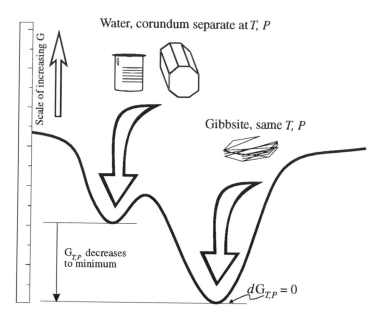

Figure 4.7: Gibbs free energy decreases in spontaneous processes at a given temperature and pressure.

### 4.4.4  Gibbs Energy Change as Useful Work

A ball in a metastable equilibrium valley (e.g., Figure 1.2) is capable of doing work as it rolls down to lower elevations (once it is pushed over the barrier). The maximum work it can do is exactly equal to the work required to push the ball back up to its metastable elevation. One way of understanding the Gibbs energy is that it is equal to the maximum amount of useful work that chemical systems can do as they change from metastable states to stable states, underlining the usefulness of the ball-in-valley analogy.

However, we must first distinguish between *total* work and *useful* work. Chemical systems undergoing change (i.e., in which reactions occur) can do various kinds of work. For instance, batteries can do electrical work. While undergoing these reactions, the chemical system invariably has some change in volume, because it is most unlikely that the reaction products would have exactly the same volume as the reactants. This change in volume $\Delta V$ takes place under some ambient pressure $P$, so that $P \, \Delta V$ work is done during the reaction regardless of whether any other kind of work is done or not—if the reaction is to take place, it cannot be avoided. This "work against the atmosphere" (or against the confining pressure, whatever it is) usually is not *useful*; it simply takes place whether we like it or not, and at atmospheric pressure it is often a rather small part of the total energy change. Although we can decide to eliminate electrical

# The Second Law:

| | |
|---|---|
| **WORDS:** | There is a system property, entropy (S), which always increases in spontaneous reactions in isolated systems (those having constant U and V). |
| **EQUATION:** | $\Delta S_{U,V} \geq 0$ |

## IMPLIES

⇓

| | |
|---|---|
| **WORDS:** | There is a system property, Gibbs free energy (G), which always decreases in spontaneous reactions in constant T,P systems. |
| **EQUATION:** | $\Delta G_{T,P} \leq 0$ |

Figure 4.8: All you need to know about the Second Law and the derivation of the Gibbs free energy function. These statements and equations can be written equally well in their molar forms (using $U$, $S$, $V$, and $G$).

work or other kinds of mechanical work from our systems, we cannot eliminate this $P \Delta V$ work (unless we consider only constant volume systems, which is not usually very practical). We will switch here from considering the total work done, $P \Delta V$, to the work done per mole of substance, $P \Delta V$ (note the change from V to $V$).

Net work other than $P \Delta V$ work can be written

$$\begin{aligned} w_{net} &= (w_{total} - w_{P \Delta V}) \\ &= (w + P \Delta V) \end{aligned}$$

and because $q \leq T \Delta S$ [equation (4.6)], it follows from the First Law ($\Delta U = q + w$) that

$$w \geq \Delta U - T \Delta S$$

Adding $P \Delta V$ to both sides,

$$w + P \Delta V \geq \Delta U - T \Delta S + P \Delta V$$

and so

$$w_{net} \geq \Delta U - T \Delta S + P \Delta V \tag{4.12}$$

Now,

$$G = U - TS + PV$$
$$dG = dU - T\,dS - S\,dT + P\,dV - V\,dP$$
$$dG_{T,P} = dU - T\,dS + P\,dV \tag{4.13}$$

and

$$\Delta G_{T,P} = \Delta U - T \Delta S + P \Delta V \tag{4.14}$$

So, combining (4.12) and (4.14),

$$w_{net} \geq \Delta G_{T,P}$$

or, the other way around,

$$\Delta G_{T,P} \leq w_{net}$$

However, don't forget that according to our sign convention, $-w$ is the work done by a system, or available from a system, so we should perhaps write

$$-\Delta G_{T,P} \geq -w_{net}$$

In other words, the net useful work available from a system cannot be greater than the decrease in $G$ that the system undergoes. For example, if a battery is doing work by lighting the bulb in a flashlight, the maximum amount of useful work it can do is given by its decrease in $G$ toward stable equilibrium, when the battery is dead (we continue this thought in §9.5.1). If the system does *no* work other than expanding or contracting against its confining pressure (no work other than $P \Delta V$ work), then $w_{net} = 0$, and

$$\Delta G_{T,P} \leq 0 \tag{4.15}$$

This result is not surprising, as it agrees with our conclusion in §4.4.2, but it does serve to link the Gibbs free energy with an intuitive concept, the available work.

> **Example**
>
> The maximum amount of work, other than the work done by the confining pressure, available from reaction (2.1) is
>
> $$
> \begin{aligned}
> \Delta_r G^\circ &= \Delta_f G^\circ_{Al_2O_3 \cdot 3H_2O(s)} - \Delta_f G^\circ_{Al_2O_3(s)} - 3\Delta_f G^\circ_{H_2O(l)} \\
> &= -2310.21 - (-1582.3) - 3(-237.129) \\
> &= -16.523 \, kJ \\
> &= -16523 \, J
> \end{aligned}
> $$
>
> By comparison, $P\Delta V$ work done by atmospheric pressure during this reaction is 1.61 J (Problem 6, Chapter 3). Don't worry about the meaning of $\Delta_f$. We'll get to that (§5.4).

### 4.4.5 Heat of Reaction

In processes at constant pressure, the work done, as we have seen, is $-P\Delta V$. Therefore the First Law can be written

$$
\Delta U = q_P - P\Delta V
$$

where $q_P$ is the heat transferred in the constant pressure process. Thus

$$
q_P = \Delta U + P\Delta V
$$

from which we can see that the heat transferred in constant pressure processes is equal to a function involving only state variables, and so it is itself a state variable. Don't forget that we went to some trouble in Chapter 3 to show that in general neither $q$ nor $w$ is a state variable; it is only in the special case of constant pressure processes that they both become state variables. Because of this, we can define a new term, enthalpy,

$$
H = U + PV \tag{4.16}
$$

which has the differential form

$$
dH = dU + P \, dV + V \, dP
$$

At constant pressure, this becomes

$$
dH_P = dU + P \, dV
$$

or

$$
\Delta H_P = \Delta U + P\Delta V \tag{4.17}
$$

and since

$$\Delta U = q_P - P \, \Delta V$$

therefore

$$\Delta H_P = q_P \qquad (4.18)$$

All we have done here is notice that, because work becomes a fixed quantity in constant pressure processes, then heat does too, by the First Law. And because constant pressure processes are so common (including all reactions carried out at atmospheric pressure, such as most biochemical reactions), it is convenient to have a state variable defined to equal this heat term. Defining enthalpy as in (4.16) accomplishes this, and we now have a "heat of reaction" term, which will be useful in all constant pressure processes.

---

**Example**

The standard heat of reaction for reaction (2.1) is

$$
\begin{aligned}
\Delta_r H^\circ &= \Delta_f H^\circ_{\text{Al}_2\text{O}_3 \cdot 3\text{H}_2\text{O}(s)} - \Delta_f H^\circ_{\text{Al}_2\text{O}_3(s)} - 3\Delta_f H^\circ_{\text{H}_2\text{O}(l)} \\
&= -2586.67 - (-1675.7) - 3(-285.83) \\
&= -53.48\,\text{kJ} \\
&= -53480\,\text{J}
\end{aligned}
$$

Again, the minus sign means heat is given out. This amount of heat would raise the temperature of a liter of water about 12°C.

---

Note that because $H$ is a state variable, $\Delta H$ is perfectly well-defined between any two equilibrium states. But when the two states are at the same pressure, $\Delta H$ becomes equal to the total heat flow during the process from one to the other, and in practice enthalpy is little used except in this context.

Processes having a negative $\Delta_r H$ ($\Delta_r H < 0$) are termed *exothermic*, and those having a positive $\Delta_r H$ are termed *endothermic*.

## 4.4.6   The Heat Capacity

An older name for the enthalpy is the "heat content." This name is somewhat discredited for good reasons, but nevertheless it helps a little in conveying the essential idea behind the next concept, the heat capacity. The molar heat capacity can be defined as the amount of heat required to raise the temperature of one mole of a substance by one degree. Of course, some substances require much more heat to do this than do others. Metals such as copper generally require less heat to warm them up than do insulators—it takes less change in their "heat content" to change their temperature. Just imagine trying to make a heating element for a stove from an insulator such as glass or ceramic.

The formal definition is

$$\frac{dH}{dT} = C_p \qquad (4.19)$$

or

$$\frac{d\Delta H}{dT} = \Delta C_p \qquad (4.20)$$

Thus heat capacity is the rate of change in $H$ with $T$. A large $C_p$ means that $H$ changes a lot for a given change in $T$; it will have a high "heat content."

## 4.5  SUMMARY

What you should know at this point is that we have defined a parameter, entropy, which can tell us which way reactions will go, but only in isolated systems. The statement defining entropy is one way of stating the Second Law of thermodynamics. We have not said what entropy *is*, physically, nor how to measure it. Combining entropy with the First Law, and using a mathematical twist of the wrist called the Legendre Transform, we then defined another parameter, the Gibbs free energy, which can tell us which way reactions go in systems at a given temperature and pressure. We also showed that the Gibbs free energy is equal to the maximum amount of useful energy or work available from such reactions, and we defined enthalpy to be equal to the heat transferred in constant pressure processes.

It is normal at this point for newcomers to this subject to be rather confused. If we think about natural processes that we would like to understand, such as occur in living plants and animals, or even simpler inorganic processes such as occur in creating our weather patterns or in erupting volcanoes, we could be forgiven for wondering what earthly use the kind of material we have considered up to now can be. We seem to have restricted ourselves to ridiculously simple cases such as balls rolling in valleys, and even though we have claimed that certain simple inorganic processes such as melting ice and polymorphic mineral changes are analogous, we haven't shown how to do anything remotely useful.

Not only is it not yet useful, but even after restricting ourselves to simple cases and claiming not to be dealing with reality but with models of reality, we have introduced at least one concept (entropy) that so far has no physical meaning and have used a level of mathematics that, although not exceeding that taught in introductory calculus courses, has physical implications that are hard to grasp. The best remedy for this is to review the material to some extent, and to plunge ahead even if it is not entirely clear. After some familiarity with practical applications is attained, some of the earlier material will become clearer, so the best approach is continuous review, in addition to assimilation of new material.

It may well seem that we have made no progress toward understanding complex processes, but this is not true. Most natural processes are so complex that we simply must start with the very simplest ones we can think of and define our terms very carefully. Our goal of finding the secret to why reactions go in one direction and not the other may seem overly simple, but it is in fact the basic concept necessary to build up an understanding of all the natural phenomena mentioned above. Of course, even when we have mastered thermodynamics, we will find that we don't have all the answers to all our questions; in fact, we will find that the things thermodynamics can tell us are fairly limited. They have, however, a level of certainty which surpasses that of most other ways of looking at the same problems, and this makes the subject an absolutely essential element of all research into problems that involve energy transfers. You may wish to know much more than thermodynamics can tell you, but you need to know what it can tell you.

## PROBLEMS

1. Calculate the change in internal energy ($\Delta_r U°$) for reaction (2.1).

2. Calculate the standard heat of reaction for reaction (2.4).

3. Calculate the maximum amount of useful work available from reactions (2.2) and (2.4) under standard conditions. This would allow you to put some numbers on the vertical axis of Figure 2.9.

4. Calculate the work of expansion against atmospheric pressure for reaction (2.4), and compare with the maximum useful work available.

5. The unit of energy used by nutritionists is the Calorie (with a capital C), which is 1000 calories, or 1 kcal. In Appendix A, we see that 1 calorie = 4.184 Joules, so 1 Cal = 4184 J. The basal metabolic rate for humans varies from about 1300 to 1800 kcal/day. Any physical activity adds to this basal energy requirement. In other words, a person at rest would use the energy in a glass of orange juice (100 Cal) in about 100/1500 days, or 1.6 hours. How long would the energy ($\Delta_r H°$) from reaction (2.1) last an average human, if it was available?

# 5

# CALORIMETRY
# AND THE THIRD LAW

## 5.1 MEASURING $\Delta G$

We have had quite enough theoretical discussion for now. Let's see how to get some numbers into our equations so as to be able to calculate something useful. Obviously, we would like most to be able to calculate $\Delta_r G$ for any reaction of interest to us, because this will tell us which way the reaction will go, assuming that pressure and temperature are fixed. For example, if we were not sure which of the two forms of carbon was the stable form at $25\,°C$, $1$ bar, we could measure $G_{graphite} - G_{diamond}$, which is $\Delta_r G$ for the reaction

$$C(diamond) = C(graphite)$$

and if this quantity was negative, then graphite would be stable, and if it was positive, diamond would be stable.

There are quite a number of ways of determining changes in Gibbs free energy, but we will discuss only the most common one here. Others are associated with determining the equilibrium constant or cell voltages, as we will see in Chapters 8 and 9. To see how changes in $G$ are measured, consider first equations (4.14) and (4.17)

$$\Delta G_{T,P} = \Delta U - T\,\Delta S + P\,\Delta V \qquad (4.14)$$

$$\Delta H_P = \Delta U + P\,\Delta V \qquad (4.17)$$

Combining these, we have

$$\Delta G_{T,P} = \Delta H - T\,\Delta S \qquad (5.1)$$

From this, we see that we can calculate $\Delta G$ for a process if we know $\Delta H$ and $\Delta S$ for that process. We know (from §4.4.3) that $\Delta H$ is simply the heat transferred to or from the closed system during the process, and we should be able to measure that. It remains to be seen how to measure $\Delta S$ and to try to get an idea of what $\Delta S$ *means* in physical terms. We will find that we can get measurements of $S$ as well as of $\Delta H$ by measuring quantities of heat. Therefore, *calorimetry*, the art and science of measuring heat flows, is the secret to determining values of $\Delta G$.

## 5.1.1 Solution Calorimetry

Heat flows can be measured in various ways. One way is to observe some process in which heat is liberated under controlled conditions, resulting in a rise in temperature, and then duplicate that temperature rise using an electrical heater. This is the principle used in the calorimeter in Figure 5.1.

This apparatus is used to measure how much heat is liberated when a known amount of solid material, such as a mineral, dissolves. Minerals are notoriously insoluble in water and so an acid, such as hydrofluoric acid (HF), is used. The method is called *solution calorimetry*. A few grams of crushed mineral are put into the *sample holder* and sealed with gold foil. The sample holder is then place in the *reaction chamber*, which is then filled with acid. A long rod reaches from the top of the sample holder through various seals to the top of the apparatus. This assembly is then sealed and placed in a vacuum chamber, which goes into a water bath. The purpose of the vacuum and water bath is to minimize the loss of heat from the reaction chamber.

When everything has settled down, the mineral sample and the acid are at the same temperature, but are separated. The long rod is then pushed down. This punctures the seal on the sample container, and the bottom also falls out, allowing the sample to mix with the acid and dissolve. The sample holder also has fins, and rotation stirs the solution and speeds up the dissolution process. The dissolution of the mineral releases heat, which raises the temperature of the acid, and the amount of temperature change is measured by a resistance thermometer, which is wrapped around the reaction vessel. The apparatus is calibrated by using an *electrical heating coil* to raise the temperature in a different experiment, but using exactly the same setup. The voltage drop across the electrical heater and the current flowing through it are known, and so the amount of heat required to raise the temperature of the calorimeter by any given amount is known exactly by turning on the heater for a short time and observing the temperature increase. By comparing the temperature change caused by the heating coil to that caused by the mineral dissolution, the heat liberated by the mineral dissolution can be determined quite precisely. A number of small corrections must be made for various heat losses in the apparatus, plus a correction to the heat measured over the temperature interval to what would have been observed if the process had occurred at a constant temperature of 25°C. These calculations require a knowledge of the *heat capacity* of the calorimeter (see below). Because we know the mass of mineral grains used, the heat of solution per mole of mineral at 25°C then can be calculated. The whole process

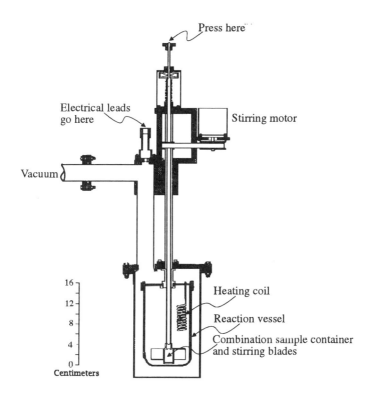

Figure 5.1: An adiabatic heat-of-solution calorimeter. The reaction vessel contains acid. Pushing down on the handle at the top punctures the upper seal and pushes out the bottom of the sample container, allowing the sample to dissolve. A thermometer is wound around the reaction vessel and records the change in temperature. No electrical leads are shown. (Simplified from Robie and Hemingway, 1972.)

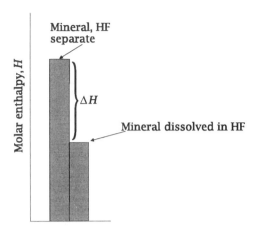

Figure 5.2: Enthalpy is a state variable of fixed but unknown value in the beginning and final equilibrium states. $\Delta H$ is obtained by measuring the heat liberated in the reaction at constant pressure.

is exacting and painstaking.

What are the meaning and use of this heat of solution? In terms of the processes we have been discussing, we have observed an irreversible reaction between a metastable state (pure acid and mineral grains, separated, at $T_1$) and a stable state (mineral dissolved in acid at $T_2$), made some measurements, and calculated from this the heat that would be released in the reaction

$$\text{mineral, HF separated} \rightarrow \text{mineral dissolved in HF} + q_{dissolution}$$

at 25°C. If the calorimeter is open to the atmosphere, then the mineral dissolution process happens at a constant pressure, and by equation (4.18), the heat measured, $q_{dissolution}$, is equal to the change in enthalpy of the system, $\Delta H$. This change is illustrated in Figure 5.2.

Of course, a heat of solution is not exactly what we wanted, although it is a $\Delta H$. We want a $\Delta H$ that is the difference between products and reactants of reactions of all kinds, such as our corundum–water–gibbsite or diamond–graphite reactions, and innumerable others. But the heat of solution technique allows us to do this. Note that because in any balanced chemical reaction, the total or bulk composition of the reactants must be exactly the same as that of the products. That's what "balanced" means—all the atoms on the left side of the reaction must appear also on the right. Therefore, if in separate experiments we dissolve the reactants and the products in the same kind of acid, we will get identical solutions. We will, however, measure different heats of solution, because the products and reactants have different structures and different energy contents. Therefore, the difference in the heats of solution must be equal to the difference in enthalpies of the products and reactants themselves.

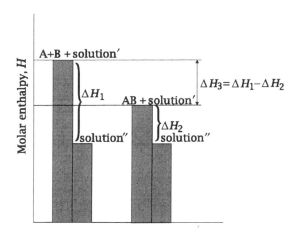

Figure 5.3: Compounds A+B together have heat of solution $\Delta H_1$. Compound AB has heat of solution $\Delta H_2$. Both processes result in solution″, so the heat of reaction A + B → AB, which cannot be carried out in a calorimeter, is $\Delta H_1 - \Delta H_2$.

To put this argument in formal terms, suppose our reaction is

$$A + B = AB$$

for example, $SiO_2 + Al_2O_3 = Al_2SiO_5$. First we dissolve the reactants, and then, in a separate experiment, we dissolve the products:

$$A + B + \text{solution}' \rightarrow \text{solution}'' + \text{heat } (\Delta H_1)$$

$$AB + \text{solution}' \rightarrow \text{solution}'' + \text{heat } (\Delta H_2)$$

As long as both solution′ and solution″ have the same compositions in both reactions, the reactions may be subtracted giving

$$A + B \rightarrow AB + \Delta H_3$$

where $\Delta H_3 = \Delta H_1 - \Delta H_2$ and is the heat of reaction ($\Delta_r H$) of the reaction A + B = AB, as shown in Figure 5.3. A, B, and AB can also refer to complex organic compounds of any kind. We need only be able to separate them into their pure forms, so as to be able to work with them.

Because of practical difficulties, the determination of $\Delta_f H°$[1] of a compound is rarely the sum of only two heats of solution, as in Figure 5.3. Quite often 10 or 15 solution reactions may have to be carried out to determine one $\Delta_f H°$, and the whole process may take several weeks.

---

[1]We get to the definition of $\Delta_f H°$ in §5.4.1. It is a special kind of $\Delta H_3$.

### 5.1.2 Entropy and Heat Capacity

Now we must consider how to measure entropy. We'll discuss what entropy *is* later. So far, all we know about entropy is that it increases in spontaneous reactions in isolated systems, and that it appears in equations such as (4.13) and (4.14). To get it into a form that would suggest a method of measurement, we combine equations (4.17) and (4.4),

$$dH = dU + P\,dV + V\,dP \qquad (4.17)$$

$$dU = T\,dS - P\,dV \qquad (4.4)$$

giving

$$dH = T\,dS + V\,dP \qquad (5.2)$$

If we choose constant pressure conditions, $dP$ becomes zero, and substituting $C_p dT$ for $dH$ [equation (4.19)], we have

$$C_p dT = T\,dS$$

or

$$dS = \frac{C_p}{T}dT \qquad (5.3)$$

Here at last we have the entropy defined in terms of something measurable, the heat capacity. Integrating (5.3), we have

$$S_{T_2} - S_{T_1} = \int_{T_1}^{T_2} \frac{C_p}{T}dT \qquad (5.4)$$

and so you see that assuming that you can get numbers for $C_p$ at a series of temperatures, you could divide each $C_p$ by its value of $T$ and evaluate the integral, giving you the difference in entropy between two temperatures.

By now you are probably accustomed to being told that we cannot know the absolute values of thermodynamic parameters, only differences. But this applies only to the internal energy, $U$, and any parameters that contain $U$, such as $H$ and $G$. Entropy is different in that we *can* get absolute values, by virtue of the Third Law of thermodynamics.

## 5.2  THIRD LAW ENTROPIES

The Third Law says that *The entropy of all pure, crystalline, perfectly ordered substances is zero at absolute zero temperature, and the entropy of all other substances is positive.* Thus the entropy of all pure, crystalline minerals such as diamond, corundum, gibbsite, and so on are all zero at $T = 0\,K$, but the entropy of, say, glass would not be zero, because it's structure is amorphous and not crystalline. If we let $T_1$ be absolute zero in equation (5.4), the entropy of minerals at any temperature, say our standard temperature of 298.15 K, is

$$S_{298} = \int_{T=0}^{T=298} \frac{C_p}{T} dT \qquad (5.5)$$

and all that is required to determine "absolute" values for the entropy of minerals is to measure their heat capacity at a series of temperatures between zero and 298.15 K and to evaluate the integral. This gives rise to another kind of calorimetry, *cryogenic*, or low-temperature calorimetry.

### 5.2.1  Cryogenic Calorimetry

A cryogenic calorimeter (Figure 5.4) is an apparatus designed for the determination of heat capacities at very low temperatures. The procedure is to cool the sample down to a temperature within a few degrees of absolute zero (a temperature of absolute zero itself is actually impossible to achieve), introduce a known quantity of heat using an electrical heating coil, and observe the resulting increase in temperature (usually a few degrees). The quantity of heat is equal to $\Delta H$, and this divided by the temperature difference gives an approximate value of $C_p$ at the midpoint of the temperature range. Corrections are then made to compensate for heat leaks, for the heat absorbed by the calorimeter, and to get exact $C_p$ values from the approximate ones. The integration of $C_p/T$ values to obtain the entropy at 298.15 K is illustrated in Figure 5.5.

## 5.3  THE PROBLEM RESOLVED

We now know how to tell which way reactions will go, not just in a theoretical way (they will decrease their Gibbs free energy at $T$, $P$), but in a practical way (how do we get this $\Delta G$?). For example, $SiO_2$ comes in several crystalline varieties (polymorphs), such as the minerals quartz and cristobalite. They have the same composition $SiO_2$, but different crystallographic structures and energy contents, and one is stable and one is metastable at 25°C, 1 bar. The problem is analogous to the diamond–graphite problem, and the reaction is

$$SiO_2^{cristobalite} = SiO_2^{quartz}$$

Which way does this reaction go at 25°C, 1 bar?

You could answer this question thermodynamically as follows:

1. Dissolve quartz in a solvent (HF acid) and measure the heat released.

2. Dissolve the same amount of cristobalite in the same amount of the same solvent and measure the heat released.

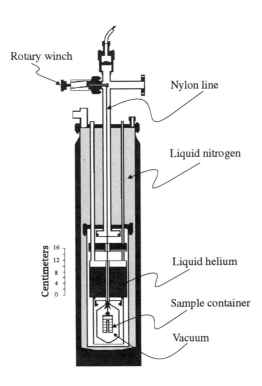

Figure 5.4: A cryogenic or low-temperature calorimeter. The sample container can be raised by the rotary winch so as to be in contact with the liquid helium reservoir for cooling to 4.2 K, or lowered into the vacuum for heating. The reentrant well in the sample container contains a heating coil. (Simplified from Robie and Hemingway, 1972.)

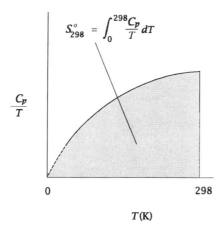

$$S_{298}^{\circ} = \int_0^{298} \frac{C_p}{T} dT$$

$\dfrac{C_p}{T}$

0                                    298

$T(K)$

Figure 5.5: A plot of $C_p/T$ vs. $T$ for a mineral. Integration gives the area under the curve, which is equal to the entropy at the upper limit of integration.

3. Because the solution after dissolution in the two cases has exactly the same composition and is identical in all respects, the difference in the two measured heat terms must be the difference in enthalpy between quartz and cristobalite, $\Delta_r H = H_{298}^{quartz} - H_{298}^{cristobalite}$ (remember, products minus reactants).

4. Measure the heat capacities of both quartz and cristobalite from near absolute zero to 298 K, and calculate $S_{298}^{quartz}$ and $S_{298}^{cristobalite}$.

5. Subtract these two entropies to give $\Delta_r S = S_{298}^{quartz} - S_{298}^{cristobalite}$.

6. Calculate $\Delta_r G = \Delta_r H - 298.15 \cdot \Delta_r S$. If this is negative, quartz is stable; if it is positive, cristobalite is stable.

You are no doubt quite sure that you don't have to do this incredible amount of work to answer such a simple question; there must be an easier way. Well, there is, but only because other people have already done this incredible amount of work, and lots more like it. In other words, you can look up the data in tables. However, you do not look up heats of solution.

The trick we have just used to get the difference in enthalpy between two minerals (that is, to dissolve them both and subtract the heats of solution) is a very useful way of determining heats of reaction, because many reactions proceed very slowly, or not at all, and so you cannot measure the heat of reaction directly in a calorimeter. You cannot measure the heat of reaction as cristobalite changes directly into quartz at 25°C, because it never does—it is a truly metastable form of $SiO_2$. However, most minerals will dissolve fairly rapidly in some kind of solvent, providing an indirect means of getting their enthalpy differences.

To have tables of data that enable you to calculate $\Delta_r H$ for any reaction, it would seem that all you need to do is tabulate heats of solution. But if you think about this for a minute or so, you find that although fine in theory, this will not work well in practice. For one thing, you would need to tabulate heats of solution for all combinations of substances that might be of interest. That is, the heats of solution of gibbsite, corundum, and water separately are not enough—you need the heat of solution of corundum + water in a 1:1 ratio. But for other reactions, you would need corundum + water in other proportions. Then you would need to be sure that the solution compositions were identical, and given the variety of solvents and concentrations of solvents used, your database would soon become very large and unmanageable. There is a better way.

## 5.4   FORMATION FROM THE ELEMENTS

### 5.4.1   Enthalpies of Formation

The idea of producing two solutions of identical composition from two different sets of starting materials so as to get $\Delta H$ between the starting materials is good, but we need something else. That something else is the notion of finding, for each mineral or other substance, the $\Delta H$ and $\Delta G$ between the substance and the elements composing it. This is done, in principle, by the method we have been discussing, that is,

crystalline compound + solution$'$ → solution$''$ + heat   ($\Delta H_1$)

component elements + solution$'$ → solution$''$ + heat   ($\Delta H_2$)

As long as both solution$'$ and solution$''$ have the same compositions in both reactions, the reactions may be subtracted giving

component elements → crystalline compound + $\Delta H_3$

where $\Delta H_3 = \Delta H_1 - \Delta H_2$ and is the heat of formation of the compound from its elements. This $\Delta H_3$ is given a special symbol, $\Delta_f H°$, where the $f$ signifies a reaction in which the compound is formed from its elements in their stable states, and the $°$ signifies a pure substance, usually at 298 K, 1 bar, but not necessarily.

This may not seem like much of an advance, but it is. Now, instead of listing innumerable heats of solution, we need list only one value of $\Delta_f H°$ for each compound. Another advantage is that although we have mentioned only one method of determining $\Delta_f H°$, dissolving both the compound and its constituent elements, there are quite a few other methods,[2] and once the $\Delta_f H°$ has been determined, they can be combined in any reactions no matter what method was used to determine them. The meaning of $\Delta_f H°$ in terms of our previous energy diagrams is shown in Figure 5.6.

---

[2]See, for example, the silica Example in §8.5.

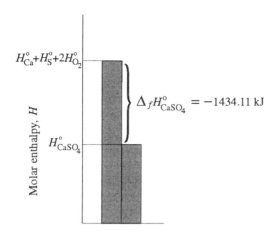

Figure 5.6: The standard enthalpy of formation of anhydrite, $CaSO_4$, is equal to the difference between the absolute enthalpy of anhydrite and the sum of the absolute enthalpies of its constituent elements. None of the absolute enthalpies is known, but the difference is known.

The change in enthalpy for *any* reaction between compounds for which there are formation-from-the-element data is given by a simple algebraic addition of these $\Delta_f H°$ terms, because in balanced reactions, the elements always cancel out. To take another example, consider the reaction

$$CaSO_4(s) \text{ (anhydrite)} + 2\,H_2O(l) = CaSO_4 \cdot 2H_2O(s) \text{ (gypsum)} \qquad (5.6)$$

Both gypsum and anhydrite occur at the Earth's surface, and it is not always clear which is the stable phase. To determine the enthalpy change in this reaction, we consider the reactions in which each phase in the reaction is formed from its elements:

$$Ca(s) + S(s) + 2\,O_2(g) = CaSO_4(s); \quad \Delta_f H°_{anhydrite} = -1434.11\,\text{kJ}$$

$$H_2(g) + \tfrac{1}{2}O_2(g) = H_2O(l); \quad \Delta_f H°_{water} = -285.830\,\text{kJ}$$

$$Ca(s) + S(s) + 3\,O_2(g) + 2\,H_2(g) \;=\; CaSO_4 \cdot 2H_2O(s);$$
$$\Delta_f H°_{gypsum} = -2022.63\,\text{kJ}$$

So for reaction (5.6) we have

$$
\begin{aligned}
\Delta_r H° &= \Delta_f H°_{gypsum} - \Delta_f H°_{anhydrite} - 2\,\Delta_f H°_{water} \\
&= -2022.63 - (-1434.11) - 2\,(-285.830) \\
&= -16.86\,\text{kJ}
\end{aligned}
$$

from which we see that the reaction between anhydrite and water to form gypsum is exothermic; that is, 16.86 kJ of heat would be released for every mole of anhydrite reacted.

It is important to realize that this heat of reaction, $\Delta_r H°$, is equal to the difference in the *absolute* enthalpies of the reactants and products—the enthalpies of the elements have nothing to do with it, because they all cancel out. Thus

$$
\begin{aligned}
\Delta_r H° &= \Delta_f H°_{gypsum} - \Delta_f H°_{anhydrite} - 2\,\Delta_f H°_{water} \\
&= H°_{CaSO_4 \cdot 2H_2O} - H°_{Ca} - H°_S - 2H°_{H_2} - 3H°_{O_2} \\
&\quad -(H°_{CaSO_4} - H°_{Ca} - H°_S - 2H°_{O_2}) \\
&\quad -2(H°_{H_2O} - H°_{H_2} - \tfrac{1}{2}H°_{O_2}) \\
&= H°_{CaSO_4 \cdot 2H_2O} - H°_{CaSO_4} - 2\,H°_{H_2O} \\
&= -16.86\,kJ
\end{aligned}
\tag{5.7}
$$

If you look carefully, you'll see that all the $H°$ terms for the elements cancel out. But if you think before you look, you'll realize that they *must* cancel out if the reaction is balanced. Otherwise there is a mistake somewhere.

Despite many statements to the contrary, the absolute enthalpies of the elements ($H°_{Ca}, H°_{O_2}$, etc.) are not assumed to be zero. There is no need to do so, because they all cancel out in balanced reactions.

## 5.4.2   Third Law Entropies

As you might expect, all the hard work of cryogenic calorimetry has already been done too, for most common substances, and the results are obtainable in tables and compilations of data. From Appendix B, we find that

$$
\begin{aligned}
S°_{CaSO_4} \text{ (anhydrite)} &= 106.7\,J\,mol^{-1}\,K^{-1} \\
S°_{CaSO_4 \cdot 2H_2O} \text{ (gypsum)} &= 194.1\,J\,mol^{-1}\,K^{-1} \\
S°_{H_2O} \text{ (water)} &= 69.91\,J\,mol^{-1}\,K^{-1}
\end{aligned}
$$

and the entropy of reaction for (5.6) is

$$
\begin{aligned}
\Delta_r S° &= S°_{gypsum} - S°_{anhydrite} - 2\,S°_{water} \\
&= 194.1 - 106.7 - 2(69.91) \\
&= -52.42\,J\,mol^{-1}\,K^{-1}
\end{aligned}
$$

**So is Gypsum or Anhydrite Stable?**

Now that we have numerical values for both $\Delta_r H°$ and $\Delta_r S°$ for reaction (5.6), it is a simple matter to calculate $\Delta_r G°$ to see which way the reaction goes. Our number for enthalpy is in kJ and that for entropy is in J, so we must convert one of them to be consistent. Converting kJ to J, we have

$$
\begin{aligned}
\Delta_r G^\circ &= \Delta_r H^\circ - T \Delta_r S^\circ \\
&= -16860 - 298.15(-52.42) \\
&= -1231 \, J
\end{aligned}
$$

which is negative; therefore, gypsum is more stable than anhydrite in the presence of water at Earth surface conditions. We repeat that what we have found is that the *assemblage* of anhydrite plus water is metastable with respect to gypsum at 25°C, 1 bar. Anhydrite by itself is not metastable, as there is no other form of $CaSO_4$ that has a lower energy.

### 5.4.3 Gibbs Energy of Formation From the Elements

Finally, although we must do the calorimetry experiments in order to calculate free energy differences, there is usually no need to use $\Delta_r H^\circ$ and $\Delta_r S^\circ$ values from tables to calculate $\Delta_r G^\circ$. Values of $\Delta_f G^\circ$ for most compounds have been calculated and are also to be found in the same tables of data, and so we can use these values directly, instead of going through the $\Delta_r H^\circ - T \Delta_r S^\circ$ calculation.

For example, $\Delta_f G^\circ$ for anhydrite can be calculated from

$$
\Delta_f G^\circ_{CaSO_4} = \Delta_f H^\circ_{CaSO_4} - T \Delta_f S^\circ_{CaSO_4}
$$

where $\Delta_f S^\circ_{CaSO_4}$ is

$$
\Delta_f S^\circ_{CaSO_4} = S^\circ_{CaSO_4} - S^\circ_{Ca} - S^\circ_S - 2 S^\circ_{O_2}
$$

Don't forget that absolute entropies are obtainable for the elements just as well as for compounds, and these numbers are available in tables of data, such as Appendix B. These numbers are

| Substance | $S^\circ$, $J \, mol^{-1} K^{-1}$ |
|-----------|-----------|
| $CaSO_4(s)$ | 106.7 |
| $Ca(s)$ | 41.42 |
| $S(s)$ | 31.80 |
| $O_2(g)$ | 205.138 |

So

$$
\begin{aligned}
\Delta_f S^\circ_{CaSO_4} &= 106.7 - 41.42 - 31.80 - 2 \times 205.138 \\
&= -376.796 \, J \, mol^{-1} K^{-1}
\end{aligned}
$$

Therefore, the Gibbs free energy of formation of anhydrite is

$$
\begin{aligned}
\Delta_f G^\circ_{CaSO_4} &= \Delta_f H^\circ_{CaSO_4} - T \Delta_f S^\circ_{CaSO_4} \\
&= -1434110 - 298.15(-376.796) \\
&= -1321768 \, J \\
&= -1321.77 \, kJ
\end{aligned}
$$

which is the number for $\Delta_f G°$ in Appendix B ($-1321.79$ kJ), within the limits of accuracy of the data.

The calculation for determining whether gypsum or anhydrite is stable is therefore a little easier—we just look up the $\Delta_f G°$ numbers instead of both the $\Delta_f H°$ and $S°$ numbers. Thus

$$
\begin{aligned}
\Delta_r G° &= \Delta_f G°_{CaSO_4 \cdot 2H_2O} - \Delta_f G°_{CaSO_4} - 2\,\Delta_f G°_{H_2O} \\
&= -1797280 - (-1321790) - 2(-237129) \\
&= -1232\,J
\end{aligned}
\tag{5.8}
$$

which is what we got before ($-1231$ J), within the limits of accuracy of the tabulated data.

It is important to realize that in equation (5.8), although we use the free energies of formation, the $G$ values for the elements all cancel out, and what we calculate is the difference between the absolute free energies of the compounds in the reaction. Free energy and enthalpy are similar in this respect [see equation (5.7)].

### 5.4.4   An Aqueous Organic Example

To emphasize that our model is just as useful for organic or biochemical processes as for mineralogical ones, let's take another look at the reaction involving amino acids we considered in §2.6.1. Equations (2.2) and (2.4) are, to repeat,

$$
C_8H_{16}N_2O_3(aq) + H_2O(l) = C_6H_{13}NO_2(aq) + C_2H_5NO_2(aq) \tag{5.9}
$$

$$
C_6H_{13}NO_2(aq) + C_2H_5NO_2(aq) = 2H_2(g) + 2NH_3(g) + 4H_2O(l) + 8C_{graphite} \tag{5.10}
$$

From the tables in Appendix C, we find the following properties:

| Substance | Formula | $\Delta_f G°$, J mol$^{-1}$ |
|---|---|---|
| leucine | $C_6H_{13}NO_2(aq)$ | $-343088$ |
| glycine | $C_2H_5NO_2(aq)$ | $-370778$ |
| leucylglycine | $C_8H_{16}N_2O_3(aq)$ | $-462834$ |
| hydrogen | $H_2(g)$ | $0$ |
| ammonia | $NH_3(g)$ | $-16450$ |
| water | $H_2O(l)$ | $-237129$ |
| graphite | $C(s)$ | $0$ |

Therefore for reaction (5.9),

$$
\begin{aligned}
\Delta_r G° &= \Delta_f G°_{leucine} + \Delta_f G°_{glycine} - \Delta_f G°_{leucylglycine} - \Delta_f G°_{water} \\
&= -343088 - 370778 - (-462834) - (-237129) \\
&= -13903\,J
\end{aligned}
$$

and for reaction (5.10)

$$
\begin{aligned}
\Delta_r G^\circ &= 2\,\Delta_f G^\circ_{H_2(g)} + 2\,\Delta_f G^\circ_{NH_3(g)} + 4\,\Delta_f G^\circ_{H_2O(l)} + 8\,\Delta_f G^\circ_{C(s)} \\
&\quad - \Delta_f G^\circ_{leucine} - \Delta_f G^\circ_{glycine} \\
&= 2\,(0) + 2\,(-16450) + 4\,(-237129) + 8\,(0) \\
&\quad - (-343088) - (-370778) \\
&= -267550\,J
\end{aligned}
$$

Thus we see that, as illustrated in Figure 2.9, both reactions have a negative "chemical energy," or $\Delta_r G^\circ$. However, to say any more about these reactions, we must emphasize a factor we have not yet mentioned, and which we cannot develop fully until Chapter 8 (§8.6).

## The Problem With Solutions

It matters not a bit whether the substances we consider are organic or inorganic, stable or metastable, as long as we have data for them. But it matters a great deal whether they are pure substances (such as gypsum, quartz, diamond, liquid water, etc.) or are dissolved in some solvent, as with all substances designated $(aq)$ in the tables. The problem is that the Gibbs energy (and all other properties) of a pure substance is a fixed and known quantity, but the Gibbs energy of a substance in solution depends on its concentration. The tabulated values of $\Delta_f G^\circ$ for $(aq)$ substances are for one particular standard concentration. Therefore, although we have calculated a negative $\Delta_r G^\circ$ for our two reactions above, they both involve at least some dissolved substances and, therefore, the conclusion that the reactions should proceed spontaneously applies only when all the $(aq)$ substances have the standard concentrations. We look more carefully at these problems in Chapters 7 and 8.

## Enzymes as Catalysts

One more thing to note about chemical reactions is that living organisms have evolved mechanisms involving enzymes that overcome the energy barriers between reactants and products for reactions required by the organism. Such reactions, therefore, proceed easily and quickly, whereas in the inorganic world, diamond persists forever in its metastable state. No organism needs to change diamond to graphite, so no enzymes exist for this reaction. Living organisms also have mechanisms that drive some reactions "uphill," or against the free energy gradient. Thus peptide bonds are formed in organisms, as well as broken. The energy required to do this is obtained ultimately from the sun, but the exact mechanisms are complex. The study of such reactions forms a large part of the science of biochemistry.

## 5.5   THE MEANING OF ENTROPY

As long as we are dealing with pure compounds, we have answered just about all our questions. We have an energy parameter, the Gibbs free energy, which always decreases in spontaneous reactions at a given $T$ and $P$, and we know how to measure this energy term—calorimetry. We know, however, that this energy term, $\Delta G$, is made up partly of a fairly comprehensible term $\Delta H$, which is just a heat flow term, and another term $\Delta S$, which is more mysterious. All we know about this one is that we defined the Second Law such that the entropy always increases in spontaneous reactions in isolated systems. The entropy is not itself an energy term, but the product of $T$ and $S$, or $T$ and $\Delta S$, is an energy term.

But what *is* entropy? We have avoided this topic until now. If we had to rely on classical thermodynamics, we would know little more than we have already said about entropy. It is a parameter, with a method of measurement, which increases in spontaneous processes, *even when no energy changes are possible*, that is, in isolated systems. We would also notice, after measuring the entropy of many substances, that the entropies of gases are relatively large, those of solids relatively small, and those of liquids somewhere in between, but we would probably not have any mental picture of what entropy represents physically.

If you look at some processes that are quite irreversible but that involve little or no energy change, such as the mixing of two gases or the spreading of a colored dye in water, you observe a driving force for processes that is quite different from the energy-drop paradigm we have been pursuing (the ball rolling downhill). There is no energy drop when gases mix, but they do so invariably and irreversibly, and this shows that another driving force for reactions is *mixing*, or an increase in "mixed-upness," which will take place if it is possible. If you think about gases as collections of countless tiny molecules zipping around with the speed of rifle bullets, but with quite a lot of space between them, you realize if two different gases are brought together, it is no more difficult to understand why they will always mix together than it is to understand why a ball will roll downhill. But this mixing process involves no energy change (at least for ideal gases), so the First Law of thermodynamics is powerless—it cannot be the basis for a thermodynamic explanation. The Second Law and entropy do provide it. Entropy can be thought of as a *degree of mixed-upness*, and increasing the randomness or mixed-upness of systems is *one* of the driving forces for spontaneous reactions.

The confusion arises because it is not the only one—the ball rolling down the hill (energy decrease) is also one. It is the two together that provide the complete answer. In some processes energy decrease is the dominant factor, and in others, mixing or entropy increase is the dominant factor. The two are brought together in the Gibbs energy equation

$$\Delta G_{T,P} = \Delta H - T \Delta S \tag{5.1}$$

es as well (warming the cards, warming the gases).

s the fact that the concept of entropy transcends the lim-
cs into statistical mechanics, information theory, and even
interest in it as universal as it is. We pursue this relation-
y and probability a little further in Chapter 12 (§12.4).


Y


s the transition from somewhat abstract theory to usable
energy and enthalpy are forms of energy, closely related
deep pond," U (Figure 3.2). A common way to measure this
calorimetry, an exacting experimental procedure in which
is carried out, such as dissolution of a solid phase, and the
bsorbed is measured. If the reaction occurred at constant
ed $q$ is a $\Delta H$, and if not, it is fairly easily converted into a
be measured by calorimetry, though of a different type,
thalpy and entropy measurements gives $\Delta G$ numbers.

e are no absolute values for $G$ or $H$ is a decided nuisance,
y around this is by using the "formation from the ele-
his means that for every compound, calorimetry is used
$\Delta G$ for the reaction in which a compound is formed from
its most stable form, and these quantities are given the
$G°$, where the subscript $f$ stands for "formation from the
uperscript ° means all substances are in their pure ref-
$H$ and $\Delta G$ for any other reaction can then be found by
° and $\Delta_f G°$ terms for compounds.

d quite a few subscripts and superscripts all at once here,
g. The logical relationships among these terms is shown
as an example. The most general term for a change in $G$
ers to any change in the Gibbs energy of any system be-
states (stable or metastable), not necessarily associated
on. Many of the equations in Chapter 4 use this "plain
the discussion is very general. A special case is the $\Delta G$
and reactants of a chemical reaction, called $\Delta_r G$, so this
the more general term $\Delta G$. A special kind of chemical
pure compounds, whose thermodynamic parameters can
so a subset of all $\Delta_r G$ values can be called $\Delta_r G°$, to in-
and reactants are in their pure reference states. Finally,
is the reaction in which a compound is formed from its
ure reference states, and this is called $\Delta_f G°$.

Here there are two factors that to
tive or negative. One is $\Delta H (= \Delta U$
heat and work in the process repr
change due to the mixing factor
ative (the process is exothermic)
and so both factors are negative (
ative. In other reactions, $\Delta H$ is 
ice), but the $\Delta S$ term is sufficien
$\Delta G$ is negative in spite of the pc
temperatures, when $T$ is large. T
is that whether or not any partic
two competing factors. System
also want to maximize their mi:
decides the issue.

## 5.5.1 Entropy in Statistic

You may not have noticed, but
thermodynamics for a moment
we referred to gases as collecti
In Chapter 2 we defined the te
atoms or molecules—tradition
about such things. You make 
substances of known composi
matical framework, and that's
$G$ of water is less than the $G$ c
entropy, you must look outsic

Statistical mechanics cons
and has developed another v
modynamics. In this view, e
of events occurring within a f
tion has led to the widespreac
thermodynamic concerns. Fo
increase its entropy, and tidy
are analogous to the mixing (
cards are not warmed or coo
equations for the calculation
justification for equating en
the use (and misuse) of this
of the most common misusε
always increases, period. T
verse is steadily increasing, 
within the universe are spo:
in. your refrigerator. Other
affected not only by mixin₂

but by energy chan;

Nevertheless, it
its of thermodynam
daily life that make:
ship between entroj

## 5.6 SUMMAR

This chapter contaiı
numbers. The Gibbs
to the "energy in the
kind of energy is by
some kind of reactio
heat ($q$) released or a
pressure, the measuı
$\Delta H$. Entropy can als«
and combining the eı

The fact that ther
but a very simple w
ments" convention. 
to measure $\Delta H$ and $\Delta$
its elements, each in
symbols $\Delta_f H°$ or $\Delta_f$
elements," and the s
erence states. The $\Delta$
combining these $\Delta_f H$

We have introduce
which can be confusiı
in Figure 5.7, using $G$
is simply $\Delta G$. This re
tween two equilibriuı
with a chemical react
$\Delta$" notation, because 
between the products
represents a subset oı
reaction involves only 
be found in tables, anc
dicate that all product
a special case of $\Delta_r G°$
elements, all in their p

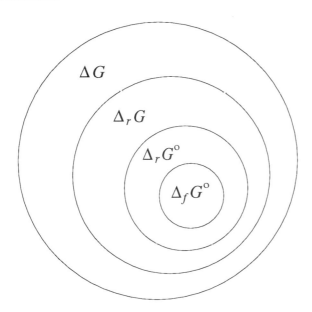

Figure 5.7: The hierarchy of $\Delta G$ terms.

# PROBLEMS

1. Calculate the entropy of formation from the elements of anorthite ($CaAl_2Si_2O_8$). Combine this with the enthalpy of formation from the tables to calculate the Gibbs free energy of anorthite. Compare with the value in the tables.

2. Calculate $\Delta_r H°$ and $\Delta_r G°$ for the reaction

$$NaAlSiO_4(nepheline) + 2\,SiO_2(quartz) = NaAlSi_3O_8(low\ albite)$$

   at 25°C, 1 bar. Is the reaction endothermic or exothermic? Which way would the reaction go under standard conditions in the absence of kinetic barriers? What actually happens if you put quartz and nepheline together?

3. (a). Calculate the value of $R$ in $J\,mol^{-1}\,K^{-1}$ from the ideal gas equation ($PV = nRT$).
   (b). Use the dimensions (Appendix A) of energy and pressure to show that $J\,bar^{-1}$ is a volume term, and calculate the conversion factor from $J\,bar^{-1}$ to $cm^3$.

4. Is magnesite ($MgCO_3$) or nesquehonite more stable in water?

5. There are six naturally occurring oxides and hydroxides of aluminum listed in Appendix B, but only four of these have complete data (corundum, boehmite, diaspore, gibbsite). Note that the compositions of these

phases differ only by the number of $H_2O$, so it is relatively easy to write reactions between them. By writing balanced reactions between these four phases, determine which one is most stable in water.

6. The origin of red-bed sandstones, in which the grains are coated with minute amounts of hematite, has long been controversial. A key question in the controversy is whether hematite is stable in water at low temperatures. Calculate whether goethite or hematite is stable in the presence of water at 25°C.

7. If you look more carefully into the origins of the Third Law (§5.2), you find that what it really says is that the entropy of all perfectly crystalline substances *is the same* at absolute zero temperature, not necessarily with a value of zero $J\,mol^{-1}\,K^{-1}$. We essentially *assign* a value of zero to this entropy. Show that this has no effect on any practical applications.

# 6

# SOME SIMPLE APPLICATIONS

We now know how to determine in which direction any chemical reaction will proceed at a given temperature and pressure, at least when all the products and reactants are pure phases. When even one of the products or reactants is a solute, that is, part of a solution, we would be stuck because we would not be able to put it into our calorimeters, and so we would not as yet be able to get data for it. We will start considering this problem in the next chapter. Before going on, however, we should explore some relationships between the concepts we have defined so far, so as to make sure we fully understand them.

## 6.1  ADDITIVITY OF STATE VARIABLES

At several points in our discussions so far, we have mentioned or assumed that we can add and subtract state variables such as $\Delta H$, for example as shown in Figure 5.3. This is perhaps obvious, but it is so fundamental that we will emphasize it here.

We said in Chapter 2 (§2.3) that a state variable is a property of a system that has a fixed value when the system is at equilibrium, whether we know the value of that property or not. For example, a mole of water at 25°C, 1 atm has a fixed but unknown enthalpy $H$, a fixed enthalpy of formation $\Delta_f H° = -285.830$ kJ mol$^{-1}$, and a fixed entropy $S° = 69.91$ J mol$^{-1}$. We also said that this means that the changes in these properties between equilibrium states depends only on the equilibrium states, and not on what happens between the time the system leaves one equilibrium state and the time it settles down in its new equilibrium state. Therefore, if two different reactions produce the same compound, we can subtract the $\Delta_r H°$, $\Delta_r G°$, $\Delta_r S°$, and so on, of these reactions to get the $\Delta$ of the combined reaction, and the properties of that compound will cancel out. For example, carbon dioxide, $CO_2$, might be produced from the oxidation of either graphite or carbon monoxide, CO:

$$C + O_2 = CO_2; \quad \Delta_r G° = -394.359 \text{ kJ}$$

85

$$CO + \tfrac{1}{2}O_2 = CO_2; \quad \Delta_r G° = -257.191 \text{ kJ}$$

Subtracting the reactions and the $\Delta_r G°$ values (reverse the second reaction, change the sign of $\Delta_r G°$, and add), we get

$$C + \tfrac{1}{2}O_2 = CO; \quad \Delta_r G° = -137.168 \text{ kJ}$$

Thus we get the properties of a reaction that is impossible to carry out experimentally from two reactions that are relatively easy do experimentally.

## 6.2  SIMPLE PHASE DIAGRAMS

The reason we are interested in knowing $\Delta_r G$ for reactions is that we can then tell which way the reaction will go, or which side is more stable at one particular $T$ and $P$. If we know how $\Delta_r G$ varies with $T$ and $P$, we might find that under some conditions $\Delta_r G$ changes sign, so that the other side is more stable. This implies that there is a boundary between regions of $T$ and $P$, with one side of the reaction stable on one side of the boundary, and the other side of the reaction stable on the other side of the boundary. A phase diagram shows which phases are stable as a function of $T$, $P$, composition, or other variables.

For example, calcium carbonate ($CaCO_3$) has two polymorphs, calcite and aragonite. Their properties (from Appendix B) are

| Formula | Form | $\Delta_f H°$ | $\Delta_f G°$ | $S°$ | $V°$ |
|---------|------|------|------|------|------|
| | | kJ mol$^{-1}$ | | J mol$^{-1}$ K$^{-1}$ | cm$^3$ mol$^{-1}$ |
| $CaCO_3$ | calcite | $-1206.92$ | $-1128.79$ | 92.9 | 36.934 |
| $CaCO_3$ | aragonite | $-1207.13$ | $-1127.75$ | 88.7 | 34.150 |

Because $\Delta_f G°_{calcite} < \Delta_f G°_{aragonite}$, we conclude immediately that calcite is the stable form of $CaCO_3$ at 25°C, 1 bar, and that aragonite is a metastable form. But what about other temperatures and pressures? Is aragonite stable at high temperature? At high pressure? How can we tell?

### 6.2.1  LeChatelier's Principle

When looking at thermodynamic data, or the results of some thermodynamic calculation, it is always a good idea to ask yourself if it makes sense, if it is reasonable. To some extent this is a matter of experience, but in another way, "making sense" means obeying LeChatelier's Principle. This simply says that *if a change is made to a system, the system will respond such as to absorb the force causing the change.* For example, if the pressure on a system is raised, the system will respond by lowering its volume, that is, by being compressed. Systems never expand as a result of increased pressure. The result of a change in temperature is less obvious, though equally certain. If the temperature of a system is raised, the enthalpy and the entropy of the system will both increase.

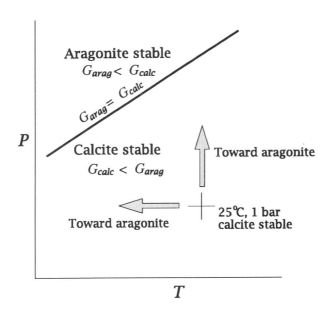

Figure 6.1: The form of the calcite-aragonite phase diagram deduced from LeChatelier's principle.

This is because of equations (4.20) and (5.3), which show that the temperature derivative of each is a simple function of $C_p$, the heat capacity, which is always positive for pure compounds.

Therefore by looking at $V°$ and $\Delta_f H°$ or $S°$ for calcite and aragonite, and assuming that the relative magnitudes of these properties do not change much with $T$ and $P$, we can tell something about their relative positions on the phase diagram. We note that $V_{aragonite} < V_{calcite}$; therefore, increasing the pressure on calcite should favor the formation of aragonite. Also, $\Delta_f H°$ and $S°$ for calcite are greater than the values for aragonite, and so raising the temperature of calcite will *not* favor the formation of aragonite. In other words, *lowering* the temperature of calcite should favor the formation of aragonite. If the stability field of aragonite lies somewhere at higher pressure and lower temperature than 25°C, 1 bar, the boundary between the two phases must have a positive slope, as shown in Figure 6.1. This is the common case for phase boundaries; it is normal for the high-pressure, lower volume side to be the lower enthalpy, lower entropy side. The most common exception to this is the ice-water transition, as shown in Figure 3.1.

In Figure 6.1 we see that a phase diagram is a kind of free energy map — it shows a $T$-$P$ region where calcite is stable ($G_{calcite} < G_{aragonite}$), and another where aragonite is stable ($G_{aragonite} < G_{calcite}$). These two regions are necessarily separated by a line where $G_{aragonite} = G_{calcite}$, the phase boundary.

## 6.2.2   The Effect of Pressure on $\Delta_r G°$

Having figured out the relationship between calcite and aragonite qualitatively, the next step is to define the stability field of aragonite, that is, to calculate the position of the phase boundary. This should be possible, because we know that

$$\partial G / \partial P = V \tag{4.10}$$

and thus

$$\partial \Delta G / \partial P = \Delta V$$

$\Delta G$ and $\Delta V$ refer to the difference in $G$ and $V$ between any two equilibrium states. In this case we are dealing with a chemical reaction between two compounds in their pure states, so we can also write

$$\partial \Delta_r G° / \partial P = \Delta_r V°$$

Integrating this equation between 1 bar and some higher pressure $P$, we have

$$\Delta_r G°_P - \Delta_r G°_{1\ bar} = \int_{1\ bar}^{P} \Delta_r V° \, dP$$

and if we assume that $\Delta_r V°$ is a constant, this becomes

$$\begin{aligned}
\Delta_r G°_P - \Delta_r G°_{1\ bar} &= \Delta_r V° \int_{1\ bar}^{P} dP \\
&= \Delta_r V° (P - 1)
\end{aligned}$$

We could use this to evaluate $\Delta_r G°_P$ at any chosen value of $P$. However, we are particularly interested in a value of $\Delta_r G°_P = \Delta_r G°_{P_{eqbm}} = 0$, that is, on the phase boundary. We know the values of

$$\begin{aligned}
\Delta_r G°_{1\ bar} &= \Delta_f G°_{aragonite} - \Delta_f G°_{calcite} \\
&= -1127.75 - (-1128.79) \\
&= 1.04 \text{ kJ} \\
&= 1040 \text{ J}
\end{aligned}$$

and

$$\begin{aligned}
\Delta_r V° &= V°_{aragonite} - V°_{calcite} \\
&= 34.150 - 36.934 \\
&= -2.784 \text{ cm}^3
\end{aligned}$$

So we can solve the equation for $P_{eqbm}$, the pressure of the calcite–aragonite equilibrium at 25°C.[1]

However, there is one little problem.

---

[1]Note that although $\Delta_f G°$ has units of J mol$^{-1}$, when we write a reaction and take a difference in $\Delta_f G°$, the units for the difference are simply joules, and similarly for other parameters, as noted in the Example in Chapter 2.

### The Units of Volume

Volumes are generally measured in cubic centimeters, milliliters, liters, and so on. But if you look at an equation such as

$$w = -P\,\Delta V$$

you see that we have a problem with our units. Work ($w$) and $P\Delta V$ are obviously energy terms (joules), but the product of $P$ in bars and $\Delta V$ in $cm^3$ is not joules. We must always convert our volumes to joules bar$^{-1}$, so that the product of $P$ and $V$ or $\Delta V$ is joules. The conversion factor (Appendix A) is

$$1\,cm^3 = 0.10\,J\,bar^{-1}$$

so now our $\Delta_r V^\circ$ is $-2.784 \times 0.1 = -0.2784\,J\,bar^{-1}$.

Now we can solve for pressure $P_{eqbm}$:

$$
\begin{aligned}
\Delta_r G^\circ_{P_{eqbm}} - \Delta_r G^\circ_{1\,bar} &= \Delta_r V^\circ (P_{eqbm} - 1) \\
0 - 1040 &= -0.2784(P_{eqbm} - 1) \\
P_{eqbm} &= 3737\ \text{bar}
\end{aligned}
$$

The relationship between $G$ and $P$ in this calculation is shown in Figure 6.2. This gives us one point on the calcite-aragonite phase boundary. We also know that the boundary has a positive slope, and so we could sketch a diagram that would be approximately right, but we really need one more piece of information—either another point on the boundary or its slope.

## 6.3 THE SLOPE OF PHASE BOUNDARIES

The phase boundary is the locus of $T$ and $P$ conditions where $\Delta_r G = 0$, i.e., where

$$G_{calcite} = G_{aragonite} \tag{6.1}$$

It follows that on the boundary,

$$dG_{calcite} = dG_{aragonite} \tag{6.2}$$

This simply says that as you move along the boundary, the change in $G_{calcite}$ has to be the same as the change in $G_{aragonite}$; otherwise you won't stay on the boundary. From equation (4.8) we have

$$dG = -S\,dT + V\,dP \tag{4.8}$$

This applies to each mineral, and combining with (6.2) gives

$$-S_{calcite}dT + V_{calcite}dP = -S_{aragonite}dT + V_{aragonite}dP$$

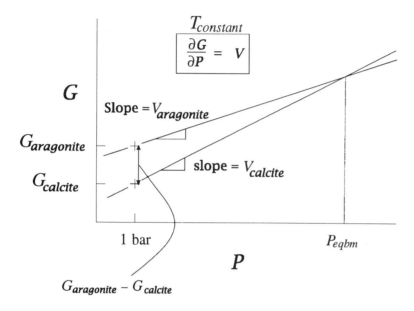

Figure 6.2:  The relationship between $G$ and $P$.  Note that we don't know individual $G$ values, so there are no numbers on the $y$-axis.  We do know $G_{calcite} - G_{aragonite}$ and the slopes of the lines (the molar volumes), and this is sufficient to solve for $P_{eqbm}$.  At $P_{eqbm}$, $G_{calcite} = G_{aragonite}$, the two phases can coexist, and we have a phase boundary.

Rearranging this gives

$$\frac{dP}{dT} = \frac{(S_{calcite} - S_{aragonite})}{(V_{calcite} - V_{aragonite})}$$

or, for any reaction

$$\frac{dP}{dT} = \frac{\Delta_r S}{\Delta_r V} \tag{6.3}$$

which gives the slope of an equilibrium phase boundary in terms of the entropy and volume changes between the phases involved in the reaction. This is called the Clapeyron equation.

Equation (5.1) says

$$\Delta G_{T,P} = \Delta H - T \, \Delta S \tag{5.1}$$

This applies to any change between two equilibrium states at the same $T$ and $P$. If those two equilibrium states have the same value of $G$, such as calcite and aragonite do on their phase boundary (6.1), then $\Delta G_{T,P} = 0$, and

$$\Delta H = T \, \Delta S \tag{6.4}$$

or

$$\frac{\Delta H}{T} = \Delta S \tag{6.5}$$

This is a useful relationship for any phase boundary, which is the usual place to find $\Delta G_{T,P} = 0$. This gives an alternative form of the Clapeyron equation,

$$\frac{dP}{dT} = \frac{\Delta H}{T \, \Delta V} \tag{6.6}$$

## 6.3.1 The Slope of the Calcite–Aragonite Boundary

We have one point on the calcite-aragonite boundary at 3737 bar, 25°C. If we assume that the $\Delta_r S$ and the $\Delta_r V$ at this $P$ and $T$ are the same as those at 1 bar, 25°C, we can calculate the slope from the data in our tables. Thus

$$
\begin{aligned}
\Delta_r S &= S_{aragonite} - S_{calcite} \\
&= 88.7 - 92.9 \\
&= -4.2 \quad J K^{-1}
\end{aligned}
$$

and

$$
\begin{aligned}
\Delta_r V &= V_{aragonite} - V_{calcite} \\
&= 34.150 - 36.934 \\
&= -2.784 \quad cm^3 \\
&= -0.2784 \quad J bar^{-1}
\end{aligned}
$$

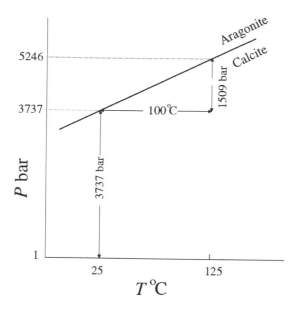

Figure 6.3: The calcite–aragonite phase diagram.

Therefore

$$\frac{dP}{dT} = \frac{\Delta_r S}{\Delta_r V}$$
$$= \frac{-4.2}{-0.2784}$$
$$= 15.09 \text{ bar/}°C$$

Therefore, to get another point on the calcite–aragonite phase boundary, we simply choose an arbitrary temperature increment, say 100°C, calculate the corresponding pressure increment, $100 \times 15.09 = 1509$ bar, and add these increments to our first point. We now have a second point at 125°C, 3737 + 1509 = 5246 bar, and we can plot the boundary as in Figure 6.3.

Keep in mind that we have assumed that the $\Delta_r S$ and $\Delta_r V$ from the tables are unchanged at all temperatures and pressures, that is, that they are constants. This is quite a good approximation for a reaction involving only solid phases such as this one, but you would not use it for reactions involving liquids, gases, or solutes. In general, all thermodynamic parameters do vary with $T$ and $P$, so phase boundaries are in principle curved and not straight as we have assumed. However, the amount of curvature is quite small in some cases, such as this one.

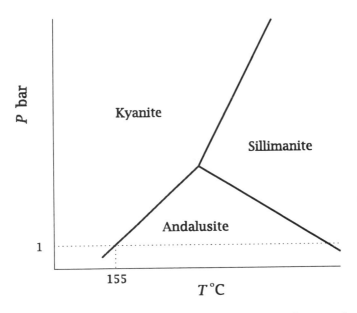

Figure 6.4: The phase diagram for the aluminum silicate polymorphs.

# 6.4 ANOTHER EXAMPLE

### 6.4.1 The Effect of Temperature on $\Delta_r G°$

To illustrate the effect of temperature on $\Delta_r G°$, we could continue with the calcite-aragonite case and try to calculate the temperature where the phase boundary crosses the 1 bar pressure line (Figure 6.3). Unfortunately, this turns out to be below absolute zero, so it is not a very useful example. As another case let's consider the polymorphs of $Al_2SiO_5$. There are three of these, kyanite, andalusite, and sillimanite. Therefore there are three two-phase boundaries, and these three boundaries meet at a single point, where $G_{kyanite} = G_{andalusite} = G_{sillimanite}$ as shown in Figure 6.4. These minerals, which form quite commonly in rocks subjected to high temperatures and pressures in the Earth's crust, are of special interest to geologists who study these rocks because the "triple point," the point where the three phase boundaries meet, is in the middle of a rather common range of $T$-$P$ conditions. If a rock contains one of these minerals, the geologist immediately has a general idea of the $T$ and $P$ conditions at the time the rock formed.

According to Figure 6.4, the kyanite-andalusite boundary crosses the 1 bar line at some elevated temperature. We should be able to calculate what this is by methods perfectly analogous to those we used for calcite-aragonite. The data we need (from Appendix B) are:

| Formula | Form | $\Delta_f H°$ | $\Delta_f G°$ | $S°$ | $V°$ |
| | | kJ mol$^{-1}$ | | J mol$^{-1}$ K$^{-1}$ | cm$^3$ mol$^{-1}$ |
|---|---|---|---|---|---|
| Al$_2$SiO$_5$ | kyanite | −2594.29 | −2443.88 | 83.81 | 44.09 |
| Al$_2$SiO$_5$ | andalusite | −2590.27 | −2442.66 | 93.22 | 51.53 |

First, we note that all seems to be well to start with, in that there is no conflict between the data and Figure 6.4. Kyanite has the lower value of $\Delta_f G°$, and so it should be the stable phase at 25°C, 1 bar, as shown in the diagram. The values of $\Delta_r S°$ and $\Delta_r V°$ would indicate that kyanite is the high-pressure phase, and that the kyanite–andalusite boundary has a positive slope, also as shown by the diagram.

To calculate the temperature of the kyanite–andalusite boundary at 1 bar, we start with equation (4.9),

$$(\partial G/\partial T)_P = -S \qquad (4.9)$$

from which we can write immediately[2]

$$(\partial \Delta_r G°/\partial T)_P = -\Delta_r S°$$

Integrating this from 298.15 K to some higher temperature $T$, we get

$$\Delta_r G_T° - \Delta_r G_{298}° = \int_{298}^{T} -\Delta_r S° \, dT \qquad (6.7)$$

and if we assume that $\Delta_r S°$ is a constant, this becomes

$$\begin{aligned} \Delta_r G_T° - \Delta_r G_{298}° &= -\Delta_r S_{298}° \int_{298}^{T} dT \\ &= -\Delta_r S_{298}° (T - 298.15) \qquad (6.8) \end{aligned}$$

Now if we let $\Delta_r G_T° = 0$, $T$ becomes $T_{eqbm}$, and we can solve for this. From the tables, $\Delta_r G_{298}° = 1220$ J and $\Delta_r S_{298}° = 9.41$ J K$^{-1}$, so

$$\Delta_r G_T° - \Delta_r G_{298}° = -\Delta_r S_{298}° \int_{298}^{T_{eqbm}} dT$$

---

[2]It is not immediately clear to many students why, if $(\partial G/\partial T)_P = -S$, we can "write immediately" $(\partial \Delta_r G°/\partial T)_P = -\Delta_r S°$, that is, why we can just stick in a $\Delta$ whenever we wish. It is because the derivative relationship can be applied to all terms of any balanced reaction. For example, if the reaction is A + 2B = C (e.g., reaction 5.6),

$$\begin{aligned} \Delta_r G° &= \Delta_f G_C° - \Delta_f G_A° - 2\Delta_f G_B° \\ &= G_C° - G_A° - 2G_B° \end{aligned}$$

and

$$\begin{aligned} (\partial \Delta_r G°/\partial T) &= (\partial G_C°/\partial T) - (\partial G_A°/\partial T) - 2(\partial G_B°/\partial T) \\ &= -S_C° + S_A° + 2S_B° \\ &= -(S_C° - S_A° - 2S_B°) \\ &= -\Delta_r S° \end{aligned}$$

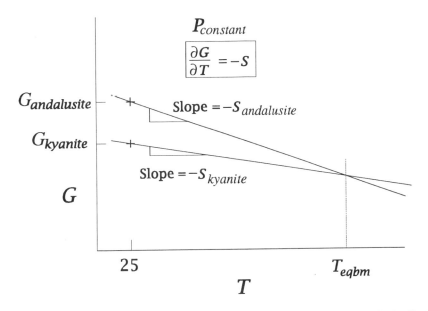

Figure 6.5: The relationship between $G$ and $T$. Note the general similarity to Figure 6.2, with the exception that the slopes are negative.

$$
\begin{aligned}
0 - 1220 &= -9.41(T_{eqbm} - 298.15) \\
T_{eqbm} &= 427.8 \text{ K} \\
&= 154.6°\text{C}
\end{aligned}
$$

The relationship between $G$ and $T$ is shown in Figure 6.5. Note that in Figure 6.2, the slope of $G$ vs. $P$ is positive, whereas in Figure 6.5, the slope of $G$ vs. $T$ is negative. This is because for pure substances $V$ is always positive, and $S$ is always positive by virtue of the Third Law. This is only true in general for pure substances; for differences or for solutes, these quantities may be negative, as we will see.

### 6.4.2  A Different Formula for $\Delta_r G_T^\circ$

Another useful way of expressing the effect of temperature on $G$ is given by expanding (6.8). Thus

$$
\begin{aligned}
\Delta_r G_T^\circ - \Delta_r G_{298}^\circ &= -\Delta_r S^\circ(T - 298.15) \\
\Delta_r G_T^\circ - (\Delta_r H_{298}^\circ - 298.15\,\Delta_r S_{298}^\circ) &= -T\,\Delta_r S_{298}^\circ + 298.15\,\Delta_r S_{298}^\circ
\end{aligned}
$$

Collecting and rearranging terms gives

$$
\Delta_r G_T^\circ = \Delta_r H_{298}^\circ - T\,\Delta_r S_{298}^\circ \tag{6.9}
$$

In other words, you can calculate $\Delta_r G°$ at some temperature $T$ using the values of $\Delta_r H°$ and $\Delta_r S°$ at 298.15 K. However, this is subject to the same restriction as before, that both $\Delta_r H°$ and $\Delta_r S°$ are not functions of temperature. Of course, both these terms always *are* functions of temperature, but often this can be neglected without introducing much error, especially if $T$ is not very different from 298 K.

Both (6.8) and (6.9) are therefore approximations, to be used with caution. There are much more accurate formulae available involving the heat capacity (see Appendix C), but these approximations will be sufficient for our purposes.

## 6.5  SUMMARY

The main idea in this chapter is to show how to calculate the effects of temperature and pressure on the Gibbs free energy of solid phases, as well as to emphasize the importance of the condition $\Delta_r G = 0$ for given reactions. We did this by considering simple phase diagrams, of the kind that are useful to geochemists thinking about reactions between minerals in the Earth, but of course the principles are applicable to a much wider variety of situations.

The examples are quite simple in the sense that possible changes in $\Delta_r S°$, $\Delta_r H°$, and $\Delta_r V°$ during changes in $T$ and $P$ are ignored. These changes will be particularly important in reactions involving gases and/or solutes, and this topic is pursued a little further in Appendix C.

## PROBLEMS

1. At what pressure will graphite be converted to diamond at 25°C? At 100°C?

2. Calculate the pressure at which jadeite is in equilibrium with low albite and nepheline at 300°C.

3. At what temperature does gibbsite break down (dehydrate) to form corundum and water? Assume a pressure of 1 bar. Actually, because the temperature is greater than 100°C and liquid water is present, the pressure will be somewhat greater than 1 bar, but this will not greatly affect the free energies.

4. Show that dolomite is stable relative to calcite plus magnesite at 25°C and 1 bar.

5. Above what temperature will $\alpha$-cristobalite become the stable form of $SiO_2$ (i.e., more stable than $\alpha$-quartz) at 1 bar?

6. Calculate the pressure and temperature of the triple point in Figure 6.4.

# 7

# SOLUTIONS AND ACTIVITIES

## 7.1  NO PURE SUBSTANCES

If the world were made of pure substances, our development of the thermodynamic model would now be complete. We have developed a method, based on measurements of heat flow, that enables predictions to be made about which way reactions will go in given circumstances. If our measurements (thermodynamic data) are accurate, and our knowledge of the system sufficient to define the problem, then the prediction will always be correct. But thermodynamic data are not always accurate, and not always available, and reactions may occur that we did not anticipate. Or, more commonly, no reaction at all may occur, even though thermodynamics says that it should. As usual, we find that although our model is relatively simple, and in a sense infallible, the real world is very complex, and the thermodynamic model usually represents only a beginning to our understanding of natural systems.

One of the reasons that the world is so complex is that pure substances are relatively rare, and strictly speaking they are nonexistent (even "pure" substances contain impurities in trace quantities). Most natural substances are composed of several components, and the result is called a *solution*. Therefore, we need to develop a way to deal with components in solution in the same way that we can now deal with pure substances—we have to be able to get numerical values for the free energies, enthalpies, and entropies of components in solutions. We will then be able to predict the outcome of reactions that take place entirely in solution, such as the ionization of acids and bases, and reactions that involve solids and gases as well as dissolved components, such as whether minerals will dissolve or precipitate. Our thermodynamic model will then be complete.

## 7.2   MEASURES OF CONCENTRATION

A number of concentration terms are used in describing solutions, and it is naturally important to be able to change from one to another. Consider a solution containing a number of components, $n_1$ moles of component 1, $n_2$ moles of component 2, and so on. If it is an aqueous solution, then water is one of the components, normally the major component.

### Mole Fraction

The mole fraction of any one component is defined as

$$X_1 = \frac{n_1}{\sum n}$$

$$X_2 = \frac{n_2}{\sum n}$$

and so on, where $\sum n$ is the total number of moles of components, $n_1 + n_2 + n_3 + \cdots$. The mole fraction is very commonly used, especially in theoretical discussions, because it is perfectly general. It is inconvenient for aqueous solutions.

### Molality

The molality ($m$) of component 1 is the number of moles of component 1 ($n_1$) per kilogram of pure solvent, usually water. A kilogram of water has $1000/18.0154 = 55.51$ moles of $H_2O$, so a solution of $n_1$ moles of component 1 in a kilogram of water has a molality $m_1$ and a mole fraction of

$$X_1 = \frac{m_1}{m_1 + 55.51}$$

or

$$m_1 = \frac{55.51 X_1}{1 - X_1}$$

Note that even if the aqueous solution contains several solutes, the molality is the number of moles of one of them in 1000 g of *pure* water.

The molality is commonly used in thermodynamic calculations because it is independent of the temperature and pressure of the solution.

### Molarity

The molarity ($M$) of component 1 is the number of moles of component 1 in 1 liter of solution (not a liter of pure solvent). This is a convenient unit in the laboratory, where solutions are prepared in volumetric flasks. It has the disadvantage that as temperature or pressure changes, the volume of the solution changes but the definition of the liter does not, and the molarity is therefore a function of temperature and pressure. We will not use this unit.

**Parts per Million**

The parts per million (ppm) of a component is the number of grams of that component in a million grams of solution. To convert to or from molality you therefore need to know the molecular weight of the component. For example, an aqueous solution that is $10^{-4}$ molal in NaCl contains $0.0001 \times 58.4428 = 0.005844$ grams of NaCl in $(1000 + 0.005844)$ grams of solution, or about $10^3$ grams of solution. Therefore there would be 5.844 grams of NaCl in $10^6$ grams of solution, or 5.8 ppm. If the solution contains a number of other solutes, they should all be included in the calculation, but it is common practice to ignore all components except the solute of interest and water.

Parts per million are used for trace levels of components. It usually does involve some approximation and has no real advantage over molality other than avoiding small numbers.

# 7.3 THE PROPERTIES OF SOLUTIONS

When we mix two substances together, sometimes they dissolve into one another, like sugar into coffee or alcohol into water, and sometimes they do not, like oil and water. In either case, if we thought about it at all, we would probably expect that the properties of the mixture or solution would be some kind of average of the properties of the two separate substances. This is more or less true for some properties, but decidedly not true for the most important one, free energy.

## 7.3.1 The Volume of Mixing

If two substances are immiscible (they do not dissolve into one another to any appreciable extent, like oil and water), obviously the volume of the two together is simply the sum of the two volumes separately. But if they are completely miscible (they dissolve into one another completely, forming a solution), this may be more or less true, but probably not exactly true. Why?

If you mix white sand and black sand together, there is no interaction or chemical reaction at all between the two kinds of sand, and the volume of the mixture is the same as the two volumes separately. If the volume of the white sand is $V_w$ and the volume of the black sand is $V_b$, the total volume is

$$V = V_w + V_b$$

It's sort of like stacking boxes as is Figure 7.1a. There is no change in total volume just because they are together. However, using total volumes usually turns out to be inconvenient. If the volume per mole of white sand is $V_w$ and of black sand is $V_b$,[1] then the total volume is

$$V = n_w V_w + n_b V_b \tag{7.1}$$

---

[1]Note the change from roman V to italic $V$, as explained in §2.4.1

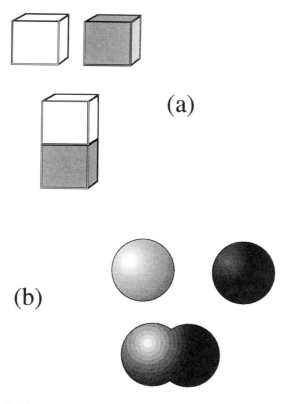

Figure 7.1: (a). There is no volume change when boxes are stacked together—they do not interact. (b). When molecules are mixed together, they may occupy less volume than they did separately.

where $n_w$ and $n_b$ are the number of moles of white and black sand in the mixture. The molar volume is defined as the total volume divided by the number of moles of all components in the system (i.e., the molar volume of pure white sand is therefore $V_w/n_w$); so if the mixture contains $n_w$ moles of white sand and $n_b$ moles of black sand, the total number of moles in the mixture is $n_w + n_b$. Dividing both sides of equation (7.1) by $n_w + n_b$, we get

$$V = X_w V_w + X_b V_b \qquad (7.2)$$

Here, $V$ is the *molar volume* of the mixture and $X$ is the *mole fraction*, where

$$X_w = \frac{n_w}{\sum n}$$
$$= \frac{n_w}{n_w + n_b} \qquad (7.3)$$

and similarly for $X_b$. These equations all simply say that the volume of the mixture is the same as the volume of the two things separately. The introduction of $n$ and $X$ is just to determine how much of each is used. If we plot molar volume against mole fraction of either component sand, we get a straight line (Figure 7.2a), called the *ideal mixing* line.

Clearly these relations do not depend on the grain size of the sands; they depend on the fact that the sands do not react in any way with each other. Each grain of white sand is indifferent to what kind of sand is next to it. Now imagine that the grain size of the sands gets smaller and smaller. Soon they get so small that you can no longer distinguish the colors— the mixture becomes gray. Imagine the grain size continuing to get smaller and smaller—right down to atomic proportions, so that instead of having a mechanical mixture of black and white sand, we have a true solution of black and white atoms. If the black and white atoms continue to have no attraction, repulsion, or chemical reaction with one another, the volume of the two together will continue to be exactly the same as the sum of the two separately. Actually, we have oversimplified a bit—it may be that the white molecules interact with each other even in the pure state, and similarly with the black molecules. If these interactions are very similar in nature, then when they are mixed together the molecules will continue to interact with each other in the same way, and the volumes will be additive. In other words, it is not necessary for there to be no molecular interactions for ideal mixing, only that white molecules react with black molecules in exactly the same way that they do with other white molecules.

But suppose that at this molecular size, white (w) and black (b) particles are attracted to one another more than to others of the same kind, perhaps even forming a new kind of particle (wb). Because of this attraction, the particles will be closer together than they would otherwise be, and the total volume of the mixture will be smaller, as shown in Figure 7.1b, and instead of getting a straight mixing line as in Figure 7.2a, the line is curved downward as in Figure 7.2b. Alternatively, if the white and black particles repel each other, the total volume will be greater, and in Figure 7.2b the curved line for the molar volume

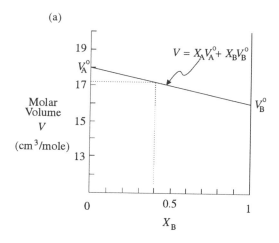

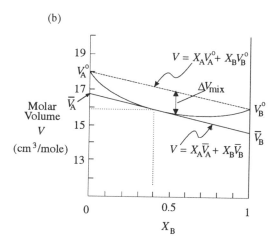

Figure 7.2: (a). The molar volume of solutions of A and B, which are completely miscible but do not attract or repel each other at the molecular level. The molar volume of pure A $(V_A^\circ)$ is 18.0 cm$^3$ mol$^{-1}$ (might be water), and that of pure B $(V_B^\circ)$ is 16.0 cm$^3$ mol$^{-1}$. The molar volume of a solution having $X_B = 0.4$ is $0.6 \times 18.0 + 0.4 \times 16.0 = 17.2$ cm$^3$ mol$^{-1}$. (b). The molar volumes of solutions of A and B, but where A and B attract each other in solution. The molar volume of a solution having $X_B = 0.4$ is reduced to 15.9 cm$^3$ mol$^{-1}$. It may still be calculated, but using $\overline{V}_A$ and $\overline{V}_B$ instead of $V_A^\circ$ and $V_B^\circ$.

of the mixture will lie above the straight line that represents no interaction. The volume change on mixing ($\Delta V_{mix}$, Figure 7.2b) caused by the attraction between A and B is the difference between the straight line and the curved line. The straight line

$$V = X_A V_A^\circ + X_B V_B^\circ \tag{7.4}$$

is called *ideal mixing* and is rarely observed. The curved line represents non-ideal mixing, the general case.

### 7.3.2  Two Kinds of Ideality

If the two components mix ideally as described above, their volumes are additive, and we have an *ideal solution*. Ideal cases are commonly used in thermodynamics because they provide a simple model to which real cases can be compared. Unfortunately, there are actually two kinds of ideal solution, and both are commonly used. We will not pursue this topic at all; we need only to mention the two types, because they result in two kinds of activity coefficients that we will introduce later.

The kind of ideal solution we have just described is called a Raoultian solution, or one that obeys Raoult's Law. In this kind of solution, a molecule of A "sees" only other molecules of A around it, even if some of those molecules are B. Molecule A interacts with B in the same way that it interacts with other A molecules, so it does not behave any differently when B is introduced. Since no two different molecules are exactly alike, this is a hypothetical situation, but it nevertheless serves as a very useful *model* that real solutions are compared to.

Another kind of ideal solution is called a Henryan solution, or one that obeys Henry's Law. In this kind of ideality, a molecule of A "sees" only molecules of B, no matter what the concentration of A. Molecules of A and B can interact in any way, strongly or weakly, attractive or repulsive, but whatever it is, it is unchanged by changing concentrations. If a molecule of A is to "see" (be affected by) only molecules of B, it must be in very low concentration, or very dilute; otherwise, molecules of A will be closely spaced and begin to affect each other. This kind of ideality is therefore what happens at "infinite dilution" (see Appendix C).

### 7.3.3  Partial Molar Volumes

Now suppose in our mixture of white and black particles that attract each other, that we are not satisfied to have the total volume or the molar volume of the mixture as a whole. We would like to know the volume of each component in the mixture, not just the combined volume. But how can this be done, when each is dispersed at the molecular level and is interacting strongly with another component? Simple. Just draw the tangent to the molar volume curve at the composition you are interested in. The intercepts of this tangent give the volumes of each component in the solution, called *partial molar volumes*, which are combined to give the total molar volume in exactly the same way as the black and white sands in equation (7.2) and Figure 7.2a.

Looking at partial molar volumes in this way, they seem to be just a sort of geometrical construct. They are defined such that they can be substituted for $V_A$ and $V_B$ in equation (7.2) in cases where mixing results in a curved line for the molar volumes; thus

$$V = X_A \overline{V}_A + X_B \overline{V}_B \qquad \text{(Figure 7.2b)} \qquad\qquad (7.5)$$

In Figure 7.2b we have shown a case where A and B are attracted to each other, and their partial molar volumes are both *less* than the volumes of the pure components ($\overline{V}_A < V_A^\circ$). If A and B repelled one another, the mixing line would lie above the straight line and the partial molar volumes would be *larger* than the pure volumes. There is no general rule for the shapes and positions of these mixing curves; they must be measured experimentally. This would be done by density measurements in the case of volume, and calorimetry in the case of enthalpy and entropy. It is quite possible for the mixing curve to be shaped such that in a certain range of composition one of the tangent intercepts is at less than zero volume—a negative partial molar volume. This is why some of the tabulated thermodynamic parameters in Appendix B are negative for some solute components. It is, of course, not possible for pure components to have a negative volume.

### The Room Analogy

But there is another way of looking at partial molar volumes which shows that they really are the volume of a mole of each component in solution. Just for a change we will switch from components A and B to a solution of salt (NaCl) in water. Consider an extremely large quantity of water — say enough to fill a large room (Figure 7.3). Now let's add enough salt to make the concentration exactly 1 molal, and adjust the volume of the solution so that the room is full and a little excess solution sticks up into a calibrated tube inserted into the ceiling. By observing changes in the level of solution in the tube, we can accurately record changes in the V of the solution in the room.

Now, when we add a mole of NaCl (58.5 g of NaCl occupying $27\,\text{cm}^3$) to the solution, the change in concentration is very small. In fact, if we can detect *any* change in concentration by the finest analytical techniques available, then our room is too small, and we must find and inundate a larger one. Eventually, we will fill a sufficiently large room with salt solution that on adding 58.5 g of NaCl we are unable to detect any change in concentration — it remains at 1.000 mole NaCl/kg $H_2O$. But although the concentration remains unchanged, the volume of course does not. The salt added cannot disappear without a trace. The level in the tube in the ceiling changes, and the $\Delta V$ seen there is evidently the volume occupied by 1 mole of NaCl in a 1 molal NaCl solution, in this case about $15\,\text{cm}^3\,\text{mol}^{-1}$ of NaCl. This is, in quite a real sense, the volume occupied

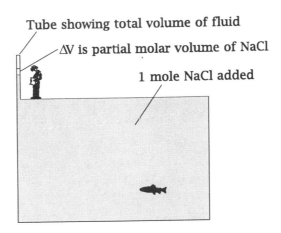

Figure 7.3: A roomful of 1 molal salt solution. The observer sees the change in volume caused by adding one mole of salt, which is the partial molar volume of salt in the 1 molal solution.

by a mole of salt in that salt solution and has a right to be thought of as a *molar volume* (just as much as 27 cm$^3$ mol$^{-1}$ is the molar volume of crystalline salt) rather than as an arbitrary mathematical construct. It is referred to as the *partial* molar volume of NaCl in the salt solution, $\overline{V}_{NaCl}$. You will be well advised to think of partial molar properties in this sense, that is, as *molar* properties of solutes in solutions of particular compositions, rather than in terms of the partial derivative that defines them mathematically (Appendix C).

Some readers will have difficulty in seeing how, on adding our salt, the concentration does not change but the total volume does. If this is the case, think of the room as containing not a solution, but nine million white tennis balls and one million black tennis balls, all mixed together. The room is full, the balls are arranged so that no space is available for another ball, and a few balls overflow into the tube in the ceiling. The total volume is the volume of ten million tennis balls. Now we add one more black tennis ball, somewhere in the middle of the room. The fractional concentration of black balls changes from $10^6/10^7$ to $(10^6 + 1)/(10^7 + 1)$, or from 0.1 to 0.10000009, a change so small it is completely negligible.[2] But the total volume has changed by the volume of one tennis ball, and this change must be reflected by the level of the balls in the tube, which will rise by the volume of one ball. We can even extend the analogy by imagining that the balls in the room are compressed by the pressure, so that when we add another ball, it becomes compressed too, and the level in the tube rises by the volume of a compressed tennis ball, not a normal (standard state) tennis ball.

It is possible that the solute added would cause the $\Delta V$ to be negative. Be-

---

[2]If you don't find it negligible, just imagine a bigger room and more tennis balls, until the change *is* negligible.

cause of very strong interactions between the solute molecules and the water molecules, the volume of liquid in the room might actually decrease on adding one mole of solute. The solute would then have a negative partial molar volume. It is of course also possible to use any concentration of solution to fill the room, even pure water. In this case, when one mole of NaCl is added to the roomful of pure water, the water remains pure—the NaCl is so dilute it cannot be detected analytically. But $\Delta V$ is still observed and is the partial molar volume of NaCl in pure water, or at *infinite dilution*.

Needless to say, measurements are not actually done in this bizarre fashion; procedures have been worked out to calculate these quantities from more normal types of laboratory measurements.

### 7.3.4 The Free Energy of Mixing

We have discussed the volume of mixtures first because it is more easily visualized than other properties, but in fact enthalpy is quite analogous to volume. The same diagrams could be used, just changing $V$ to $H$. But free energy is different. You should not be surprised at this, because we spent considerable effort looking for an energy parameter that would always decrease in a spontaneous process, and two substances dissolving into one another is a perfect example of a spontaneous process. Therefore, we cannot use Figure 7.2a for free energies—the $G$ of a solution will never lie on a straight line between the $G$ of the pure components, and it will never lie above that line; it will always fall below, as in Figure 7.2b and Figure 7.4.

But there are two differences between Figures 7.4 for $G$ and 7.2b for $V$. One difference is that despite appearances, the molar free energy curve is actually asymptotic to each vertical axis, whereas the volume curve is not. This is not terribly important. The other is that there are no numbers on the $y$-axis in Figure 7.4 as we do not have absolute values for $G$. The free energy of pure B has been placed arbitrarily 2000 J above that for A, but this was simply artistic license. In general $G_A^\circ \neq G_B^\circ$, but there is no way to know the difference in absolute $G$ values of different compounds, and so Figure 7.4 is rather hypothetical. But don't worry; all the essential information is available on a real diagram, obtained by plotting values of the *difference* between $(X_A G_A^\circ + X_B G_B^\circ)$ and $G$ of the solution, labeled $\Delta G_{mix}$ in Figure 7.4. This is shown in Figure 7.5.

### 7.3.5 Partial Molar Free Energies

In Figure 7.4 the partial molar free energies are given by the tangent intercepts, as in Figure 7.2. The only difference here is that the partial molar free energies (the Gibbs energy of a mole of dissolved component) are given a special symbol, $\mu$, and a special name, the chemical potential. Because the mixing curve (the line giving the $G$ of the solution in Figure 7.4) is concave downward, the chemical potentials of A and B necessarily lie below the corresponding values of $G_A^\circ$ and $G_B^\circ$, the molar free energies of the pure compounds. That is, $\mu_A < G_A^\circ$ and

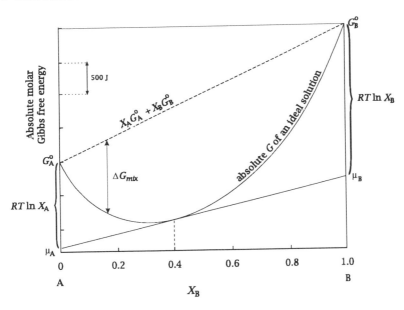

Figure 7.4: The molar free energy ($G$) of an ideal solution of A and B. The absolute free energies are unknown, but B has been placed arbitrarily 2000 J above A. $\Delta G_{mix}$ can be calculated and results in the curve joining $G_A^\circ$ and $G_B^\circ$. The left and right intercepts of the tangent to this curve give $\mu_A$ and $\mu_B$ in the solution for which the tangent is drawn ($X_B = 0.4$).

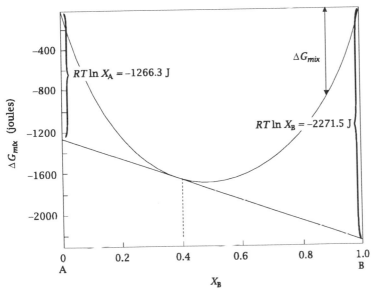

Figure 7.5: Free energy of mixing, from Figure 7.4.

similarly for B. The fact that this is so provides the reason why A and B form a solution. The reaction is

$$n_A A + n_B B = (n_A A, n_B B)_{solution} \qquad (7.6)$$

and the free energy change for this reaction is $\Delta G_{mix}$ (Figure 7.4), which is negative. During this reaction, both A and B lower their free energy contents— $G_A^\circ$ is lowered to $\mu_A$ and $G_B^\circ$ is lowered to $\mu_B$. If the mixing line lies *above* the straight line joining $G_A^\circ$ and $G_B^\circ$, then $\Delta G_{mix}$ would be positive, the dissolution reaction (7.6) would not be spontaneous, and no solution would form—A and B would be immiscible, like oil and water.

We will be using the symbol $\mu$ frequently, and you must be sure about its meaning. We went on at great length in §7.2.3 about how the partial molar volume (say $\overline{V}_A$) is the volume of a mole of dissolved A, usually quite different from (and possibly smaller or greater than) the volume of a mole of pure A. In exactly the same way, the partial molar free energy ($\overline{G}_A$) is the free energy per mole of dissolved A, which is always less than the free energy per mole of pure A. The only difference (besides the fact that partial molar free energy gets a special symbol $\mu$) is the familiar problem that we have absolute values for $V$ but not for $G$, and so we have absolute values for $\overline{V}$ but not for $\mu$. Therefore, while you may see statements like $\overline{V}_{NaCl} = 15 \, cm^3 \, mol^{-1}$, values for $\mu$ should appear only in difference terms, such as $\Delta\mu$, $\mu - \mu^\circ$, or $\mu - G^\circ$.

## 7.4   THE RELATION BETWEEN COMPOSITION AND FREE ENERGY

There is one aspect of Figure 7.4 we have not yet mentioned. That is the fact that, as shown by the labels on the diagram,

$$\left.\begin{array}{rcl} \mu_A - G_A^\circ & = & RT \ln X_A \\ \mu_B - G_B^\circ & = & RT \ln X_B \end{array}\right\} \qquad (7.7)$$

We will not derive this equation here (see Appendix C), but if you think about it you'll see that it explains why we said the curve in Figure 7.5 is asymptotic to each vertical axis.

When $X_B = 0.4$ and $X_A = 0.6$, $R = 8.31451 \, J \, deg^{-1} \, mol^{-1}$, and $T = 298.15 \, K$, equations (7.7) give

$$\mu_A - G_A^\circ = -1266.3 \, J$$

and

$$\mu_B - G_B^\circ = -2271.5 \, J$$

This says that a mole of A has 1266.3 J less in the dissolved state than in the pure state, and this is the "thermodynamic explanation" for why A dissolves in B.

Check these numbers against the positions of $\mu_A$, $\mu_B$, $G_A^\circ$, $G_B^\circ$ in Figures 7.4 and 7.5, and note that the same information about $\mu_A$ and $\mu_B$ at any concentration is available from the experimentally derivable Figure 7.5, not just our hypothetical diagram in Figure 7.4.

Note especially that equations (7.7) provide a relationship between the concentration of a solution component and its free energy. This is an important milestone. Equations (4.9) and (4.10) showed how free energy varies with $T$ and $P$, respectively (expanded upon in Figures 6.2, 6.5); now we can see how free energy varies with concentration of something in solution.

If we can calculate the free energy of solids, liquids, gases, and solutes over a range of $T$, $P$, and composition $(X)$, we have just about solved all our problems, in principle. Basically, from here on we will be amplifying and coming to grips with practical matters.

## 7.4.1 Ideal and Nonideal Mixing

In §7.1.1 we said that when A-A, A-B, and B-B interactions are all about the same, we get a straight line relationship [equation (7.5)] called ideal mixing, whereas the more general case, when there is some sort of unequal interaction between the compounds, is nonideal mixing. In Figure 7.4 we have drawn the free energy of mixing for the ideal case, where A and B do not interact at all, and where the volume of mixing follows the straight line relationship [equation (7.5)]. What about the more general case of non-ideal mixing? In this case, $\mu_A$ and $\mu_B$ are changed, but the relations (7.7) are very convenient, so we introduce a *correction factor* for the nonideal case, such that we can continue to use the convenient equations. The correction factor is $\gamma_R$ (the R stands for Raoult, and means that the correction factor is a measure of the deviation from Raoultian ideality), and equations (7.7) become

$$\left.\begin{array}{rcl} \mu_A - G_A^\circ &=& RT \ln X_A \, \gamma_{R_A} \\ \mu_B - G_B^\circ &=& RT \ln X_B \, \gamma_{R_B} \end{array}\right\} \tag{7.8}$$

When the correction factor is 1.0, the compounds are mixing ideally. These correction factors can be measured or calculated theoretically, or they can be ignored (i.e., assumed to be 1.0).

## 7.4.2 Tabulated Free Energies of Solutes

Because the free energy of dissolved compounds changes with concentration $(X)$, we must choose one concentration for the purposes of tabulating data. This has been chosen to be 1 molal, but for reasons we won't discuss here (see Appendix C), the 1 molal reference state is chosen to be a Henryan kind of ideal solution, that is, one that has the properties of a very dilute solution. For example, for component NaCl, the quantity $\mu_{NaCl} - G_{NaCl(s)}^\circ$ is determined experimentally for the ideal 1 molal solution,[3] and this is added to $\Delta_f G^\circ_{\text{NaCl}(s)}$ to give the free energy of formation of aqueous NaCl.

Tabulated formation-from-the-element values for aqueous solutes are therefore similar to those for solids—they represent a difference between the property for the solute and the sum of the properties of its constituent elements. The difference is that it is for a concentration of 1 molal. Tabulated properties for aqueous ions ($Na^+$, $Cl^-$, etc.) have a little more complicated meaning, but are also for 1 molal concentration. All complications cancel out in balanced reactions anyway, so we won't go into this any further.

## 7.4.3  Reactions in Solution

Reactions involving dissolved compounds, either alone or with solids and gases, are different in an important way from reactions involving pure compounds only, such as pure solids. To see why, consider two reactions, one between pure compounds and one between dissolved components.

The first is a reaction in which all products and reactants are pure substances, the kind of reaction we have been considering up to now. It is

$$NaAlSiO_4(s) + 2\,SiO_2(s) = NaAlSi_3O_8(s) \tag{7.9}$$

The second is a reaction in which all products and reactants are dissolved in water and are capable of changing their concentration:

$$H_2CO_3(aq) = HCO_3^-(aq) + H^+(aq) \tag{7.10}$$

The temperature and pressure are normal, 298.15 K and 1 bar. As usual, we want to know which way each reaction will go. Reaction (7.9) presents no problem. We look up the values of $\Delta_f G^\circ$ for each compound, and calculate $\Delta_r G^\circ$:

$$
\begin{aligned}
\Delta_r G^\circ &= \Delta_f G^\circ_{\text{NaAlSi}_3\text{O}_8} - \Delta_f G^\circ_{\text{NaAlSiO}_4} - 2\,\Delta_f G^\circ_{\text{SiO}_2} \\
&= -3711.5 - (-1978.1) - 2(-856.64) \\
&= -20.12\,\text{kJ}
\end{aligned}
$$

We see that the reaction as written is spontaneous; $NaAlSiO_4$ (nepheline) and $SiO_2$ (quartz) at 1 bar pressure should react together to form $NaAlSi_3O_8$ (albite). If the reaction does proceed (thermodynamics doesn't tell us whether it will or not, only that the energy gradient favors it), then nepheline and quartz get used up during the reaction. However, while being used up, they do not change their free energies. The reaction should actually proceed as long as any reactants are left. When either the nepheline or the quartz is used up completely, the reaction must stop. This reaction can be represented graphically as in Figure 7.6. Here we use bars to represent the magnitude of the combined free energy

---

[3]How can you experiment with an ideal solution, which does not exist? Simple—you experiment with real solutions over a range of compositions, then *extrapolate* your findings to zero concentration, or infinite dilution. Some solute properties at infinite dilution have the ideal 1 molal value.

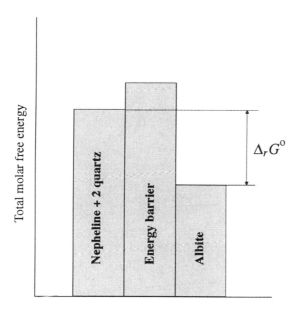

Figure 7.6: Molar free energies when all products and reactants are pure compounds. The free energy of reaction is given by $\Delta_r G°$ because all products and reactants are in their reference states, and this does not change during the reaction until one of the reactants disappears.

of the products and of the reactants. The difference in the height of the bars represents $\Delta_r G°$, the driving force for the reaction. The middle bar represents an activation energy barrier that prevents the reaction from occurring. It is put there to form a link with the discussion in Chapter 2, but thermodynamics is unable to calculate the size of this barrier, or anything whatever about it. Nevertheless it is often there, and is the reason that one of the states is metastable. The point here is that the size of the bars does not change during the reaction, if it proceeds, because none of the products or reactants changes in any way—only the amounts present change. The value of $\Delta_r G°$ never goes to zero.

Reaction (7.10) is different. We can start off the same way, by looking up the values of $\Delta_f G°$ for each compound;

$$
\begin{aligned}
\Delta_r G° &= \Delta_f G°_{HCO_3^- (aq)} + \Delta_f G°_{H^+ (aq)} - \Delta_f G°_{H_2CO_3 (aq)} \\
&= -586.77 + 0 - (-623.109) \\
&= 36.339 \, kJ
\end{aligned}
$$

This is positive, and so the reaction goes spontaneously to the left. So far, so good. But as soon as the reaction starts, the concentrations of $H^+$ and $HCO_3^-$ start to decrease, the concentration of $H_2CO_3$ starts to increase, and the free energies of all three change, as shown in Figure 7.7. All we can say from the tabulated data is that if all three aqueous species were present at a concentration of 1 molal (actually an *activity* of 1.0; see below), the reaction would start to go to the left. But suppose we are interested in some other concentrations? And what happens to the reaction after it starts? Because the solutes can change their concentrations and their free energies, the situation is quite different from the "all pure substances" situation.

## 7.5  ACTIVITIES

Reaction (7.10) is the first example we have seen of a reaction that can stop because it reaches equilibrium, rather than because one of the reactants is used up. In reaction (7.10) it seems reasonable, though we have not demonstrated it, that because the concentrations and free energies of $H^+$ and $HCO_3^-$ are decreasing, and that of $H_2CO_3$ is increasing, that sooner or later they will reach a balance, where

$$
\mu_{H^+} + \mu_{HCO_3^-} = \mu_{H_2CO_3} \tag{7.11}
$$

so that $\Delta_r \mu = 0,$[4] and the reaction will stop because there is no driving force in either direction. To complete our model, we need to develop a method of dealing with reactions like this, which will allow us to determine the conditions for which the reaction reaches equilibrium. This method will have to use the known variation of free energy with composition [equation (7.8)], and allow

---

[4]$\Delta_r \mu = 0$ is just another way of writing $\Delta_r G = 0$, but is more appropriate for aqueous solutes.

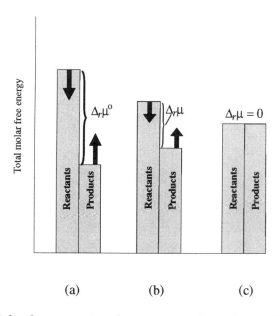

(a)                    (b)                    (c)

Figure 7.7: Molar free energies of reactants and products of a reaction between aqueous solutes. (a). A hypothetical starting condition, represented by the numbers in the tables of data (note superscript ° on $\Delta_r\mu°$, indicating standard conditions for all products and reactants). (b). Either the reaction in (a) after it has proceeded for some time, or a beginning state where the reactants and products are not all at 1 molal (no superscript °). The point is that, as shown by the arrows, the free energy of the reactants decreases and that of the products increases during the reaction. (c). Sooner or later, a state of equilibrium is reached, when the free energies of reactants and products are equal.

us to use solute concentrations of any value. The result is the equilibrium constant, the subject of the next chapter.

However, before we derive the equilibrium constant, we must generalize our formula (7.8) relating composition and free energy. We said that in the general nonideal case, $\mu - G°$ is not given exactly by $RT \ln X$, and so we add a correction factor $\gamma_R$, which can be measured or in many cases calculated theoretically, to give $\mu - G° = RT \ln X\gamma_R$. The term $X\gamma_R$ is actually an *activity* and is just one of many activity terms that can be used in this expression. The generalized equation relating free energy and composition is thus

$$\mu_i - G_i° = RT \ln a_i \qquad (7.12)$$

where $a_i$ is the *activity* of component $i$ in a solution. Life would be considerably simpler if the only activity term we used was $X\gamma_R$, but in fact we trade some simplicity for convenience, and we have different activity terms for different kinds of solutions.

The following are some rules that will be useful in relating concentrations in various kinds of solution to free energy differences. The rules define various activity terms.

## Solid and Liquid Solutions

You may not have realized that there is such a thing as a solid solution, but they are very common in the mineral world. The mineral olivine, for example, which commonly occurs in certain kinds of rocks, has a composition that can be written $(Mg, Fe)_2SiO_4$. The (Mg,Fe) means that one of the atomic positions in the mineral lattice can be occupied by either Mg or Fe, and the composition of the mineral can change all the way from almost pure $Mg_2SiO_4$ to almost pure $Fe_2SiO_4$, depending on the ratio of Mg to Fe in the mineral. Olivine is therefore called a solid solution, because its composition is variable, just as the composition of a solution of salt in water is variable. The fact that olivine has a fixed three-dimensional structure with either Mg or Fe in one of the structural positions, whereas water has no fixed structure, is irrelevant in this case. The fact is that both need a compositional variable such as $X_{NaCl}$ or $X_{Mg_2SiO_4}$ to describe their composition.

For solid and liquid solutions, we use

$$a_i = X_i \gamma_{R_i} \qquad (7.13)$$

where $a_i$ is the activity of $i$, $X_i$ is the mole fraction of $i$, and $\gamma_{R_i}$ is the activity coefficient of $i$. Subscript $R$ means that the activity coefficient is a correction factor relative to the Raoultian type of ideal solution.

Note that for a pure compound $i$, $a_i = 1$ because $X_i = 1$, and $\gamma_{R_i} = 1$ by definition. It follows from (7.8) that $\mu_i = G_i°$ for a pure substance. This just means that we can use either symbol for pure substances, but it is important to remember that $a = 1$ for pure substances.

## Gaseous Solutions

Mixtures of gases are also solutions. To describe the composition of components in a gaseous solution, we could use mole fraction, but more commonly we use the partial pressure, which is analogous. If a gaseous solution has a total pressure $P_{total}$ and the gaseous components have mole fractions $X_1$, $X_2$, $X_3...$, the partial pressures of these components are defined as

$$\left.\begin{aligned}
P_1 &= X_1 \cdot P_{total} \\
P_2 &= X_2 \cdot P_{total} \\
&\vdots \qquad\qquad \vdots \\
\text{etc.}
\end{aligned}\right\} \tag{7.14}$$

and the activities of the components are

$$\left.\begin{aligned}
a_1 &= P_1 \, y_{f_1} \\
a_2 &= P_2 \, y_{f_2} \\
&\vdots \qquad\quad \vdots \\
\text{etc.}
\end{aligned}\right\} \tag{7.15}$$

where $P_i$ is the partial pressure of component $i$ and $y_f$ is another kind of activity coefficient. The term $P_i \, y_{f_i}$ is given a special name, the *fugacity* of $i$ ($f_i$), and $y_{f_i}$ is a special kind of activity coefficient, the *fugacity coefficient*. Fugacity has units of bars.

## Aqueous Solutions

The unit of concentration most commonly used for aqueous solutions is molality, so the activity for aqueous species is

$$a_i = m_i \, y_{H_i} \tag{7.16}$$

where $m_i$ is the molality of $i$ and $y_{H_i}$ is another kind of activity coefficient, this time based on the Henryan kind of ideal solution.

Don't worry about all these activity coefficients ($y_R$, $y_f$, $y_H$). They are included here for completeness, but in a great many cases, and certainly at the introductory level, we may completely neglect them, that is, we assume they have values of 1.0. This is sufficiently accurate in many situations.

## 7.5.1  Summary of Activity Rules

For solid and liquid solutions:

$$a_i = X_i \, y_{R_i} \qquad (7.17)$$

For gaseous solutions:

$$a_i = P_i \, y_{f_i} \qquad (7.18)$$
$$\phantom{a_i} = f_i \qquad (7.19)$$

For aqueous solutions:

$$a_i = m_i \, y_{H_i} \qquad (7.20)$$

## 7.5.2  What Is the Activity?

The activity is a term in the thermodynamic model that allows us to relate the concentration of something in solution to its Gibbs free energy. Its defining equation is (7.12), which is always true. It takes different forms [equations (7.17)–(7.20)], depending on the kind of solution. Think of it as a concentration, because that is the purpose it serves. It is in all cases a concentration multiplied by a factor that accounts for the difference between an ideal solution and a real solution.

### A Problem With Activity Units?

If you look at the units of the terms in this equation, you see that $\mu$ and $G°$ and $RT$ all have units of energy; that is, $J\,mol^{-1}$. This means that for the equation to balance dimensionally, $a_i$ must be dimensionless—it must have no units. This is evidently true for $a_i = X_i \, y_{R_i}$, because $X$ is dimensionless, and all the $y$ terms are dimensionless. However, $a_i = P_i \, y_{f_i}$ and $a_i = m_i \, y_{H_i}$ seem to present a problem, as both $P$ and $m$ have units.

If we took the time to develop the subject in more detail (see Appendix C), we would see that in each case the activity is actually a ratio, with the denominator equal to one. For example, the strict definition of the activity of species $i$ in an aqueous solution is

$$a_i = \frac{(m_i \, y_{H_i})}{(m_i \, y_{H_i})°} \qquad (7.21)$$

and in a gaseous solution is

$$a_i = \frac{f_i}{f_i°} \qquad (7.22)$$

where the denominator refers to a reference state and is always set to 1 (1 $m$, or 1 bar). Therefore, $a_i$ is seen to be actually dimensionless (the units of $m$ or

$f$ cancel out), and our "rules" for the use of activity [equations (7.16)-(7.19)] are fine because they are abbreviations.

## 7.5.3 Standard State Values of Thermodynamic Properties

From now on, when we refer to the tabulated or reference state properties of substances, we mean the pure substance for solids and liquids, the pure gas at 1 bar for gases, and for aqueous solutes a solution having Henryan ideality at a concentration of 1 molal. These are technically speaking the *standard states* chosen for the various compounds. They are the values that are tabulated in reference tables, such as Appendix B, and are signified by putting the superscript ° on the property symbol, for example, $\mu°$, $S°$. They are always tabulated for a temperature of 298.15 K and a pressure of 1 bar, but they can easily be calculated for other temperatures and pressures. An example of this is given in Appendix C. The essential thing about the standard state is therefore not its $T$ and $P$, though these are most often 25°C and 1 bar, but its physical state (pure solid or liquid; ideal $1m$ solute, etc.).

Note too that when we write reactions such as

$$CH_4(g) = CH_4(aq)$$

as we have been from the beginning, the $(g)$, $(aq)$, $(s)$, and $(l)$ following each formula refers not only to the physical state of the substance in real life, in this case methane in a gas phase in equilibrium with methane in an aqueous phase, but indicates what standard state is used in applying thermodynamics. For example, the reactions

$$H_2S(g) = H^+ + HS^-$$

and

$$H_2S(aq) = H^+ + HS^-$$

give totally different equilibrium constants (to be discussed in the next chapter) with different meanings. The system you are interested in may be entirely aqueous and may not even have a gas phase, but if you use data from the tables for $H_2S(g)$ instead of for $H_2S(aq)$, you will calculate an equilibrium constant in which $a_{H_2S}$ refers to the partial pressure of $H_2S$ in a gas phase, which *would* be in equilibrium with the ions, if there was a gas phase, which is probably not what you wanted. Many compounds have data for several different standard states, so pick the right one for the situation.

Note too that we do not write $(aq)$ after ions, because it goes without saying that they are all aqueous, $(aq)$, and only one standard state is ever used for them. We are not dealing with ionized plasmas.

## 7.6 SUMMARY

This chapter introduces the concepts necessary to use thermodynamics on solutions and their components. It is first necessary to have a firm grasp of the meaning of the molar volume and molar energies and so on of pure compounds, such as water. Then you find that solutions (solid, liquid, or gas solutions) can also have molar properties. And finally you see how the molar property of a solution is "split up" into the partial molar properties of its individual components, so that these components can be used by themselves in chemical reactions.

The problems in understanding this material are made more difficult in the case of molar enthalpy and molar free energy because of the now familiar problem of not having absolute values of $H$ and $G$ for any substances, as well as the necessity of always using differences when measurements are involved. It is important to remember that substances do have absolute, finite values of $H$, $G$, and $\mu$—we just don't know what they are. We do use these terms (the absolute quantities) in discussion, and in diagrams without units on the axis.

Finally, we have introduced the activity, which is rather like a chameleon, in that it changes its form depending on its surroundings. This can cause confusion, and often does. At all times, however, it is used in place of a concentration term and forms a direct link between the concentration of a substance in solution and its free energy. Its chameleonlike habit of changing its form comes about because we use different reference states for solids and liquids, gases, and solutes, but if you follow the "rules" for its use, you will have no problems. A deeper look into its origins reveals it to be always dimensionless, as are activity coefficients (Appendix C).

## PROBLEMS

1. The equation for $\Delta G_{mix}$ in Figure 7.5 is

$$\Delta G_{mix} = RT(X_A \ln X_A + X_B \ln X_B).$$

   Use calculus to show that this has a minimum at $X_A = X_B = 0.5$.

2. Calculate $\Delta G_{mix}$ for $X_B = 0.4$, and compare with Figures 7.4 and 7.5.

3. If the molality of solute A is 0.05 and its activity coefficient ($y_H$) is 0.8, what can you say about the chemical potential of the solute?

4. The composition of the air we breathe is tabulated below.

| Gas | Percent by volume |
|-----|-------------------|
| $N_2$ | 78.084 |
| $O_2$ | 20.946 |
| Ar | 0.934 |
| $CO_2$ | 0.035 |
| $CH_4$ | 0.00017 |
| $H_2$ | 0.00005 |

Calculate the partial pressure and fugacity of each gas, assuming the atmosphere is an ideal gas. Atomic weights are not required.

5. If a sample of air was compressed to 100 bar, what would be the fugacity of methane in the gas? Assume it behaves as an ideal gas ($y_f = 1.0$).

6. If the fugacity coefficient of methane at 100 bar is actually 0.95, what is its fugacity?

# 8

# THE EQUILIBRIUM CONSTANT

## 8.1 REACTIONS AT EQUILIBRIUM

In the last chapter, we discussed the idea that chemical reactions not only go one way or the other (our main problem), but they can stop going for two reasons. Either one of the reactants is used up, or the reaction can reach an equilibrium state, with all products and reactants present in a balanced condition. The second possibility is the subject of this chapter—how much can we predict about this balanced state of equilibrium?

In Chapter 4 we defined the molar Gibbs energy, $G$, which always decreases in spontaneous reactions ($\Delta G < 0$). In Chapter 6, we used the fact that a reaction at equilibrium (e.g., calcite $\rightleftharpoons$ aragonite) does not go either way ($\Delta G = 0$) to calculate the $T$ and $P$ of equilibrium between phases. The expression $\Delta G = 0$ expresses a balance between the free energies of calcite and aragonite, that is, $G_{calcite} = G_{aragonite}$ (§6.2.1). If there is more than one reactant or product, the same relationship must hold (the $G$ of reactants and products are equal), but each side is now a sum of $G$ terms, and the $G$ terms for solutes are properly written as $\mu$ rather than $G$. We have already seen an example of this, in equation (7.11). Of course, not all products and reactants need be solutes. For example, the reaction

$$SiO_2(s) + 2H_2O = H_4SiO_4(aq) \qquad (8.1)$$

shows what happens when quartz dissolves in water. Molecules of $SiO_2$ dissolve and combine with water molecules to form the solute species $H_4SiO_4$. This dissolution process continues until the solution is saturated with silica, and then stops. The system is then at equilibrium, because

$$\mu_{H_4SiO_4} = \mu_{SiO_2} + 2\mu_{H_2O} \qquad (8.2)$$

If we added some $H_4SiO_4$ to this solution it would then be supersaturated, $\mu_{H_4SiO_4}$ would be greater than its equilibrium value, and the reaction would tend to go to the left, precipitating quartz.[1]

## 8.2   THE MOST USEFUL EQUATION IN THERMODYNAMICS

To find out what we can say about this balanced equilibrium state when several solutes and other phases are involved, let's consider a general chemical reaction

$$aA + bB = cC + dD \tag{8.3}$$

where A, B, C, and D are chemical formulae, and $a$, $b$, $c$, $d$ (called stoichiometric coefficients) are any numbers (usually small integers) that allow the reaction to be balanced in both composition and electrical charges, if any. When this reaction reaches equilibrium,

$$c\mu_C + d\mu_D = a\mu_A + b\mu_B$$

and

$$\begin{aligned} \Delta_r\mu &= c\mu_C + d\mu_D - a\mu_A - b\mu_B \\ &= 0 \end{aligned} \tag{8.4}$$

By our definition of activity, equation (7.12),

$$\mu_A = \mu_A^\circ + RT \ln a_A$$

$$\mu_B = \mu_B^\circ + RT \ln a_B$$

$$\mu_C = \mu_C^\circ + RT \ln a_C$$

and

$$\mu_D = \mu_D^\circ + RT \ln a_D$$

Substituting these expressions into (8.4), we get

$$\begin{aligned} \Delta_r\mu &= c\mu_C + d\mu_D - a\mu_A - b\mu_B \\ &= c(\mu_C^\circ + RT \ln a_C) + d(\mu_D^\circ + RT \ln a_D) \\ &\quad -a(\mu_A^\circ + RT \ln a_A) - b(\mu_B^\circ + RT \ln a_B) \\ &= (c\mu_C^\circ + d\mu_D^\circ - a\mu_A^\circ - b\mu_B^\circ) + RT \ln a_C^c + RT \ln a_D^d \\ &\quad -RT \ln a_A^a - RT \ln a_B^b \\ &= \Delta_r\mu^\circ + RT \ln \left( \frac{a_C^c \, a_D^d}{a_A^a \, a_B^b} \right) \end{aligned}$$

---

[1]At the risk of becoming repetitious, we note that it is in our model reaction that quartz precipitates.  In real life, something else might happen—nothing might precipitate, or some other $SiO_2$ phase such as silica gel might precipitate.

There may be any number of reactants and products, and so to be completely general we can write

$$\Delta_r \mu = \Delta_r \mu^\circ + RT \ln \prod_i a_i^{v_i} \tag{8.5}$$

where $i$ is an index that can refer to any product or reactant, $v_i$ refers to the stoichiometric coefficients of the products and reactants, with $v_i$ positive if $i$ is a product, and negative if $i$ is a reactant. $\prod_i$ (or $\prod_i$) is a symbol meaning "product of all $i$ terms," which means that all the $a_i^{v_i}$ terms are to be multiplied together (much as $\sum_i a_i$ would mean that all $a_i$ terms were to be added together). So in our case, the $v$ terms are $c$, $d$, $-a$, and $-b$, and

$$\prod_i a_i^{v_i} = a_C^c \, a_D^d \, a_A^{-a} \, a_B^{-b}$$

$$= \frac{a_C^c \, a_D^d}{a_A^a \, a_B^b}$$

In the general case, $\prod_i a_i^{v_i}$ is given the symbol $Q$, so (8.5) becomes

$$\Delta_r \mu = \Delta_r \mu^\circ + RT \ln Q \tag{8.6}$$

We must be perfectly clear as to what (8.6) means. In Figure 8.1 (a variation of Figure 7.7) are pictured the possible relationships between the free energies of the products and reactants in reaction (8.3).

First, the term $\Delta_r \mu^\circ$ refers to the difference in free energies of products and reactants when each product and each reactant, whether solid, liquid, gas, or solute, is in its pure reference state. This means the pure phase for solids and liquids [e.g., most minerals, $H_2O(s)$, $H_2O(l)$, alcohol, etc.], pure gases at 1 bar [e.g., $O_2(g)$, $H_2O(g)$, etc.], and dissolved substances [solutes, e.g., $NaCl(aq)$, $Na^+$, etc.] in ideal solution at a concentration of 1 molal. Although we do have at times fairly pure solid phases in our real systems (minerals such as quartz and calcite are often quite pure), we rarely have pure liquids or gases, and we never have ideal solutions as concentrated as 1 molal. Therefore, $\Delta_r \mu^\circ$ usually refers to quite a hypothetical situation. It is best not to try to picture what physical situation it might represent, but to think of it as just the difference in numbers that are obtained from tables.

$\Delta_r \mu$, on the other hand, is the difference in free energy of reactants and products as they actually occur in the system you are considering, which may or may not have reached equilibrium. The activities in the $Q$ term (the concentrations, fugacities, mole fractions, etc. of the products and reactants) change during the reaction as it strives to reach equilibrium and at any particular moment result in a particular value of $\Delta_r \mu$. Thus $\Delta_r \mu^\circ$ is a number obtained from tables that is independent of what is happening in the real system you are considering, but $\Delta_r \mu$ and $Q$ are linked together—whatever activities (think *concentrations*) are in $Q$ will result in a certain value of $\Delta_r \mu$.

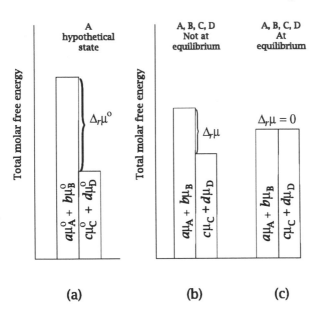

Figure 8.1: Possible free energy relationships in the reaction $a$A + $b$B → $c$C + $d$D.

If it makes more sense, you can write equation (8.6) as

$$\Delta_r \mu - \Delta_r \mu^\circ = RT \ln Q \qquad (8.7)$$

which means that whatever terms are in $Q$ control how different the chemical potentials ($\Delta_r \mu$) are from their standard tabulated values ($\Delta_r \mu^\circ$). When all activities in $Q$ are 1.0, then there is no difference, $\Delta_r \mu = \Delta_r \mu^\circ$.

We are especially interested in the value of $Q$ when our systems reach equilibrium, that is, when the product and reactant activities have adjusted themselves spontaneously such that $\Delta_r \mu = 0$. In this state, the $\prod_i a_i^{\nu_i}$ term is called $K$, instead of $Q$, and (8.6) becomes

$$0 = \Delta_r \mu^\circ + RT \ln K$$

or

$$\Delta_r \mu^\circ = -RT \ln K \qquad (8.8)$$

Standard states usually refer to pure substances (except for the aqueous standard states) in which $\mu = G$, so this equation is often written

$$\boxed{\Delta_r G^\circ = -RT \ln K} \qquad (8.9)$$

This equation has been called, with some reason, the most useful in chemical thermodynamics, and it certainly merits the most careful attention. Most important is the fact that the activity product ratio ($K$) on the right-hand side is independent of variations in the system composition. Its value is controlled completely by a difference in standard (tabulated) state free energies ($\Delta_r G°$) and so is a function only of the temperature and pressure. It is a constant for a given system at a given temperature or temperature and pressure and is called the *equilibrium constant*. Its numerical value for a given system is not dependent on the system actually achieving equilibrium, or in fact even existing. Its value is fixed when the reacting substances are chosen. The left-hand side refers to a difference in free energies of a number of different physical and ideal states, which do not represent any real system or reaction. The right-hand side, on the other hand, refers to a single reaction that has reached equilibrium, or more exactly, to the activity product ratio that would be observed if the system had reached equilibrium.

The great usefulness of equation (8.9) lies in the fact that knowledge of a few standard state free energies allows calculation of an indefinite number of equilibrium constants. Furthermore, these equilibrium constants are very useful pieces of information about any reaction. If $K$ is very large, it shows that a reaction will tend to go "to completion," that is, mostly products will be present at equilibrium, and if $K$ is small, the reaction hardly goes at all before enough products are formed to stop it. If you are a chemical engineer designing a process to produce some new chemical, it is obviously of great importance to know to what extent reactions should theoretically proceed. The equilibrium constant, of course, will never tell you whether reactants will actually react, or at what rate; there may be some reason for reaction kinetics being very slow. It indicates the activity product ratio at equilibrium, not whether equilibrium is easily achievable.

---

Finally, <u>NEVER</u> write equation (8.9) as

$$\Delta_r G = -RT \ln K$$

---

that is, omitting the superscript °, because doing so indicates a complete lack of understanding of the difference between $\Delta_r G$ and $\Delta_r G°$ and is just about grounds for failing any course in this subject. Let's go over it once more. $\Delta_r G°$ (or $\Delta_r \mu°$) is the difference in free energy between products and reactants when they are all in their reference states (pure solids and liquids, solutes at ideal 1 molal, gases at 1 bar), determined directly from the tables. Products and reactants are virtually never at equilibrium with each other under these conditions ($\Delta_r G°$ or $\Delta_r \mu°$ never becomes equal to zero). $\Delta_r G$ (or $\Delta_r \mu$) is the difference in free energy between products and reactants in the general case (when at least one of the products or reactants is *not* in its reference state) and becomes equal to zero when the reaction reaches equilibrium. $\Delta_r G$ cannot be used in place of $\Delta_r G°$ in (8.9) because this would mean, among other things, that every reaction at equilibrium ($\Delta_r G = 0$) would have an equilibrium constant of 1.0.

## 8.2.1    A First Example

Let's calculate the equilibrium constant for reaction (7.10),

$$H_2CO_3(aq) = HCO_3^-(aq) + H^+(aq)$$

First we write, as before,

$$\Delta_r G° = \Delta_f G°_{HCO_3^-} + \Delta_f G°_{H^+} - \Delta_f G°_{H_2CO_3}$$

Getting numbers from the tables, we find

$$
\begin{aligned}
\Delta_r G° &= -586.77 + 0 - (-623.109) \\
&= 36.339 \text{ kJ} \\
&= 36399 \text{ J}
\end{aligned}
$$

The fact that this number is positive is not as significant as in our previous examples. In this case it means that *if* $H_2CO_3$, $HCO_3^-$, and $CO_3^{2-}$ were all present in an ideal solution, and each had a concentration of 1 molal, the reaction would go to the left. This hypothetical situation is not of much interest. We want the value of $K$.

Inserting this result in equation (8.9), we get

$$
\begin{aligned}
\Delta_r G° &= -RT \ln K \\
36339 &= -(8.3145 \times 298.15) \ln K
\end{aligned}
$$

so

$$
\begin{aligned}
\ln K &= -36339/(8.3145 \times 298.15) \\
&= -14.659 \\
\text{or} \qquad K &= 4.30 \cdot 10^{-7} \\
&= 10^{-6.37}
\end{aligned}
$$

If you don't like dealing with natural logarithms, you can use the conversion factor $\log x = \ln x/2.30259$ (Appendix A). This gives

$$
\begin{aligned}
\log K &= -36339/(2.30259 \times 8.3145 \times 298.15) \\
&= -6.37
\end{aligned}
$$

directly.

This means that when these three aqueous species are at equilibrium,

$$\frac{a_{HCO_3^-} \cdot a_{H^+}}{a_{H_2CO_3}} = 10^{-6.37}$$

This is the answer to our question in §7.3.3 ("what happens to the reaction after it starts?"). The reaction continues until the ratios of the activities of the products and reactants equals the equilibrium constant, in this case $10^{-6.37}$. It doesn't matter what the starting activities were, and individual activities at equilibrium can be quite variable. In other words the values of $a_{H_2CO_3}$ and of $(a_{HCO_3^-} \cdot a_{H^+})$ are not determined, nor are the values of $a_{HCO_3^-}$ or $a_{H^+}$ individually; only the ratio expressed by $K$ is fixed. In specific cases, the values of these individual activities are determined by the bulk composition of the solution, but this is another topic. For now, we are content to determine $K$. In this case $K$ is the ionization constant for carbonic acid, $H_2CO_3$. It is a very small number, meaning that carbonic acid is a weak acid.

## 8.3 SPECIAL MEANINGS FOR *K*

Equilibrium constants are also sometimes equal to system properties of interest, such as vapor pressures, solubilities, phase compositions, and so on. This is because quite often it can be arranged that all activity terms drop out (are equal to 1.0) except the one of interest, which can then be converted to a pressure or composition.

### 8.3.1 *K* Equal to a Solubility

#### Quartz-Water Example

In our quartz-water example [equation (8.1)], the equilibrium constant expression is

$$K = a_{H_4SiO_4} / (a_{SiO_2} a_{H_2O}^2) \qquad (8.10)$$

At this point the expression is perfectly general, valid for any conditions, and $K$ is calculable from equation (8.9) if we know the free energies of the three species in their reference states. However, if we are dealing with pure quartz dissolving into pure water, then by our definitions (§7.4.1) $a_{SiO_2} = 1$ and $a_{H_2O} = 1$. Therefore

$$
\begin{aligned}
K &= a_{H_4SiO_4} \\
&= (m_{H_4SiO_4} \, y_{H_4SiO_4}) \\
&= m_{H_4SiO_4} \quad \text{assuming } y_{H_4SiO_4} = 1.0
\end{aligned}
$$

This shows that assuming $y_{H_4SiO_4}$ is 1.0, which happens to be an excellent approximation in this case, we can calculate the concentration of silica ($m_{H_4SiO_4}$) in equilibrium with quartz, that is, the solubility of quartz.

Following our routine, we write for the reaction as written

$$\Delta_r G^\circ = \Delta_f G^\circ_{H_4SiO_4} - \Delta_f G^\circ_{SiO_2} - 2 \Delta_f G^\circ_{H_2O}$$

Then, getting numbers from the tables,

$$\Delta_r G^\circ = -1307.7 - (-856.64) - 2(-237.129)$$

$$= \quad 23.198 \, \text{kJ}$$
$$= \quad 23198 \, \text{J}$$

Then

$$\Delta_r G^\circ \quad = \quad -RT \ln K$$
$$23198 \quad = \quad -(8.3145 \times 298.15) \ln K$$

so

$$\log K \quad = \quad -23198/(2.30259 \times 8.3145 \times 298.15)$$
$$= \quad -4.064$$

Thus the molality of $SiO_2$ in a solution in equilibrium with quartz is about $10^{-4.064}$, or about 5.2 ppm.[2]

Note the strangeness of what we are doing here. On the left-hand side [of (8.9)] we enter the free energies of the reactants and products, which in this case includes $\Delta_f G^\circ$ of $H_4SiO_4$ at a concentration of one molal (its concentration in its reference state) and on the right-hand side calculated its equilibrium concentration, only a few ppm. Remember what we said in deriving the equilibrium constant—the left-hand side consists of tabulated reference state data; it has nothing to do with real systems or with equilibrium. But from these data, equilibrium ratios and sometimes compositions can be calculated.

## 8.3.2  $K$ Equal to Fugacity of a Volatile Species

The next example is the same in principle. Consider the reaction

$$6 \, Fe_2O_3(s) = 4 \, Fe_3O_4(s) + O_2(g) \tag{8.11}$$

for which the equilibrium constant is

$$K = \frac{a_{Fe_3O_4}^4 \, a_{O_2}}{a_{Fe_2O_3}^6}$$

If the reaction involves pure hematite $Fe_2O_3$ and pure magnetite $Fe_3O_4$, then $X_{Fe_2O_3} = 1$ and $X_{Fe_3O_4} = 1$, so $a_{Fe_2O_3} = 1$ and $a_{Fe_3O_4} = 1$. Therefore

$$K \quad = \quad a_{O_2}$$
$$= \quad f_{O_2}$$
$$= \quad P_{O_2} y_f$$

Assuming that the activity coefficient $y_f$ is 1.0, again in this case an excellent approximation, we can calculate the partial pressure of oxygen in a gas phase in equilibrium with the two minerals hematite and magnetite.

---

[2]If the molality of $H_4SiO_4(aq)$ is $x$, then the molality of $SiO_2(aq)$ is also $x$, as there is 1 mole of $SiO_2$ in each.

Following the routine, we write

$$
\begin{aligned}
\Delta_r G^\circ &= 4\,\Delta_f G^\circ_{Fe_3O_4} + \Delta_f G^\circ_{O_2} - 6\,\Delta_f G^\circ_{Fe_2O_3} \\
&= 4(-1015.4) + 0 - 6(-742.2) \\
&= 391.6 \text{ kJ} \\
&= 391600 \text{ J}
\end{aligned}
$$

$$
\begin{aligned}
\Delta_r G^\circ &= -RT \ln K \\
391600 &= -(8.3145 \times 298.15) \ln K
\end{aligned}
$$

$$
\begin{aligned}
\log K &= -391600/(2.30259 \times 8.3145 \times 298.15) \\
&= -68.40
\end{aligned}
$$

So the oxygen fugacity in equilibrium with hematite and magnetite at 25°C and 1 bar is $10^{-68.40}$ bar. This is an incredibly small quantity, which would have absolutely no significance if it were simply a partial pressure, unconnected to thermodynamics. A partial pressure of this magnitude would be produced by one molecule of oxygen in a volume larger than that of a sphere with a diameter of the solar system.[3] However, it is in fact a parameter in the thermodynamic model, just as valid as any other part of the model. It can be used, for example, to calculate other parameters that might be more easily measurable. For example, the reaction

$$CH_4(g) + O_2(g) = CO_2(g) + 2\,H_2(g) \tag{8.12}$$

is one that you might be interested in if you were studying the bottom muds in Figure 2.1c. The equilibrium constant for this is

$$
\begin{aligned}
\Delta_r G^\circ &= -394.359 + 2(0) - (-50.72) - 0 \\
&= -343.639 \text{ kJ} \\
&= -343639 \text{ J}
\end{aligned}
$$

$$
\begin{aligned}
\log K &= 343639/(2.30259 \times 8.3145 \times 298.15) \\
&= 60.203
\end{aligned}
$$

which means that at equilibrium,

$$\frac{f_{CO_2} \cdot f_{H_2}^2}{f_{CH_4} \cdot f_{O_2}} = 10^{60.203}$$

---

[3]See §9.11.

Now $10^{60.203}$ is just as ridiculous as $10^{-68.40}$ in a sense. But if we insert the value $f_{O_2} = 10^{-68.40}$ into this expression, we get

$$\frac{f_{CO_2}}{f_{CH_4}} f_{H_2}^2 = 10^{60.203} \cdot 10^{-68.40}$$

$$= 10^{-8.20}$$

which begins to look a little more reasonable. This tells you something about how the $CO_2/CH_4$ ratio varies with $f_{H_2}$. For example, you could say that according to the thermodynamic model, if $f_{O_2}$ is controlled by hematite-magnetite, the $CO_2$ and $CH_4$ fugacities (partial pressures) are equal when $f_{H_2}$ is $10^{-4.1}$ bar, and this might in fact be a measurable quantity in the muds.

The point is that by writing a few reactions and using thermodynamics, your thoughts about what might be happening in the bottom muds or any other environment take shape in a controlled fashion—controlled, that is, by the implied hypothesis of chemical equilibrium. Your system may not be at complete equilibrium, but your model is, because that is a good place to start. And the fact that one of your thermodynamic parameters, such as $f_{O_2}$, turns out to be impossibly small or large does not make it ridiculous; it just means you won't be able to measure it directly, and you might want to concentrate on other parameters to which your impossible one is connected by the model.

## 8.4   $K$ IN SOLID–SOLID REACTIONS

It should be evident by now that the equilibrium constant is most useful in reactions between dissolved substances, those that change their activities during the reaction. Reactions of the other kind, between pure substances that do not change their activities during the reaction [e.g., reaction (7.9)] have no need of an equilibrium constant because in general they do not reach an equilibrium; they proceed until one of the reactants disappears. But what happens if you *do* calculate $K$ for such a reaction—what does it mean? Let's do this for reaction (7.9) and see what happens.

$$\Delta_r G° = -20.12 \text{ kJ}$$

$$= -20120 \text{ J}$$

$$\log K = 20120/(2.30259 \times 8.3145 \times 298.15)$$

$$= 3.52$$

$$K = 3349$$

As usual, this means that *at equilibrium*,

$$\frac{a_{NaAlSi_3O_8}}{a_{NaAlSiO_4} a_{SiO_2}^2} = 10^{3.52}$$

If we in fact have pure nepheline, pure albite and pure quartz involved in the reaction, then we come up with the same answer as before. The activities of $NaAlSi_3O_8$, $NaAlSiO_4$, and $SiO_2$ are 1.0 by our definitions, so the ratio $(a_{NaAlSi_3O_8}/a_{NaAlSiO_4}a_{SiO_2}^2)$ is fixed at 1.0 and can never be equal to 3349—the three pure minerals can never reach equilibrium at $25°C$, 1 bar. But suppose the minerals are not pure—suppose they are solid solutions. In briefly discussing solid solutions in §7.4, we gave the example of olivine, which has the composition $(Mg, Fe)Si_2O_4$ and can be considered to be a *solution* of $Mg_2SiO_4$ and $Fe_2SiO_4$ in the same crystal structure. Therefore, if the mole fraction of $Mg_2SiO_4$ in olivine is 0.5, we can say

$$a_{Mg_2SiO_4}^{olivine} = X_{Mg_2SiO_4}^{olivine} \gamma_R$$
$$= 0.5$$

assuming $\gamma_R$ is 1.0. In other words, the activity of $Mg_2SiO_4$ *can* be less than 1.0, even though there is a pure mineral with this composition. The activity is less than 1.0 when $Mg_2SiO_4$ is *not* pure, but occurs as a component of a solid solution. Both $NaAlSiO_4$ and $NaAlSi_3O_8$ form solid solutions with other compounds, so that $a_{NaAlSiO_4}$ and $a_{NaAlSi_3O_8}$ could easily be less than 1.0, if these species were present in solid solutions rather than in the pure minerals. In this particular case, having $a_{NaAlSi_3O_8}$ less than 1.0 would not help to achieve equilibrium. Equilibrium could only be achieved by lowering $a_{NaAlSiO_4}$ or $a_{SiO_2}$. For example, in the presence of pure nepheline and pure albite, $a_{SiO_2}$ would have to be 0.0173 to achieve equilibrium. This of course could not happen if quartz was present, but $a_{SiO_2}$ might be controlled in some other way, such as by the amount of dissolved $SiO_2$ in a solution that is undersaturated with quartz. To calculate this $SiO_2$ concentration, the reaction would be written using $SiO_2(aq)$ rather than $SiO_2(s)$ (see Problem 6 at the end of this chapter).

There is an important lesson here. When we write a chemical reaction, we look up a value of $\Delta_f G°$ for each chemical formula. The values of $\Delta_f G°$ are determined for those chemical species in very particular states—pure solids, ideal 1 molal solution, and so on. If our reaction is concerned with those species in those particular states, then the result is directly applicable to our problem— the value of $\Delta_r G$ is the same as the value of $\Delta_r G°$, and the reaction accordingly will go or not go. This case basically arises only when dealing with pure solids. When dealing with solutions, $\Delta_r G°$ is only a starting point. The reacting species are never in their reference states and have values of free energy that add up to $\Delta_r G$, not $\Delta_r G°$. The chemical formulae in our reactions represent species in some kind of solution, and we deal with these solutions with our activity terms, which are basically concentrations.

Reactions between solid phases such as (7.9) are in principle no different from any other kind of reactions, such as (7.10). The only difference is that there is in fact such a thing as relatively pure albite and quartz, and the like, to which the numbers in the tables apply directly, and we are sometimes interested in reactions between these pure compounds. In principle, however, each chemical formula in a chemical reaction, whether $Mg_2SiO_4(s)$ or $HCO_3^-(aq)$, can and usually does occur in a solution of some kind, with an activity controlled by its concentration.

## 8.5  CHANGE OF $K$ WITH TEMPERATURE

To get the effect of temperature on $K$, assuming as before (Chapter 6) that $\Delta_r H°$ and $\Delta_r S°$ are constants (not affected by temperature), we need only combine equations (8.9) and (6.9),

$$\begin{aligned} \Delta_r G° &= -RT \ln K \\ &= \Delta_r H°_{298} - T \Delta_r S°_{298} \end{aligned}$$

so

$$\ln K = \frac{-\Delta_r H°_{298}}{RT} + \frac{\Delta_r S°_{298}}{R} \qquad (8.13)$$

or

$$\log K = \frac{-\Delta_r H°_{298}}{2.30259\,RT} + \frac{\Delta_r S°_{298}}{2.30259\,R} \qquad (8.14)$$

As both $\Delta_r H°$ and $\Delta_r S°$ are assumed constant, this can be rewritten

$$\log K = a(1/T) + b$$

where $a$ and $b$ are constants, which is an equation in the form $y = ax + b$, meaning that $\log K$ is a linear function of $1/T$. An example of this is shown in Figure 8.2.

In Figure 8.2a are shown some solubility data for quartz, measured at a constant pressure of 1000 atm. As discussed in §8.3.1, these numbers can be interpreted as values of the equilibrium constant for the quartz dissolution reaction. The same data plotted as $\log m_{SiO_2(aq)}$ vs. $1/T$, where $T$ is in Kelvins, are shown in Figure 8.2b. Obviously this shows a good linear correlation, indicating that $\Delta_r H°$ does not change greatly over the temperature range of 25 to 300°C.

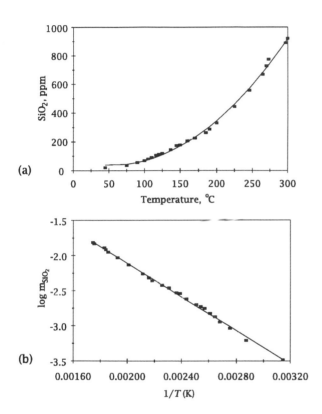

Figure 8.2: (a). The solubility of quartz in water as a function of temperature at a pressure of 1000 bar. (From Morey et al., 1962.) (b). The same data converted to $\log m_{SiO_2(aq)}$ and plotted vs. the reciprocal of absolute temperature.

**Example**

A plot of $\log K$ vs. $1/T$ can be used to obtain an estimate of $\Delta_r H°$ for the reaction for which $K$ is the equilibrium constant. According to the authors (Morey et al., 1962), the slope of the line in Figure 8.2b (fitted by the method of least squares) is $-1180\,$K, and so from (8.14),

$$\frac{\Delta_r H°}{2.30259\,R} = -(-1180)$$

and

$$
\begin{aligned}
\Delta_r H° \;&=\; 1180 \times 2.30259 \times 8.3145 \\
&=\; 22590\,\text{J} \\
&\approx\; 22.6\,\text{kJ}
\end{aligned}
$$

However, although the data may appear to be quite linear, confirming a constant $\Delta_r H°$ and $\Delta_r S°$, you must realize that a gentle curvature can easily be obscured by small random experimental errors, and even a gentle curvature implies a significant change in slope and of $\Delta_r H°$. In this case, a theoretical treatment of these and other data (contained in program SUPCRT92, Johnson et al., 1992) shows that $\Delta_r H°$ can vary (at 1 kbar) from 35.2 kJ at 25°C to 23.2 kJ at 300°C, while retaining an excellent fit to the data. The assumption of constant $\Delta_r H°$ is not suitable for accurate work, but is often useful nonetheless.

## 8.5.1   Another Example

As an example of the effect of $T$ on $K$, as well as some of the other points we have made, consider the reaction

$$CaCO_3(s) + SiO_2(s) = CaSiO_3(s) + CO_2(g) \tag{8.15}$$

This is an important reaction at high temperatures, when granites intrude limestones at depth in the Earth, but we will consider it at low temperatures and 1 bar pressure.

First, we get the equilibrium constant, as usual,

$$
\begin{aligned}
\Delta_r G° \;&=\; \Delta_f G°_{CaSiO_3} + \Delta_f G°_{CO_2} - \Delta_f G°_{CaCO_3} - \Delta_f G°_{SiO_2} \\
&=\; -1549.66 + (-394.359) - (-1128.79) - (-856.64) \\
&=\; 41.411\ \text{kJ} \\
&=\; 41411\ \text{J}
\end{aligned}
$$

$$
\begin{aligned}
\Delta_r G° \;&=\; -RT \ln K \\
41411 \;&=\; -(8.3145 \times 298.15) \ln K
\end{aligned}
$$

$$
\begin{aligned}
\log K_{298} \;&=\; -41411/(2.30259 \times 8.3145 \times 298.15) \\
&=\; -7.25
\end{aligned}
$$

This means, as usual, that

$$\frac{a_{CaSiO_3} a_{CO_2}}{a_{CaCO_3} a_{SiO_2}} = 10^{-7.25}$$

and because all the solid phases are pure, their activities are all 1.0, and we write

$$a_{CO_2} = 10^{-7.25}$$
$$= f_{CO_2}$$

Of course, both $CaSiO_3$ (wollastonite) and $CaCO_3$ (calcite) often form solid solutions and in natural situations might have activities less than 1.0, as discussed above. However, we are interested here in the pure phases.

The calculated $f_{CO_2}$ of $10^{-7.25}$ can be thought of as meaning that if calcite, wollastonite, and quartz were at equilibrium with a gas phase having a pressure of 1 bar at 25°C, the partial pressure of $CO_2$ in that gas would be about $10^{-7.25}$ or $5.6 \times 10^{-8}$ bar. As long as the three minerals remain pure and at equilibrium, the equilibrium constant will continue to be equal to $f_{CO_2}$, and so we can calculate the temperature at which the $CO_2$ pressure (fugacity) will reach 1 bar by calculating the change in $K$ with $T$.

To do this, we will first get another expression for the effect of $T$ on $K$ that will be more convenient. From (8.13) you can see that the slope of the graph of $\ln K$ as a function of $1/T$ is $-\Delta_r H°/R$, which is to say that at a temperature of 298 K,

$$\frac{d \ln K}{d(1/T)} = -\frac{\Delta_r H°_{298}}{R}$$

Integrating this between 298 K and $T$, we get

$$\int_{298}^{T} d \ln K = -\frac{\Delta_r H°_{298}}{R} \int_{298}^{T} d(1/T)$$

and so

$$\ln K_T - \ln K_{298} = -\frac{\Delta_r H°_{298}}{R} \left( \frac{1}{T} - \frac{1}{298.15} \right)$$

or

$$\ln K_T = \ln K_{298} - \frac{\Delta_r H°_{298}}{R} \left( \frac{1}{T} - \frac{1}{298.15} \right) \qquad (8.16)$$

or

$$\log K_T = \log K_{298} - \frac{\Delta_r H°_{298}}{2.30259 R} \left( \frac{1}{T} - \frac{1}{298.15} \right) \qquad (8.17)$$

By substituting terms, you can easily show that these are equivalent to our previous equations, (8.13) and (8.14). Remember, they are valid only for constant $\Delta_r H°$ and $\Delta_r S°$.

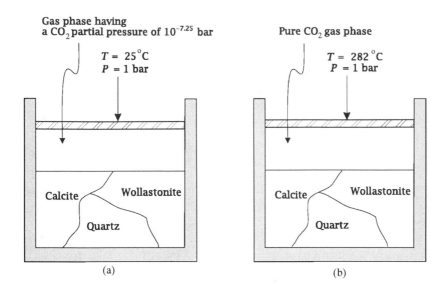

Figure 8.3: The fugacity of $CO_2$ in equilibrium with calcite, wollastonite, and quartz.

Now we need $\Delta_r H^\circ$ for reaction (8.15). This is

$$
\begin{aligned}
\Delta_r H^\circ &= \Delta_f H^\circ_{CaSiO_3} + \Delta_f H^\circ_{CO_2} - \Delta_f H^\circ_{CaCO_3} - \Delta_f H^\circ_{SiO_2} \\
&= -1634.94 + (-393.509) - (-1206.92) - (-910.94) \\
&= 89.411 \text{ kJ}
\end{aligned}
$$

If we want to calculate the temperature $T$ at which $f_{CO_2}$ reaches 1 bar while in equilibrium with calcite, quartz, and wollastonite, then $K_T = 1$, $\log K_T = 0$, and using our value of $\Delta_r H^\circ$, we get

$$
0 = -7.25 - \frac{89411}{2.30259 \times 8.3145} \left( \frac{1}{T} - \frac{1}{298.15} \right)
$$

from which $T = 555$ K or about $282°C$. The meaning of these calculations is illustrated in Figure 8.3.

## A Common Error

> You must remember that you <u>CANNOT</u> calculate $\ln K$ at $T$ from
>
> $$\Delta_r G^\circ_{298} = -RT \ln K \qquad (8.9)$$

where $\Delta_r G°$ comes from the normal tables, simply by changing $T$ from 298.15 to some other value. In this equation (8.9), $\Delta_r G°$ and $K$ must refer to the same temperature. If you want $K$ at some temperature other than 25°C, *first* get $\Delta_r G°$ at that new temperature from (6.9) or some other method, *then* get $K$ from (8.9), using your new value of $\Delta_r G_T°$ in place of $\Delta_r G_{298}°$. Of course, this procedure has essentially been done for you in equations such as (8.14) and (8.17).

## 8.6 THE AMINO ACID EXAMPLE AGAIN

Let's write equation (5.9) one more time.

$$C_8H_{16}N_2O_3(aq) + H_2O(l) = C_6H_{13}NO_2(aq) + C_2H_5NO_2(aq) \tag{8.18}$$

or

$$\text{leucylglycine} + \text{water} = \text{leucine} + \text{glycine} \tag{8.19}$$

$$
\begin{aligned}
\Delta_r G° &= \Delta_f G°_{leucine} + \Delta_f G°_{glycine} - \Delta_f G°_{leucylglycine} - \Delta_f G°_{water} \\
&= -13903 \, J
\end{aligned}
$$

We now know that our calculation of this $\Delta_r G°$ (§5.4.4), the reaction in which a peptide bond between two amino acids is broken, was only a beginning. The value of $-13903\,J$ means that if all reactants and products had unit activity (leucine, glycine, and leucylglycine had concentrations of 1 molal, and water was pure), the reaction would start to go to the right; leucylglycine would start to break down to leucine and glycine. But we note again the fundamental difference between this reaction between dissolved compounds, and reaction (7.9) between solid compounds. Repeating (7.9) here,

$$NaAlSiO_4(s) + 2\,SiO_2(s) = NaAlSi_3O_8(s) \tag{8.20}$$

$$
\begin{aligned}
\Delta_r G° &= \Delta_f G°_{NaAlSi_3O_8} - \Delta_f G°_{NaAlSiO_4} - 2\,\Delta_f G°_{SiO_2} \\
&= -20.12 \, kJ
\end{aligned}
$$

The value of $\Delta_r G°$ of $-20120\,J$ means that reaction (8.20) will also go to the right. But this reaction will continue to go (strictly, it *should* continue to go, according to our model) until either $NaAlSiO_4(s)$ or $SiO_2(s)$ is used up. Thus $NaAlSiO_4(s)$ and $SiO_2(s)$ *are not stable together*—one of them must disappear.

This is not the case with leucylglycine. We cannot say that leucylglycine is not stable in water—what happens to it depends entirely on its concentration and on the concentrations of other things in solution such as leucine and glycine. The unit activities are only a starting point, and a very unrealistic one at that. The next step is to calculate the equilibrium constant for (8.18).

$$\Delta_r G° = -RT \ln K$$

$$-13903 \quad = \quad -(8.3145 \times 298.15)\ln K$$

$$
\begin{aligned}
\log K_{298} \quad &= \quad 13903/(2.30259 \times 8.3145 \times 298.15) \\
&= \quad 2.436
\end{aligned}
$$

Thus

$$\frac{a_{leucine}\, a_{glycine}}{a_{leucylglycine}\, a_{water}} = 10^{2.436}$$

The activity (mole fraction) of water in biochemical systems is usually close to 1.0, so we see that although leucylglycine is not "unstable" in water, its concentration at equilibrium must be quite a bit less than that of its constituent amino acids. For example, if leucine and glycine had concentrations of say $10^{-3}\, m$ (activities of $10^{-3}$), the equilibrium activity of leucylglycine would be $10^{-8.436}$ (concentration $10^{-8.436}\, m$). So with concentrations of $10^{-3}$, $10^{-3}$, and $10^{-8.436}$, leucine, glycine, and leucylglycine would not react at all, but would be at equilibrium. In fact, with a concentration of leucylglycine of less than $10^{-8.436}$, the reaction as written would go to the left—leucylglycine would form from the two amino acids. So remember this—unless the reaction consists only of pure phases,

| *you cannot tell which way the reaction will go by looking at $\Delta_r G°$.* |

You can always tell which way the reaction will go by looking at $\Delta_r G$.

   Look at equations (8.7) and (8.8) one more time. When leucine, glycine, leucylglycine, and water all have unit activities, (8.7) becomes

$$
\begin{aligned}
\Delta_r\mu \quad &= \quad \Delta_r\mu° + RT\ln Q \\
-13903 \quad &= \quad -13903 + RT\ln\left(\frac{1 \times 1}{1 \times 1}\right)
\end{aligned}
$$

In other words, $\Delta_r\mu$ is the same as $\Delta_r\mu°$; the driving force for the reaction can be obtained directly from the tables, as for solid–solid reactions. When products and reactants have reached equilibrium,

$$
\begin{aligned}
\Delta_r\mu \quad &= \quad \Delta_r\mu° + RT\ln K \\
0 \quad &= \quad -13903 + RT\ln\left(\frac{10^{-3} \times 10^{-3}}{10^{-8.436} \times 1}\right)
\end{aligned}
$$

Now the $\ln K$ term exactly balances the $\Delta_r\mu°$ term, and the driving force for the reaction is zero. If $a_{leucylglycine} < 10^{-8.436}$, the driving force ($\Delta_r\mu$) becomes positive.

### 8.6.1   Peptides Favored at Higher Temperatures

To round out our discussion of this reaction, let's calculate the effect of temperature on the equilibrium constant in reaction (8.18). From Appendix B we find the following data:

| Substance | Formula | $\Delta_f H°$, J mol$^{-1}$ | $S°$, J mol$^{-1}$ K$^{-1}$ |
|---|---|---|---|
| leucine | $C_6H_{13}NO_2(aq)$ | $-632077$ | 215.48 |
| glycine | $C_2H_5NO_2(aq)$ | $-513988$ | 158.32 |
| leucylglycine | $C_8H_{16}N_2O_3(aq)$ | $-847929$ | 299.16 |
| water | $H_2O(l)$ | $-285830$ | 69.91 |

$$\begin{aligned}
\Delta_r H° &= \Delta_f H°_{leucine} + \Delta_f H°_{glycine} - \Delta_f H°_{leucylglyine} - \Delta_f H°_{water} \\
&= -632077 - 513988 - (-847929) - (-285830) \\
&= -12306 \text{ J}
\end{aligned}$$

Suppose we wanted the value of $K$ at 100°C. Equation (8.17) then becomes

$$\begin{aligned}
\log K_T &= \log K_{298} - \frac{\Delta_r H°_{298}}{2.30259\,R}\left(\frac{1}{T} - \frac{1}{298.15}\right) \\
&= 2.436 - \frac{-12306}{2.30259 \times 8.31451}\left(\frac{1}{373.15} - \frac{1}{298.15}\right) \\
&= 2.00
\end{aligned}$$

Alternatively, by calculating $\Delta_r S°$, you could use equation (6.9) first, then equation (8.9). Thus

$$\begin{aligned}
\Delta_r S° &= S°_{leucine} + S°_{glycine} - S°_{leucylglyine} - S°_{water} \\
&= 215.48 + 158.32 - 299.16 - 69.91 \\
&= 4.73 \text{ J}
\end{aligned}$$

Then

$$\begin{aligned}
\Delta_r G°_{373} &= \Delta_r H°_{298} - T\,\Delta_r S°_{298} \\
&= -12306 - 373.15 \times 4.73 \\
&= -14071 \text{ J}
\end{aligned}$$

from which

$$\begin{aligned}
\log K_{373} &= \frac{-\Delta_r G°_{373}}{2.30259\,RT} \\
&= -\frac{-14071}{2.30259 \times 8.31451 \times 373.15} \\
&= 1.97
\end{aligned}$$

There will often be a small discrepancy in $\log K$ calculated in different ways, as here (2.00 vs. 1.97), because of slight inconsistencies in the data. In other words, to get answers that are exactly the same no matter which way the calculation is done, the data in the tables for each compound must satisfy the relation

$$\Delta_f G° = \Delta_f H° - 298.15 \times \Delta_f S°$$

Because enthalpy, entropy,and free energy data come from different experiments, using a variety of methods, this relation is often not satisfied exactly in the tabulated data.

The interesting aspect of this calculation of $K$ is that according to the data, leucylglycine (and perhaps all peptide bonds in proteins) becomes *more stable* as temperature increases. Thus for the same concentrations of leucine and glycine ($10^{-3}m$) as before, we find the leucylglycine concentration is $10^{-8.0}$ $m$ at 100°C, compared to $10^{-8.436}$ $m$ at 25°C. That is, its concentration is more than doubled. This result is quite interesting to those scientists trying to figure out how life could have begun in the early days of the Earth, 3.5 billion years ago. The fact that increasing temperatures do not impair but in fact aid the bonding of simple amino acids, the building blocks of life, has led to thoughts that perhaps life began when the oceans were at higher temperatures, or in particular locations (volcanic environments) where heat was available.

This result is typical of the value of thermodynamics. It does not and cannot tell you how life began, but it can tell you which processes are possible and which impossible, and what the effects of changing the constraints on your system will be. This guides the development of scientific ideas in an essential way and provides a universally agreed-upon bedrock from which to start. However, it is up to you to think of the processes to ask thermodynamics about, and this is the creative part of science.

## 8.7  SUMMARY

This chapter contains a sudden increase in the amount of practical, usable material. If you ever have occasion to use thermodynamics in a practical situation, it will very likely involve the use of the equilibrium constant.

The molar Gibbs free energy of a dissolved substance changes with the concentration of the substance. The activity is a dimensionless concentrationlike term that is used to give the free energy in a particular state, in terms of its difference from its value in some reference state [equation (7.12)]. When a reaction has reached equilibrium, the activities of the various products and reactants can have a variety of values individually, but their ratio, as expressed in the equilibrium constant $K$, has a fixed and calculable value.

The equilibrium constant is calculated from numbers (free energies) taken from tables of standard data (derived experimentally, as discussed in Chapter 5). These standard data give the term $\Delta_r \mu°$ or $\Delta_r G°$, which is a constant for a given $T$ and $P$. It has nothing to do with whether your system or reaction has reached equilibrium ($\Delta_r \mu = 0$) or not. However, it can be used to calculate $K$, which gives the ratio of product and reactant activities your reaction will have if it ever reaches equilibrium.

The superscript ° therefore has considerable significance. It should not be omitted or inserted carelessly in your calculations.

# PROBLEMS

1. Write an equation for the reaction of methane and oxygen to give carbon dioxide and liquid water. Calculate the equilibrium constant. Given the partial pressures of $CO_2(g)$ and oxygen (Problem 4, Chapter 7), what is the equilibrium amount of methane in the atmosphere? Compare with the amount in the table. Why the discrepancy?

2. The standard free energy change for the reaction

$$2Cu(s) + \frac{1}{2}O_2(g) = Cu_2O(s)$$

as a function of temperature is sometimes given as

$$\Delta_r G^\circ \text{ (J)} = -168600 + 75.729\,T(\text{K})$$

What are the $\Delta_r H^\circ$ and $\Delta_r S^\circ$ of this reaction? (You can tell this directly from the equation given, or you can work it out from the tables, or both). What is the $f_{O_2}$ in equilibrium with Cu and $Cu_2O$ at $600°C$?

3. Calculate the vapor pressure of water at 25 and $100°C$. The reaction is simply

$$H_2O(l) = H_2O(g)$$

4. Write a reaction for the dehydration of diaspore to form corundum and gaseous water (water vapor). Calculate the equilibrium constant. If the fugacity of water in the atmosphere is controlled by evaporation as in Problem 3, which way will the reaction go?

5. Calculate the solubility of amorphous silica (i.e., the concentration of $H_4SiO_4$) in a solution having a water activity of 0.9.

6. Calculate the molality of silica ($m_{H_4SiO_4}$ or $m_{SiO_2(aq)}$) in water that equilibrates with nepheline and albite at $25°C$, 1 bar. Do the same for $100°C$, 1 bar.

7. Explain why coexisting nepheline and albite are said to "buffer" the concentration of silica in a coexisting solution.

8. In a recent large-scale experiment, $CO_2(g)$ produced from the oxidation of organic material in soils seemed to be far less than expected. Then someone suggested that perhaps cement blocks, which contain $Ca(OH)_2$ (*portlandite*) and were enclosing the experiment, were soaking up $CO_2(g)$ in the form of calcite.

   (a) Show that this is indeed possible from the reaction

   $$Ca(OH)_2(portlandite) + CO_2(g) = CaCO_3(calcite) + H_2O(l)$$

(b) What would be the equilibrium partial pressure of $CO_2(g)$ in a room containing both calcite and portlandite?

(c) Show that in the presence of liquid water, lime $[CaO(s)]$ would not be expected in the cement blocks.

(d) What would be the partial pressure of water at equilibrium with lime (CaO) and portlandite?

(e) What is the equilibrium constant of the reaction in part (a) at 150°C? At what temperature would the reaction direction be reversed, i.e., the temperature above which calcite could not form from portlandite and $CO_2(g)$ at atmospheric pressure?

9. One hypothesis for the formation of lead ores, in which the lead occurs as galena (PbS), is that hot, saline solutions carrying both lead ions ($Pb^{2+}$) and sulfate ions ($SO_4^{2-}$), circulating in the crust of the Earth, pass through a rock unit containing methane $[CH_4(g)]$, produced by the heating of organic material in the sedimentary rocks. The reaction could be written

$$Pb^{2+} + SO_4^{2-} + CH_4(g) = PbS(s)\downarrow + CO_2(g) + 2H_2O(l)$$

(a) Calculate the equilibrium constant for this reaction at 150°C.

(b) Consider the solution as it enters the methane-bearing rock. If the fugacity of both methane and $CO_2$ is 1 bar, the sulfate ion concentration is 0.001 molal, and the lead ion concentration is $10^{-6}$ molal, will galena precipitate? Assume that activity coefficients are 1.0 and that the water is sufficiently pure to have unit activity.

(c) After doing this calculation, you realize that probably the figure of $10^{-6}$ molal for lead refers to the *total* lead in solution, not the lead ion itself. In a saline solution, most of the lead will be carried as $PbCl_4^{2-}$, and the relation of this to the lead ion is

$$Pb^{2+} + 4Cl^- = PbCl_4^{2-}$$

If the solution is 3 molal in NaCl ($a_{Cl^-} = 3$), and the activity of $PbCl_4^{2-}$ is $10^{-6}$, what is $a_{Pb^{2+}}$? Does this change your previous answer? Do the calculation for 25°C, not 150°C. $\Delta_f G°$ for aqueous $PbCl_4^{2-}$ is $-557560\,J\,mol^{-1}$ at 25°C.

(d) Then it occurs to you that with so much lead and sulfate in solution, maybe anglesite (PbSO₄) should precipitate. Test this idea by calculating the solubility product for anglesite, assuming that the sulfate activity remains at 0.001. What lead ion activity would be required to precipitate anglesite? Use 25°C.

10. Calculate the vapor pressure (i.e., the $f_{S_2}$) of orthorhombic sulfur at 50°C.

11. Calculate the solubility product of gibbsite. What would be the trivalent aluminum concentration of a solution in equilibrium with gibbsite at a *pH* of 6.0?

12. If a solution having a zinc concentration of 0.173 $m$ at a $pH$ of 5.0 was in contact with quartz, would willemite $[Zn_2SiO_4(s)]$ precipitate? A relevant reaction would be

$$2\,Zn^2 + SiO_2(quartz) + 2\,H_2O(l) = Zn_2SiO_4(s) + 4\,H^+$$

13. Calculate the fugacity of oxygen in equilibrium with solid Ca and solid CaO (lime). Why is pure solid Ca never found in the Earth's crust?

14. Suppose that meteoric water that is percolating down through a bauxite deposit comes to equilibrium with gibbsite and kaolinite. What would be the silica concentration in the water?

15. At another bauxite deposit close to the one in the last question, the groundwater in the bauxite is also in contact with granite boulders containing quartz and feldspars. The weathering of these boulders results in a silica $[SiO_2(aq)]$ concentration in the groundwater of 100 ppm. Would you expect to find kaolinite or gibbsite in the bauxite? Or is there another aluminum oxide or hydroxide you would expect?

16. When carbonate rocks are metamorphosed, dolomite $[CaMg(CO_3)_2]$ and quartz $[SiO_2]$ often react to form diopside $[CaMg(SiO_3)_2]$ and carbon dioxide. Write a balanced reaction for this process, and show whether this reaction should proceed at 25°C or not. If the three minerals were at equilibrium together at 25°C, what would be the fugacity of $CO_2$? At what temperature would $f_{CO_2}$ become equal to 1 bar? Sketch $G$ vs. $T$ for this system, showing two curves, one for diopside $+ CO_2$ and one for dolomite $+$ quartz.

17. Calculate the solubility of $H_2S$ in water if $f_{H_2S} = 1$ bar, at 25°C and at 100°C.

18. (a) Calculate the concentration of silica in the ocean, assuming that it is controlled by the solubility of radiolaria shells, which are made of amorphous silica.

   (b) The $SiO_2$ concentration in sea-water is actually about 7 ppm. Use this number to calculate the Gibbs free energy of formation of $H_4SiO_4(aq)$ and of $SiO_2(aq)$ from the elements, assuming equilibrium with amorphous silica.

19. (a) Calculate the first and second ionization constants of aqueous hydrogen sulfide $[H_2S(aq)]$. Combine these to get the equilibrium constant for the reaction
$$H_2S(aq) = 2\,H^+ + S^{2-}$$

   (b) Calculate this equilibrium constant directly. It should be identical.

(c) Use the solubility product of PbS and this equilibrium constant to calculate the lead ion content of a solution saturated with hydrogen sulfide at 25°C, 1 bar ($m_{H_2S(aq)}$ = 0.1) and a *pH* of 4.0. Assume all activity coefficients are 1.0.

20. Calculate the ionization constant of acetic acid at 25 and 100°C.

21. Calculate the proportions (mole fractions) of an equilibrium mixture of $NO_2$ and $N_2O_4$ gases at 25°C, 1 bar.

# 9

# REDOX REACTIONS

## 9.1 WHY STUDY REDOX?

Normal seawater contains about 2660 ppm (0.028 $m$) of sulfur in the form of sulfate ($SO_4^{2-}$). Sulfur in this form has a valence of +6, meaning that it has six fewer electrons per atom than has native sulfur, which exists (though not in the ocean) as a yellow crystalline solid. In some parts of the ocean, however, sulfur exists in the form of dissolved hydrogen sulfide, $H_2S(aq)$. Sulfur in this form has a valence of $-2$, meaning that it has two extra electrons compared to native sulfur, and in this form it is a deadly poison. Those parts of the ocean containing this electron-rich form of sulfur contain no living organisms other than a few kinds of bacteria. Obviously, the number of electrons that each sulfur atom has is not a question of interest only to atomic physicists. Changing sulfate-sulfur to $H_2S$-sulfur or viceversa involves transferring electrons from one to the other, and this electron transfer is the basic element of redox (reduction-oxidation) reactions.

Many naturally occurring elements in addition to sulfur show similar variations in their number of electrons, with similar large differences in their chemical properties. It would be difficult to overemphasize the importance to us of these variations in valence, or numbers of electrons per atom. Biochemistry, for example, is in large part a study of redox reactions. Because natural environments show great variability in their redox state, we need to develop some kind of measurement, an index, which will be useful in characterizing these redox states, much as we use $pH$ as a measurement or index to characterize the acidity of various states, or temperature as a measurement or an index of the hotness of states. In this chapter we develop two such indexes of redox state.

## 9.2   ELECTRON TRANSFER REACTIONS

You may not have noticed it, but we have considered two kinds of reactions in previous chapters. In some, such as

$$SiO_2(s) + 2H_2O = H_4SiO_4(aq) \tag{8.1}$$

all elements on the right side have the same number of electrons that they have on the left side—there is no change in valence of any element. In others, such as

$$CH_4(g) + O_2(g) = CO_2(g) + 2H_2(g) \tag{8.12}$$

there is such a change. For example, the carbon in $CH_4$ is $C^{4-}$, and the carbon in $CO_2$ is $C^{4+}$. Each carbon atom in methane that changes to a carbon atom in carbon dioxide must get rid of 8 electrons—it is *oxidized*. Where do the electrons go? Obviously, they go to the other actors in the reaction. Oxygen in $O_2$ has a valence of zero ($O^0$), while in carbon dioxide it is $-2$ ($O^{2-}$), so in changing from $O_2$ to $CO_2$, two oxygens gain 4 electrons. The other 4 electrons go to hydrogen, which has a valence of $+1$ ($H^+$) in methane and zero in hydrogen gas. Both oxygen and hydrogen are *reduced*, if the reaction goes from left to right as written. Similarly in reaction (8.11),

$$6\,Fe_2O_3(s) = 4\,Fe_3O_4(s) + O_2(g) \tag{8.11}$$

we see that all of the iron atoms in $Fe_2O_3$ are ferric iron ($Fe^{3+}$), while one out of three iron atoms in $Fe_3O_4$ is ferrous iron ($Fe^{2+}$). The iron is partially reduced, while some oxygen in $Fe_2O_3$ is oxidized to $O_2(g)$—there is a transfer of electrons from iron to oxygen, or from oxygen to iron, depending on which way the reaction goes. Without such electron transfers, these and many other reactions, including many necessary to life processes, could not proceed.

## 9.3   THE ROLE OF OXYGEN

Both of our examples involve oxygen, which is the most common *oxidizing agent* in natural systems. In the presence of oxygen, many elements are oxidized (lose electrons, gain in valence), while oxygen is reduced. You need only think of rusty nails, green staining on copper objects, and burning logs to realize the truth of this. The process of oxidation obviously takes its name from the fact that oxygen is the premier oxidizing agent, but it is actually defined in terms of electron loss, or increase in valence. In other words, the electrons need not come from or go to oxygen; many redox reactions take place without oxygen.

Consider, for example, what happens when you put a piece of iron in a solution of copper sulfate (Figure 9.1). After a while you see the characteristic color of metallic copper forming on the surface of the iron, and the iron gradually crumbles and eventually disappears. Metallic copper precipitates, and iron

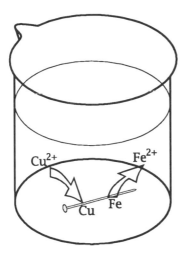

Figure 9.1: An iron nail in a solution of copper sulfate.

dissolves. The reaction is essentially

$$Cu^{2+} + Fe \rightarrow Cu + Fe^{2+} \tag{9.1}$$

We need not include the sulfate, because it is not involved in this process—being negatively charged, the $SO_4^{2-}$ ions provide an overall charge balance in the solution. In this example, copper is reduced and iron is oxidized, without the aid of oxygen. (Of course, if we wait long enough, and our solution is open to the atmosphere, oxygen dissolved in the solution will eventually oxidize the copper and the ferrous iron.)

What does thermodynamics tell us about this reaction? Looking up the data from Appendix B, we find

$$
\begin{aligned}
\Delta_r G^\circ &= \Delta_f G^\circ_{Cu} + \Delta_f G^\circ_{Fe^{2+}} - \Delta_f G^\circ_{Cu^{2+}} - \Delta_f G^\circ_{Fe} \\
&= 0 + (-78.90) - 65.49 - 0 \\
&= -144.390\,\text{kJ} \\
&= -144390\,\text{J}
\end{aligned}
$$

This means that with both metallic copper and iron present, and cupric and ferrous ions present at 1 molal concentration (and acting ideally), the reaction would proceed spontaneously, as observed. But more interestingly,

$$
\begin{aligned}
\Delta_r G^\circ &= -RT \ln K \\
-144390 &= -(8.3145 \times 298.15) \ln K
\end{aligned}
$$

$$
\begin{aligned}
\log K &= 144390/(2.30259 \times 8.3145 \times 298.15) \\
&= 25.295
\end{aligned}
$$

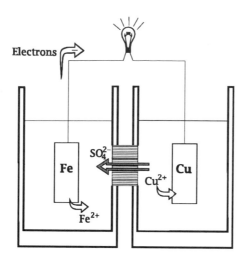

Figure 9.2: An electrolytic cell. Iron dissolves on one side, and copper precipitates on the other. A porous liquid junction allows sulfate to migrate between the solutions. The cell reaction is identical to the reaction in Figure 9.1.

Thus

$$K = \frac{a_{Fe^{2+}}}{a_{Cu^{2+}}}$$
$$= 10^{25.295}$$

This means that to reach equilibrium, the activity (~concentration) of $Fe^{2+}$ would have to be enormously greater than that of the $Cu^{2+}$, so the reaction will always proceed as written, and cannot be made to go the other way (at least, not by simply adjusting the ion concentrations).

## 9.4   A SIMPLE ELECTROLYTIC CELL

Obviously in this case, and in most natural circumstances, the electron transfer takes place on a molecular level, and we have no control over it. However, what if we could separate the iron and copper, and have the electrons travel through a wire from one to the other? Well, why not? In Figure 9.2, the beaker has been divided into two parts. In one we have a solution of ferrous ions and sulfate ions in contact with a piece of iron (an iron electrode), and in the other we have a solution of cupric ions and sulfate ions in contact with a piece of copper (a copper electrode). Considered separately (and in the absence of oxygen, which would oxidize both electrodes), nothing at all happens. But if we connect a wire between the electrodes, a current begins to flow, because reaction (9.1) wants to occur, and now it can. Iron dissolves, forming more

$Fe^{2+}$ in solution, and the electrons, instead of attaching themselves to some immediately adjacent copper ions, must travel through the wire before being able to do that. That is, when they get to the copper electrode, they jump onto some immediately adjacent $Cu^{2+}$ ions, causing copper to precipitate. If the two cells are completely separate, a positive charge would soon build up in the iron solution and a negative charge in the copper solution, stopping the reaction; so we have to provide some kind of connection (a *liquid junction*), which allows sulfate ions to migrate from one solution to the other.

By separating the two parts of the redox reaction, we have caused a current to flow through a wire. We have a simple battery, which we could use to power a light bulb, or do other useful things.

## 9.4.1 The Cell Voltage

An electrical circuit delivering direct current such as we have described obviously must have a voltage difference between the two electrodes. If we leave the wire connected to both electrodes, the cell will continue to operate until the iron all dissolves, or until there are no more copper ions to react. That is, the battery will run down, and whatever voltage we had at the beginning decreases to zero during the experiment. The voltage of the cell at any particular point is measured by attaching a voltmeter (which has an extremely high resistance) or a potentiometer (which imposes a voltage in the circuit equal and opposite to the cell voltage) across the electrodes, instead of a piece of wire. Either way, the current flow is stopped, and the voltage measured under equilibrium (no current flow in either direction) conditions. This equilibrium state is, of course, different from the equilibrium state reached when the cell "runs down" and can react no longer. It is a higher energy state that is prevented from reaching a lower energy state by some constraint, in this case an opposing voltage or a high resistance. It is in fact another example of a metastable equilibrium state.

What determines the magnitude of this cell voltage, when it is not zero? Intuitively, we would suspect that it depends a lot on what metals we use to make the electrodes, and if we thought a bit more, we might think that the concentrations of the ions in solution would have an effect too. This is exactly the case, and we must develop an equation relating the voltage to the activities of the reactants and products in the cell reaction.

## 9.4.2 Half-Cell Reactions

An oxidation reaction cannot take place without an accompanying reduction reaction—the electrons have to go somewhere—but it is convenient to nonetheless split cell reactions into two complementary "half-cell" reactions. In our copper–iron case, these half-cells are

$$\text{Oxidation}: \quad Fe(s) = Fe^{2+} + 2e \tag{9.2}$$

$$\text{Reduction}: \quad Cu^{2+} + 2e = Cu(s) \tag{9.3}$$

where $e$ is an electron, and the cell reaction is the sum of these, equation (9.1).

We also imagine that each half-cell reaction has a half-cell voltage associated with it, and that the cell voltage is the sum of the two half-cell voltages. If we had these half-cell potentials, or voltages,[1] we could tabulate them and mix and match electrodes to calculate the potential of any cell we wanted, much as we tabulate free energies of compounds so as to be able to calculate the $\Delta_r G^\circ$ of any reaction. Of course, half-cell voltages cannot be measured, just as the $G$ of compounds cannot be measured, but we can get around this, just as we did with $\Delta_f G^\circ$.

With free energies, we tabulate the difference between the $G$ of a compound and the sum of the $G$s of its constituent elements. The elements and their free energies always cancel out in balanced reactions. With electrodes, we measure and tabulate the difference of every electrode against a standard electrode, and the potential of this standard electrode cancels out in balanced cell reactions. The standard electrode chosen is the *standard hydrogen electrode*, or SHE (Figure 9.3). So if we measure both the copper electrode and the iron electrode and many others in separate experiments against the SHE, we will then be able to calculate the potential of any cell from these tabulated values. Although we can tabulate potentials for every kind of electrode, we wouldn't want to tabulate potentials for every conceivable concentration of product and reactant ions; that's very inefficient. It would be better to tabulate the potentials of each electrode for some standard conditions and to have an equation that could give the potential of the electrode (and of cells constructed from electrodes) at any particular concentrations we are interested in [see equation (9.14)].

### 9.4.3  Standard Electrode Potentials

A cell for measuring the potential of the copper–SHE cell is shown in Figure 9.3. The hydrogen electrode is a device for using the reaction

$$2H^+ + 2e \rightleftharpoons H_2(g) \tag{9.4}$$

as an electrode. The idea of using a gas as an electrode seems rather bizarre at first. It is accomplished by bubbling hydrogen gas over a specially treated piece of platinum (platinum coated with fine-grained carbon). The platinum serves simply as a source or sink of electrons; the reaction between hydrogen gas and hydrogen ions takes place at the surface of the platinum and is catalyzed by the carbon.

In order that the hydrogen electrode always give the same potential, the activities of both the hydrogen gas and the hydrogen ion must always be the same. These have been standardized at $f_{H_2(g)} = 1$ bar, and $a_{H^+} = 1$, that is, a gas pressure of one bar, and an acid concentration of about 1 molal. A hydrogen

---

[1]We use the terms *voltage* and *potential* synonymously here. Strictly speaking, voltage is just a particular kind of potential.

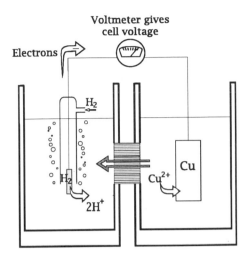

Figure 9.3: An electrolytic cell for measuring the potential of the copper electrode against the standard hydrogen electrode. If the activities of all ions and elements are 1.0, the cell voltage is the standard cell voltage, $\mathcal{E}°$. The direction of electron flow is indicated for reduction at the copper electrode. When a measurement is made, no current flows.

electrode operating under these conditions, the standard hydrogen electrode, is assigned a half-cell potential of zero volts by convention.

The half-cell reactions for the cell in Figure 9.3 are

$$\text{Oxidation:} \quad H_2(g) = 2H^+ + 2e \tag{9.5}$$

$$\text{Reduction:} \quad Cu^{2+} + 2e = Cu(s) \tag{9.6}$$

and the sum of these is the cell reaction

$$Cu^{2+} + H_2(g) = Cu(s) + 2H^+ \tag{9.7}$$

As we mentioned, the cell voltage depends on the activities of all the ions and compounds in the cell reaction; in this case it depends not only on the hydrogen gas pressure and $a_{H^+}$, but on $a_{Cu}$ and $a_{Cu^{2+}}$ as well. "Standard conditions" is defined as $a = 1$ for all products and reactants in the cell reaction, and so if the hydrogen electrode is operating under SHE conditions ($a_{H_2(g)} = 1$; $f_{H_2(g)} = 1$ bar), the copper electrode is pure Cu ($a_{Cu(s)} = 1$) and the cupric ion concentration and activity coefficient are adjusted to give $a_{Cu^{2+}} = 1$, the cell voltage will be the *standard* cell voltage, $\mathcal{E}°$

## 9.5  THE NERNST EQUATION

### 9.5.1  Work Done by Cells

In lighting the light bulb or running a small motor, our cell in Figure 9.2 is doing work. We have already seen, in Chapter 4, how much work can be done by a chemical reaction, but you may have forgotten this because we have put so much emphasis on reactions that do *no* work, other than the minimum necessary $P\Delta V$ work. But here we have a chemical reaction that is certainly doing $P\Delta V$ work (the $Fe^{2+}$ and Cu in the cell reaction will have a slightly different molar volume than the Fe and $Cu^{2+}$, and so some work is done against the atmospheric pressure on the solution), but in addition to this, it is doing work in lighting the bulb. According to §4.4.2, the maximum amount of work we can get from our cell reaction is given by the $\Delta_r G$ of that reaction, and when $\Delta_r G$ decreases to zero, we reach stable equilibrium and can get no more work from the cell.

The electrical work $w$ required to move a charge of $\mathcal{F}$ coulombs through a potential difference $\mathcal{E}$ volts is

$$-w = \mathcal{F}\mathcal{E}$$

(joules = coulombs $\times$ volts)

where $\mathcal{F}$ is the charge per mole of electrons, so if $n$ is the number of electrons appearing in the reaction as written, there are $n\mathcal{F}$ coulombs of charge, and the work is

$$- w = n\mathcal{F}\mathcal{E} \tag{9.8}$$

Because some convention must be adopted to know whether the voltage $\mathcal{E}$ is positive or negative, you may see equation (9.8) [as well as (9.9) and (9.10), below] written without the minus sign in some references. The conventions we have adopted (§9.6.1) require the minus sign.

This electrical work is by definition (Chapter 4) the $\Delta G$ associated with the process, as long as the electrical work is the only non-$P\Delta V$ work done. Therefore for any process in which $n\mathcal{F}$ coulombs are moved through a potential difference $\mathcal{E}$,

$$\Delta G = -n\mathcal{F}\mathcal{E} \tag{9.9}$$

or

$$\Delta G^\circ = -n\mathcal{F}\mathcal{E}^\circ \tag{9.10}$$

for standard state conditions. As applied to electrochemical cells, these equations are more properly $\Delta\mu = -n\mathcal{F}\mathcal{E}$ and $\Delta\mu^\circ = -n\mathcal{F}\mathcal{E}^\circ$ because many of the individual free energy terms refer to constituents in solution and hence are partial molar terms. These equations connect electrochemistry to the world of thermodynamics. They allow us to calculate the voltage that will be observed in any cell for which we know the cell reaction and the $\Delta_r G$ or the $\Delta_r G^\circ$.

## 9.5.2 Relation Between Cell Activities and Voltage

Consider the general cell reaction

$$b\text{B} + c\text{C} = d\text{D} + e\text{E} \tag{9.11}$$

Let's say that this reaction reaches equilibrium with an external measuring system, giving cell voltage $\mathcal{E}$. If operated under standard conditions, it would give cell voltage $\mathcal{E}^\circ$, and the corresponding free energies of reaction are

$$\Delta_r\mu = -n\mathcal{F}\mathcal{E} \tag{9.12}$$

and

$$\Delta_r\mu^\circ = -n\mathcal{F}\mathcal{E}^\circ \tag{9.13}$$

From equation (8.6) we have

$$\Delta_r\mu = \Delta_r\mu^\circ + RT\ln Q \tag{8.6}$$

where

$$Q = \prod_i a_i^{\gamma_i}$$

$$= \frac{a_E^e a_D^d}{a_B^b a_C^c}$$

Recall from Chapter 8 that this activity term is referred to as $Q$ rather than $K$ because it refers to a metastable equilibrium. Substitution of equations (9.12) and (9.13) gives

$$\mathcal{E} = \mathcal{E}^\circ - \frac{RT}{n\mathcal{F}}\ln Q$$

or

$$\mathcal{E} = \mathcal{E}^\circ - 2.30259\frac{RT}{n\mathcal{F}}\log Q \tag{9.14}$$

This is the Nernst equation, after the physical chemist W. Nernst, who derived it at the end of the last century. As above, $n$ is the number of electrons transferred in the cell reaction [2 in reaction (9.7)], $\mathcal{F}$ the Faraday of charge, $R$ the gas constant, and $T$ the temperature (in Kelvins). The constant 2.30259 is used to convert from natural to base 10 logs. At 25°C the quantity 2.30259 $RT/\mathcal{F}$ has the value 0.05916, which is called the Nernst slope. The importance of (9.14) is that it allows calculation of the potentials of cells having nonstandard state concentrations (i.e., real cells) from tabulated values of standard half-cell values or tabulated standard free energies.

Equation (9.11) could also be considered to represent a half-cell reaction, except that the electron is not shown. So evidently we could use the Nernst equation to calculate half-cell potentials if we knew what value to assign the chemical potential of an electron. It turns out, of course, that because the

electrons always cancel out in balanced reactions, we could assign any value
we like to the electron free energy and it would make no difference to our
calculated cell potentials. The easiest value to assign is zero, and that is what
is done. Therefore, the Nernst equation is used to calculate both half-cell and
cell potentials.

## 9.6  SOME NECESSARY CONVENTIONS

In our discussion so far, we have skipped lightly over some points which are
not important to a general understanding, but which if neglected will result in
getting the wrong answers in calculations. For example, we said (§9.4.2) that the
cell voltage is the *sum* of the two half-cell voltages. Actually, it is a little more
complicated. Because half-cell and cell reactions may be written forwards or
backwards, and a voltage by itself is not obviously positive or negative, there has
to be a set of rules to keep things straight. Unfortunately, there is more than one
set of rules. We present here the rules set out by the IUPAC (International Union
of Pure and Applied Chemistry), which are followed by most people today. Be
warned, however, that several geochemical sources use a different set of rules.

### 9.6.1  The IUPAC Rules

Some of these points have been already discussed. We include all the rules here
for completeness.

1. Cell reactions are written such that the left-hand electrode supplies electrons to the outer circuit (i.e., oxidation takes place), and the right-hand
   electrode accepts electrons from the outer circuit (i.e., reduction takes
   place).

2. The cell potential is given by

$$\mathcal{E} = \mathcal{E}_{\text{right electrode}} - \mathcal{E}_{\text{left electrode}}$$

   that is,

$$\mathcal{E} = \mathcal{E}_{\text{reduction electrode}} - \mathcal{E}_{\text{oxidation electrode}}$$

3. The cell potential is related to the Gibbs free energy by

$$\Delta_r G = -n\mathcal{F}\mathcal{E}$$

4. The electrode potential of a half-cell is equal in magnitude and sign to the
   potential of a cell formed with the electrode on the right and the standard
   hydrogen electrode ($\mathcal{E}° = 0$) on the left.

5. Standard half-cell reactions are tabulated and calculated as reductions, for example,

$$Zn^{2+} + 2e = Zn(s) \qquad \mathcal{E}^{\circ} = -0.763 \, V$$

However, the half-cell potential is a sign-invariant quantity, that is,

$$Zn(s) = Zn^{2+} + 2e \qquad \mathcal{E}^{\circ} = -0.763 \, V$$

6. For the reaction

$$bB + cC = dD + eE$$

the Nernst expression is

$$\mathcal{E} = \mathcal{E}^{\circ} - \frac{RT}{n\mathcal{F}} \ln \left( \frac{a_E^e a_D^d}{a_B^b a_C^c} \right)$$

7. In view of item 5, the Nernst expression for a half-cell is given by

$$\mathcal{E} = \mathcal{E}^{\circ} - \frac{RT}{n\mathcal{F}} \ln \left( \frac{\text{reduced form}}{\text{oxidized form}} \right)$$

## 9.6.2 Examples

Let's calculate the potential of the cell in Figure 9.2. By convention 5, both half-cells are written and calculated as reductions, no matter what is happening in the real cell. Thus, for the copper half-cell,

$$Cu^{2+} + 2e = Cu(s)$$

$$
\begin{aligned}
\Delta_r G^{\circ} &= \Delta_f G^{\circ}_{Cu(s)} - \Delta_f G^{\circ}_{Cu^{2+}} \\
&= 0 - 65.49 \, kJ \\
&= -65490 \, J \\
&= -n\mathcal{F}\mathcal{E}^{\circ}
\end{aligned}
$$

Because two electrons are involved, $n = 2$, so

$$
\begin{aligned}
\mathcal{E}^{\circ}_{Cu \, half-cell} &= -\frac{-65490}{2 \times 96485} \\
&= 0.339 \, V
\end{aligned}
$$

For the iron half-cell,

$$Fe^{2+} + 2e = Fe(s)$$

$$
\begin{aligned}
\Delta_r G^{\circ} &= \Delta_f G^{\circ}_{Fe(s)} - \Delta_f G^{\circ}_{Fe^{2+}} \\
&= 0 - (-78.90) \, kJ \\
&= 78900 \, J \\
&= -n\mathcal{F}\mathcal{E}^{\circ}
\end{aligned}
$$

$$\mathcal{E}^{\circ}_{\text{Fe half-cell}} = -\frac{78900}{2 \times 96485}$$
$$= -0.409 \, \text{V}$$

For the complete cell,

$$\mathcal{E}^{\circ}_{\text{Cu-Fe cell}} = \mathcal{E}^{\circ}_{\text{reduction half-cell}} - \mathcal{E}^{\circ}_{\text{oxidation half-cell}}$$
$$= \mathcal{E}^{\circ}_{\text{Cu half-cell}} - \mathcal{E}^{\circ}_{\text{Fe half-cell}}$$
$$= 0.339 - (-0.409)$$
$$= 0.748 \, \text{V}$$

Note that if we wrote the cell backwards,

$$Cu(s) + Fe^{2+} = Cu^{2+} + Fe(s)$$

both half-cell reactions would still be written and calculated as reductions, but the cell voltage would now be

$$\mathcal{E}^{\circ}_{\text{Cu-Fe cell}} = \mathcal{E}^{\circ}_{\text{reduction half-cell}} - \mathcal{E}^{\circ}_{\text{oxidation half-cell}}$$
$$= \mathcal{E}^{\circ}_{\text{Fe half-cell}} - \mathcal{E}^{\circ}_{\text{Cu half-cell}}$$
$$= -0.409 - (0.339)$$
$$= -0.748 \, \text{V}$$

Thus the signs of both $\Delta_r G^{\circ}$ and $\mathcal{E}^{\circ}$ of the complete cell reaction depend on how the cell is written, but the signs of the half-cell reactions do not.

These are the standard potentials. Suppose the cell is operating under non-standard (real) conditions. Let's say $a_{Cu^{2+}}$ is not 1 but 0.1, and $a_{Fe^{2+}}$ is 1.5. Using the Nernst expression for the complete cell [reaction (9.1)] at 25°C,

$$\mathcal{E} = \mathcal{E}^{\circ} - 2.30259 \frac{RT}{n\mathcal{F}} \log Q$$
$$= \mathcal{E}^{\circ} - \frac{0.05916}{n} \log \frac{a_{Fe^{2+}}}{a_{Cu^{2+}}}$$
$$= 0.748 - \frac{0.05916}{2} \log \frac{1.5}{0.1}$$
$$= 0.748 - 0.0348$$
$$= 0.713 \, \text{V}$$

So the real cell with these concentrations would have a potential of 0.713 V, rather than 0.748 V.

The Nernst expression can also be used for the half-cells. Thus for the Cu half-cell,

$$\mathcal{E} = \mathcal{E}^{\circ} - 2.30259 \frac{RT}{n\mathcal{F}} \log Q$$

$$
\begin{aligned}
&= \mathcal{E}^{\circ} - \frac{0.05916}{n} \log \frac{a_{Cu}}{a_{Cu^{2+}}} \\
&= 0.339 - \frac{0.05916}{2} \log \frac{1}{0.1} \\
&= 0.339 - 0.0296 \\
&= 0.309 \, V
\end{aligned}
$$

and for the Fe half-cell,

$$
\begin{aligned}
\mathcal{E} &= \mathcal{E}^{\circ} - 2.30259 \frac{RT}{n\mathcal{F}} \log Q \\
&= \mathcal{E}^{\circ} - \frac{0.05916}{n} \log \frac{a_{Fe}}{a_{Fe^{2+}}} \\
&= -0.409 - \frac{0.05916}{2} \log \frac{1}{1.5} \\
&= -0.409 + 0.0052 \\
&= -0.404 \, V
\end{aligned}
$$

The cell potential is then

$$
\begin{aligned}
\mathcal{E}_{Cu-Fe \ cell} &= \mathcal{E}_{reduction \ half-cell} - \mathcal{E}_{oxidation \ half-cell} \\
&= \mathcal{E}_{Cu \ half-cell} - \mathcal{E}_{Fe \ half-cell} \\
&= 0.309 - (-0.404) \\
&= 0.713 \, V
\end{aligned}
$$

as before.

There are certainly other ways to do these calculations and have them come out right. However, all conventions have their good and bad points, and the IUPAC conventions are the most commonly used.

### 9.6.3 $\mathcal{E}^{\circ}$ as a Source of $\Delta_f G^{\circ}$

We now know all we need to know about electrochemical cells, how their potentials are calculated from free energies, and how half-cell reactions can be combined into complete cell reactions. Electrochemical cells are of course of great practical importance in the form of batteries and fuel cells. In thermodynamics, the relationship between cell potential and free energy is often used the other way around. That is, cell potentials are one of the most accurate and useful sources of information about free energies of reactions and dissolved substances. Not all free energy data come from calorimetry. Most data for both $\Delta_f G^{\circ}_{Fe^{2+}}$, $\Delta_f G^{\circ}_{Cu^{2+}}$, and other ionic species come from the measurement of $\mathcal{E}^{\circ}$ in cells such as shown in Figure 9.3.

# 9.7   MEASURING REDOX CONDITIONS

So far we have only considered cells that we might construct ourselves, in the laboratory. For the scientist interested in natural environments, this is background information. What we really want to know is how to characterize natural environments as being either reducing or oxidizing on some numerical scale. Natural environments don't normally have electrodes sticking out of them, so what is the connection?

In the absence of electrodes and voltages, the redox state of a solution is characterized by the relative concentrations of reduced and oxidized ionic or molecular species in the solution. Thus in a solution containing Fe ions, the solution is relatively reduced if there are more $Fe^{2+}$ ions than $Fe^{3+}$ ions, and viceversa. In a solution containing carbon species, the solution is relatively reduced if there are more $CH_4$ molecules than $CO_2$ molecules. In a solution containing sulfur, the solution is relatively reduced if there are more $H_2S$ molecules than $SO_4^{2-}$ ions, and viceversa. And so on. For those elements that have more than one valence state in natural environments, the two (or more) states will be present in various ratios, depending on whether the environment is reducing or oxidizing. In a solution containing all of these, the $a_{Fe^{2+}}/a_{Fe^{3+}}$, $a_{CH_4}/a_{CO_2}$, and $a_{H_2S}/a_{SO_4^{2-}}$ ratios will all be different, but each will be controlled by the same factors ($T$, $P$, and the bulk composition of the solution), and so each one should give us the same index of redox conditions, if equilibrium prevails.

But what is this index? We will consider two commonly used ones, $Eh$ and $f_{O_2}$.

## 9.7.1   Redox Potential, $Eh$

Suppose you have a sample solution that contains both ferrous and ferric ions. The ferrous/ferric ratio is a measure of how reduced/oxidized the sample solution is. For any change in this ratio, some reaction involving electron transfer must take place. What we must do is insert an electrode that will supply/absorb these electrons. In other words, we need an electrode that responds to the ferrous/ferric ratio, that is, that has a half-cell potential that varies with this ratio. We could insert this electrode in the solution, connect it to a SHE, and the measured cell potential would depend on the ferrous/ferric ratio in the solution. Finding such an electrode is easier than you might think.

Note that in the Cu and Fe electrodes we have considered, the "reduced form" in both cases is the metal, Cu or Fe. The "oxidized form" is an ion in solution, $Cu^{2+}$ or $Fe^{2+}$, and the electrode, being made of the metal, is a necessary part of the half-cell. However, in the SHE, both the reduced form ($H_2$) and the oxidized form ($H^+$) are in the solution (one as a gas phase); neither is part of the electrode. The platinum electrode itself is nothing but a source or sink for electrons. We are now considering another case where both the reduced form ($Fe^{2+}$) and the oxidized form ($Fe^{3+}$) are in the solution, and so all we have to do

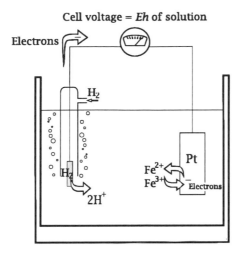

Figure 9.4: How to measure the *Eh* of a solution containing both ferrous and ferric ions. The direction of electron flow is indicated for reduction at the platinum surface. When a measurement is made, no electrons are flowing.

is provide a source and sink for electrons. All we need is a piece of platinum, as shown in Figure 9.4.

Suppose the solution is quite reduced, with $a_{Fe^{2+}}/a_{Fe^{3+}} = 10$. The half-cell reactions are (both written as reductions)

$$2H^+ + 2e = H_2(g) \qquad \mathcal{E}^\circ = 0.0\,V$$

and

$$Fe^{3+} + e = Fe^{2+} \qquad \mathcal{E}^\circ = 0.769\,V$$

The complete cell is

$$Fe^{3+} + \tfrac{1}{2}H_2(g) = Fe^{2+} + H^+$$

The Nernst equation may be written for either the complete cell or the Fe half-cell, giving the same answers. Thus

$$
\begin{aligned}
\mathcal{E} = Eh &= \mathcal{E}^\circ - \frac{0.05916}{1} \log \frac{a_{Fe^{2+}}}{a_{Fe^{3+}}} \\
&= 0.769 - 0.5916 \times \log(10) \\
&= 0.710\,V
\end{aligned}
$$

So the *Eh* of this relatively reduced solution is 0.710 V. If the solution is quite oxidized, with $a_{Fe^{2+}}/a_{Fe^{3+}} = 0.1$,

$$
\begin{aligned}
\mathcal{E} = Eh &= \mathcal{E}^\circ - \frac{0.05916}{1} \log \frac{a_{Fe^{2+}}}{a_{Fe^{3+}}} \\
&= 0.769 - 0.05916 \times \log(0.1) \\
&= 0.828\,V
\end{aligned}
$$

and the $Eh$ of this more oxidized solution is 0.828 V.

To summarize, you may think of $Eh$ of a solution as either a cell potential or a half-cell potential. It is the potential of a cell having one electrode that responds reversibly to a redox couple (such as $Fe^{2+}/Fe^{3+}$) or couples in the solution and the SHE as the other electrode. Or it is the half-cell potential of an electrode responding reversibly to a redox couple or couples in the solution. You may use any kind of electrode as the other side of the cell, as long as you correctly deduce the half-cell potential of the electrode that is responding to conditions in your solution. Half-cell potentials are *defined* in terms of the SHE by our IUPAC conventions.

Therefore, we have an index of redox conditions, only assuming that reduced and oxidized species in solution can readily exchange electrons at a platinum surface. There are some practical difficulties in this respect, as discussed below.

### 9.7.2   Redox Couples Other Than Iron

But suppose our sample solution in Figure 9.4 contains not only the ferrous/ferric redox couple, but also $Mn^{2+}$–$Mn^{4+}$, $H_2S$–$SO_4^{2-}$, $CH_4$–$CO_2$, and others. Even if all these redox couples are at equilibrium, each will have a different activity ratio. To which does our platinum electrode respond? In theory, it responds to all of them simultaneously, and all result in the same $Eh$. Each couple has a different activity ratio, but each also has a different value of $\mathcal{E}°$, and the resulting $Eh$ for the solution must be the same for each redox couple if they are at equilibrium, and if each reacts with the platinum electrode.

### 9.7.3   Some Practical Difficulties

#### No Unique $Eh$ Values

However, in practice the platinum electrode does not respond to all redox couples equally. In fact, the only ones it responds to well are Fe and Mn, and then only if concentrations are high enough. Sulfur, carbon, and many other redox pairs simply do not give up or take up electrons easily at the platinum surface, so measured $Eh$ values primarily reflect the ferrous/ferric ratio in solutions. Of course, if equilibrium prevails, this should be enough—the activity ratios of all other couples could be calculated if the ferrous/ferric (and $pH$) ratio is known. However, at Earth surface conditions in natural environments, equilibrium often does not prevail, and the only way to really know the activities of many redox pairs is to analyze the solution for both parts of the pair.

Natural solutions that have not internally equilibrated therefore cannot be said to have a unique $Eh$. Nevertheless, $Eh$ measurements are often useful in a qualitative sense and may be quite accurate in some situations, such as acid Fe-rich mine drainage systems. But even if $Eh$ measurements were not useful at all, the concept of $Eh$ is firmly established in the literature of natural environments, where it is used primarily in discussing *models* or hypothetical situations.

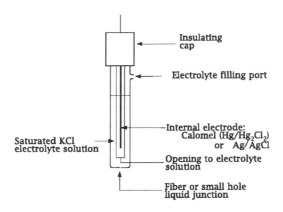

Figure 9.5: A silver/silver chloride reference electrode.

**Reference Electrodes**

Another practical difficulty is that the hydrogen electrode is a rather delicate apparatus, really only suitable for laboratory use. How do we get it into the field, to take $Eh$ measurements in natural environments? We don't. Other electrodes can be designed that have fixed potentials, independent of what solution they are put into, called reference electrodes. Their potentials with respect to the SHE can be measured, so that field measurements of cells composed of a platinum electrode and a reference electrode can be made, and the readings corrected to what they would have been had a SHE been used, giving the $Eh$.

Two common reference electrodes are the calomel ($Hg/Hg_2Cl_2$) and the silver/silver chloride ($Ag/AgCl$) electrodes. In both, the activities of all active parts of the electrode are fixed, so the electrode potential is fixed at a given temperature. For example, the $Ag/AgCl$ electrode half-cell reaction is

$$AgCl + e = Ag(s) + Cl^-$$

Both $Ag$ and $AgCl$ are present as solid phases so $a_{Ag} = a_{AgCl} = 1$, and these are immersed in a solution saturated with solid KCl, which fixes $a_{Cl^-}$ at a constant value (Figure 9.5). With all reactants and products of the half-cell reaction having fixed activities, $\mathcal{E}^\circ_{Ag/AgCl \; half-cell}$ has a fixed value, which is 0.222 V. That is, a cell composed of a $Ag/AgCl$ electrode on one side and the SHE on the other will record a constant potential of 0.222 V. When used as a reference electrode, immersed in a solution containing ferrous and ferric ions and connected to a platinum electrode, for example, the cell reaction is

$$Fe^{3+} + Ag(s) + Cl^- = Fe^{2+} + AgCl$$

and the standard cell potential of this cell is

$$\mathcal{E}^\circ = \mathcal{E}^\circ_{Fe^{2+}/Fe^{3+}} - \mathcal{E}^\circ_{Ag/AgCl}$$

$$= \quad -0.404 - 0.222$$
$$= \quad -0.626\,V$$

whereas if measured against SHE, $\mathcal{E}^{\circ}_{Fe^{2+}/Fe^{3+}}$ is $-0.404\,V$. $Eh$ measurements are defined as observed cell voltages using the SHE as reference, so if Ag/AgCl is used instead, $0.222\,V$ must be *added* $(-0.626 + 0.222 = -0.404)$ to the observed readings. For calomel reference electrodes, $0.268\,V$ must be added.

## 9.8   $Eh$–$pH$ Diagrams

The only other intensive variable of comparable significance in aqueous systems is $pH$. It too is a function of the bulk composition at a given $T$ and $P$, but both are closely related to a large number of important reactions. Therefore, it proves natural to use both as variables in diagrams of systems at fixed $T$ and $P$, and $Eh$–$pH$ diagrams have become a standard method of displaying and interpreting geochemical data. In the following sections we outline the theoretical basis for calculating these diagrams.

In this section, we will calculate portions of a simple $Eh$–$pH$ diagram for the system Mn–$H_2O$. This illustrates most of the problems encountered in calculating such diagrams. If you wish to add components such as $CO_2$ or $H_2S$, the methods are similar, and details are provided by Garrels and Christ (1965).

### 9.8.1   General Topology of $Eh$–$pH$ Diagrams

First, let us examine the completed $Eh$–$pH$ diagram for Mn–$H_2O$–$O_2$ in Figure 9.6. There are typically four different types of boundaries shown on these diagrams. The top line, labeled $O_2/H_2O$, represents conditions for water in equilibrium with $O_2$ gas at 1 atm. Above this line, a $P_{O_2}$ greater than 1 atm is required for water to exist, so that because the diagram is drawn for a pressure of 1 atm, water is not stable above this line. Similarly, the bottom line $H_2O/H_2$ represents conditions for water in equilibrium with $H_2$ gas at 1 atm. Below this line, $P_{H_2}$ values greater than 1 atm are required for water to exist, that is, at 1 atm water is not stable. Therefore, the water stability field is between these two lines. The second type of boundary separates the stability fields of minerals or solid phases such as hausmannite ($Mn_3O_4$) and pyrochroite [$Mn(OH)_2$]. These are true phase boundaries: hausmannite is thermodynamically unstable below the hausmannite/pyrochroite boundary and pyrochroite is unstable above it. Thus these first two kinds of boundary represent thermodynamic stability fields for different substances. Notice that on this diagram they all have the same slope (equal to the Nernst slope). The remaining two kinds of lines are not stability boundaries at all but refer to concentrations of dissolved ions. For example, the vertical lines within the pyrochroite stability field represent contoured solubilities of pyrochroite as $Mn^{2+}$ concentrations running from $10^{-1}$ to $10^{-6}\,m$. Finally, the dashed boundaries between aqueous species, such as that between

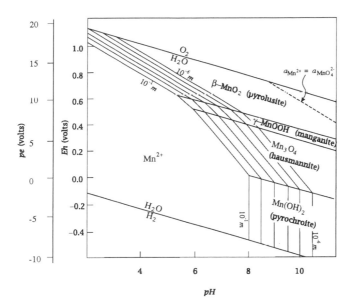

Figure 9.6: *Eh-pH* and *pe-pH* relations in the system Mn-$H_2O$-$O_2$ at 25°C, 1 bar. $Mn^{2+}$ activities and stability fields of Mn-oxide minerals are included. The *pe* and *Eh* axes are related by the formula *pe* = 5040*Eh*/*T*.

$Mn^{2+}$ and $MnO_4^{2-}$, indicate where the activities of the two species are exactly equal. To the right of this line, $Mn^{2+}$ remains present in the solution, but at a lower activity than $MnO_4^{2-}$, and vice versa.

## 9.8.2  Sample Calculations

It is possible to look up half-cell potentials for many reactions in physical chemistry textbooks and compilations of electrochemical data. However, it is usually a better procedure to choose free energy data and to use those to calculate *Eh* and $\mathcal{E}°$ for the reactions of interest.

As the first example, we will calculate the two boundaries for the stability field of water.

For the boundary $H_2O(l) - O_2(g)$, the half-cell reduction reaction is

$$4H^+ + O_2(g) + 4e = 2H_2O(l) \tag{9.15}$$

for which $n = 4$. Using the tabulated $\Delta_f G°$ for water (all others being zero), we find

$$
\begin{aligned}
\Delta_r G° &= 2(-237.129) \\
&= -474.258 \text{ kJ} \tag{9.16}
\end{aligned}
$$

Because

$$\Delta_r G^\circ = -n \mathcal{F} \mathcal{E}^\circ \tag{9.10}$$

then

$$\begin{aligned}
\mathcal{E}^\circ &= -(-474258/4 \times 96485) \\
&= 1.23 \, \text{V}
\end{aligned}$$

From the Nernst equation (9.14)

$$Eh = \mathcal{E}^\circ - 2.30259 \frac{RT}{n\mathcal{F}} \log[1/(f_{O_2} \cdot a_{H^+}^4)]$$

Setting $f_{O_2} = 1$ atm, and recalling that $pH = -\log a_{H^+}$, gives the equation for the boundary in terms of $Eh$ and $pH$.

$$Eh = 1.23 - 0.0592 pH \tag{9.17}$$

For the boundary $H_2(g) - H_2O(l)$,

$$2H^+ + 2e = H_2(g)$$

$$\begin{aligned}
\Delta_r G^\circ &= 0 \\
&= -n\mathcal{F}\mathcal{E}^\circ
\end{aligned}$$

and

$$Eh = 0 - 2.30259 \frac{RT}{n\mathcal{F}} \log(f_{H_2}/a_{H^+}^2)$$

or, with $f_{H_2} = 1$ atm, and $n = 2$,

$$Eh = -0.0592 pH \tag{9.18}$$

For boundary $Mn(OH)_2 - Mn_3O_4$,

$$Mn_3O_4(c) + 2H_2O(l) + 2H^+ + 2e = 3Mn(OH)_2(c)$$

$$\begin{aligned}
\Delta_r G^\circ &= -94140 \, \text{J} \\
&= -n\mathcal{F}\mathcal{E}^\circ
\end{aligned}$$

$$\begin{aligned}
\mathcal{E}^\circ &= -(-94140/2 \times 96485) \\
&= 0.488 \, \text{V}
\end{aligned}$$

and, with $n = 2$,

$$Eh = 0.488 - 0.0592 pH \tag{9.19}$$

For solubility of $Mn_3O_4$ as $Mn^{2+}$,

$$Mn_3O_4(c) + 8H^+ + 2e = 3Mn^{2+} + 4H_2O$$

$$\Delta_r G^\circ = -352711 \text{ J}$$

$$\begin{aligned} \mathcal{E}^\circ &= 352711/(2 \times 96485) \\ &= 1.828 \text{ V} \end{aligned}$$

$$Eh = \mathcal{E}^\circ - \frac{0.0592}{2} \log \left( \frac{a_{Mn^{2+}}^3}{a_{H^+}^8} \right)$$

or

$$Eh = 1.828 - 0.237pH - 0.0887 \log a_{Mn^{2+}} \qquad (9.20)$$

This is plotted for selected values of $Mn^{2+}$ activity ranging from $10^{-1}$ to $10^{-6}$ in Figure 9.6.

For equal activity contour of $Mn^{2+}$ and $MnO_4^{2-}$,

$$MnO_4^{2-} + 8H^+ + 4e = Mn^{2+} + 4H_2O$$

$$\Delta_r G^\circ = -672369 \text{ J}$$

$$\begin{aligned} \mathcal{E}^\circ &= 672369/4 \times 96485 \\ &= 1.742 \text{ V} \end{aligned}$$

$$Eh = 1.742 - 0.1182pH - 0.0148 \log(a_{Mn^{2+}}/a_{MnO_4^{2-}})$$

and where the activities of both aqueous species are equal, this reduces to

$$Eh = 1.742 - 0.1182pH \qquad (9.21)$$

This boundary lies at high $Eh$ and $pH$ and is illustrated in Figure 9.6.

### 9.8.3  *pe–pH* Diagrams

There is a second way of calculating the same kinds of diagrams using the alternative variable $pe$ rather than $Eh$. Because this is another way of doing exactly the same thing, it could be argued that the new variable $pe$ is unnecessary and redundant. However, the $pe$ and $Eh$ scales differ numerically, and $pe$ calculations are now used about as frequently as $Eh$. It is thus worthwhile discussing the use of this second variable.

The idea was to have an analogy between $pH$, which refers to hydrated protons, and $pe$, which would refer to hydrated electrons. Like the proton, the electron is assigned a standard $\Delta_f G^\circ = 0$. Like $pH$, $pe$ is defined in terms of activity:

$$pH = - \log a_{H^+}$$

$$pe = - \log a_e \qquad (9.22)$$

In fact the "$p$" notation is now widely used for various quantities. For example, equilibrium constants are sometimes given in terms of $pK$, where

$$pK = -\log K$$

In other words if $K = 10^{-6.37}$, then $pK = 6.37$.

Consider a half-cell reaction

$$aA + ne = bB$$

where, at equilibrium

$$K = \frac{a_B^b}{a_A^a\, a_e^n}$$

and

$$
\begin{aligned}
\log K &= \log \frac{a_B^b}{a_A^a\, a_e^n} \\
&= \log \frac{a_B^b}{a_A^a} - n \log a_e \\
&= \log \frac{a_B^b}{a_A^a} + n\, pe
\end{aligned}
$$

so

$$pe = \frac{1}{n}\log K - \frac{1}{n}\log \frac{a_B^b}{a_A^a}$$

When A and B are in their standard states of unit activity, the last term drops out, and we define $pe°$ as

$$pe° = \frac{1}{n}\log K$$

Combining equations, we get

$$
\begin{aligned}
pe &= pe° - \frac{1}{n}\log \frac{a_B^b}{a_A^a} \\
&= pe° - \frac{1}{n}\log Q 
\end{aligned}
\qquad (9.23)
$$

## 9.8.4 Comparison of $pe$ and $Eh$

There is a simple, linear relationship between the two variables $pe$ and $Eh$, which is evident in comparing the two defining equations:

$$Eh = \mathcal{E}° - 2.30259\frac{RT}{n\mathcal{F}}\log Q \qquad (9.14)$$

$$pe = pe^\circ - \frac{1}{n}\log Q \qquad (9.23)$$

Hence

$$
\begin{aligned}
pe &= Eh(\mathcal{F}/2.30259RT) \\
&= 5040\,Eh/T \\
&= Eh/0.05916 \text{ at } 298.15 \text{ K} \qquad (9.24)
\end{aligned}
$$

Thus, if you prefer, you can calculate $Eh$ and convert to $pe$ with equation (9.24), or vice versa. The relationship between the two scales is illustrated in Figure 9.6, where oxidation potential is plotted as both $Eh$ and $pe$. An $Eh$-$pH$ diagram looks exactly like a $pe$-$pH$ diagram, except that the $Y$ axis is shifted by the factor $5040/T$.

## 9.9  OXYGEN FUGACITY

As mentioned in §9.7, there are two methods in common use to represent the same fundamental variable—the oxidation state of a system. It is time to look at the second method, oxygen fugacity. This is a convenient parameter because any reaction that involves a change in oxidation state (any redox reaction) can be written so as to include oxygen as a reactant or product, whether or not oxygen is actually involved in the reaction. We gave an example in Chapter 8 (§8.3.2) of the oxidation of magnetite (in which some of the iron occurs as $Fe^{3+}$ and some as $Fe^{2+}$) to hematite (in which all of the iron occurs as $Fe^{3+}$). This reaction often occurs in systems which contain no oxygen molecules at all. Nevertheless, the calculated $f_{O_2}$ ($10^{-68.40}$) for magnetite–hematite equilibrium is a perfectly valid thermodynamic parameter, and the redox reaction involving $O_2(g)$

$$6\,Fe_2O_3(s) = 4\,Fe_3O_4(s) + O_2(g)$$

is simpler than the equivalent reaction in $Eh$ mode:

$$3\,Fe_2O_3(s) + 2\,H^+ + 2e = 2\,Fe_3O_4(s) + H_2O$$

We will discuss the relative merits of the two methods a little more further on.

### 9.9.1  Calculation of Oxygen Fugacity—$pH$ Diagrams

Because $f_{O_2}$ and $Eh$ are both indicators of the same thing — oxidation state — it is possible to draw $\log f_{O_2}$-$pH$ diagrams that are analogous to the $Eh$-$pH$ calculations we have outlined above. To illustrate this we will construct a $\log f_{O_2}$-$pH$ diagram for the same Mn-$H_2O$-$O_2$ system at 25°C already described. The completed digram is shown in Figure 9.7 and should be compared with the analogous $Eh$-$pH$ diagram of Figure 9.6. The two diagrams are similar except that most of the phase boundaries on the $Eh$-$pH$ diagram have the Nernst slope,

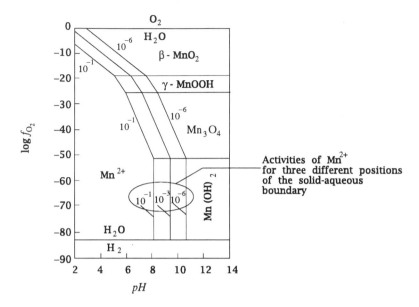

Figure 9.7: $\log f_{O_2}$-$pH$ diagram for the system $Mn$-$H_2O$-$O_2$ at 25°C, 1 atm.

whereas those on the $\log f_{O_2}$-$pH$ diagram have zero slope. Lines on $Eh$-$pH$ diagrams quite typically have nonzero slopes because hydrogen ions and electrons are so commonly involved in half-cell reactions. Reactions balanced with oxygen instead of electrons require $H^+$ ions much less frequently, and reactions that contain no hydrogen ions have zero slope.

The method of calculating $\log f_{O_2}$-$pH$ boundaries is illustrated with three examples. All other boundaries are derived in the same way. Our examples include the boundaries for water stability and for coexisting minerals, as well as the aqueous solubility contours of a mineral. Notice that half-cell reactions are not involved in these calculations.

For the water stability boundaries, the dissociation reaction of water is

$$2H_2O = O_2(g) + 2H_2(g) \qquad (9.25)$$

To calculate the equilibrium constant for this reaction,

$$
\begin{aligned}
\Delta_r G^\circ &= -2(-237129) \\
&= 474258 \, J \\
&= -RT \ln K
\end{aligned}
$$

$$
\begin{aligned}
K &= 10^{-83.1} \\
&= f_{H_2}^2 f_{O_2} \qquad (9.26)
\end{aligned}
$$

The upper boundary occurs at 1 atm $O_2(g)$ pressure or $\log f_{O_2} = 0$. The lower boundary is at 1 atm $H_2(g)$ pressure; from the equilibrium constant (9.26), this corresponds to $\log f_{O_2} = -83.1$. As noted before, water can exist under conditions outside of these boundaries, but only if the pressure of oxygen or hydrogen is greater than 1 atm.

For the boundary $Mn(OH)_2 - Mn_3O_4$,

$$\tfrac{1}{2}O_2(g) + 3Mn(OH)_2 = Mn_3O_4 + 3H_2O \tag{9.27}$$

$$\Delta_r G° = -143093 \text{ J}$$
$$= -RT \ln(1/f_{O_2}^{\frac{1}{2}})$$

Hence

$$\log f_{O_2} = -50.14 \tag{9.28}$$

For solubility of $Mn_3O_4$ as $Mn^{2+}$,

$$3H_2O(l) + 3Mn^{2+} + \tfrac{1}{2}O_2(g) = Mn_3O_4 + 6H^+ \tag{9.29}$$

$$\Delta_r G° = 114223 \text{ J}$$
$$= -RT \ln\left( \frac{a_{H^+}^6}{f_{O_2}^{\frac{1}{2}} a_{Mn^{2+}}^3} \right)$$

or

$$\log f_{O_2} = 40 - 12pH - 6\log a_{Mn^{2+}} \tag{9.30}$$

This is plotted for selected values of $Mn^{2+}$ activity ranging from $10^{-1}$ to $10^{-6}$ in Figure 9.7.

### 9.9.2 Interrelating Eh, pH, and Oxygen Fugacity

The obvious similarity between the $Eh-pH$ and $\log f_{O_2}-pH$ diagrams of Figures 9.6 and 9.7 suggests that it should be possible to convert directly from one set of coordinates to the other. This can be done using the half-cell reaction (9.15),

$$4H^+ + O_2(g) + 4e = 2H_2O(l) \tag{9.15}$$

and its related Nernst equation,

$$Eh = \mathcal{E}° - \frac{RT}{n\mathcal{F}} \ln[1/(f_{O_2} a_{H^+}^4)] \tag{9.31}$$

This equation can be used to interrelate the three variables $Eh$, $pH$, and $f_{O_2}$. The calculations are shown in §9.8.2, leading to an equation (9.17) for the $Eh$

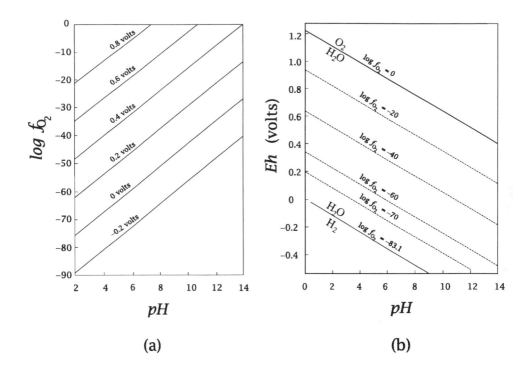

Figure 9.8: (a). $Eh$ contours (in volts) on a $\log f_{O_2}$-$pH$ diagram. (b). $\log f_{O_2}$ contours on an $Eh$-$pH$ diagram. Calculated for 25°C, 1 atm.

of solutions having an $f_{O_2}$ of 1 atm. However, instead of 1 atm, we can just as easily insert any other $f_{O_2}$ value, leading to the more general relation

$$Eh = 1.23 + 0.0148 \log f_{O_2} - 0.0592\, pH \qquad (9.32)$$

Figure 9.8 shows $Eh$ contours calculated from (9.32) drawn on a $\log f_{O_2}$-$pH$ diagram, and $\log f_{O_2}$ contours on an $Eh$-$pH$ diagram. The controlling factor is the oxidation state of the system, which in turn is controlled by the bulk composition, and $f_{O_2}, f_{H_2}, Eh, pe$, and all other related variables are simply different ways of quantifying the same thing.

## 9.10  REDOX REACTIONS IN ORGANIC CHEMISTRY

In inorganic reactions we think of H as always having valence of $+1$ ($H^+$) and oxygen as always having valence $-2$ ($O^{2-}$) and so we have no difficulty in seeing that carbon in $CH_4$ is $C^{4-}$, and carbon in $CO_2$ is $C^{4+}$. But what is the oxidation state of carbon in octane, $C_8H_{18}$, or in acetic acid, $CH_3COOH$, or in an enzyme containing hundreds of carbon, hydrogen, nitrogen, phosphorous, and other atoms? Applying our "normal" thinking to octane, we get a "nominal" valence for carbon of $18/8 = 2.25$. But how can an atom have a fractional valence? In acetic acid we get $(-4 + 4)/2 = 0$. But one carbon in acetic acid is bonded to three hydrogens and one carbon, while the other is bonded partly to O and OH. Surely there is some difference in the electron contributions of the two carbons. And in complex entities like large proteins, the idea of counting up carbons and other atoms to get some nominal oxidation state becomes silly.

As pointed out by Helgeson (1991), it is quite possible to assign nominal charge contribution of $-1$ for each C-H bond, zero for each C-C bond, and $+1$ for each C-O, C-S, or C-N bond, and arrive at fractional valences that give a consistent accounting of electron transfers in reactions, regardless of the actual extent to which the shared electrons in the covalent bonds are transferred among the carbon atoms. The fractional valences reflect some average charge on the carbon atoms, which, in the case of the alkanes, changes from $-4$ in $CH_4$ to close to $-2$ as $n$ gets large in $C_nH_{2(n+1)}$. That acetic acid is actually more oxidized than octane is readily seen by writing the reaction

$$2\,C_8H_{18} + 9\,O_2 = 8\,CH_3COOH + 2\,H_2O$$

In fact, any transfer of oxygen or hydrogen in a balanced reaction signifies oxidation-reduction. This can be verified by using a nominal valence scheme as outlined above, but it is always true. Therefore, in looking at reactions between complex molecules, biochemists do not calculate nominal charges to determine if a reaction is an oxidation or a reduction; they simply observe whether oxygen or hydrogen is involved. For example, nicotinamide adenine dinucleotide (NAD) exists in two distinct forms, $NAD^+$ and $NADH + H^+$. The reaction $NAD^+ + H_2 \rightarrow NADH + H^+$ is a reduction, and the reverse reaction is an oxidation. The reaction may involve some molecule much more complex than $H_2$, but this doesn't matter. We know this reaction is a reduction as written simply because NADH contains one extra hydrogen. Reactions involving electron transfers (redox reactions) are absolutely fundamental to all life processes.

# 9.11 SUMMARY

On a broad scale the Earth shows a large range of redox conditions, from the highly reduced Ni-Fe core through various silicate layers up into the zone of free water and eventually into the oxygen-rich atmosphere. Therefore, an indicator of the redox state is among the more important of the variables manipulated by geochemists. Like *pH*, it is an important parameter because it is intimately linked to a large number of reactions of interest to anyone trying to understand the Earth, but again like *pH*, it is actually no more or less fundamental than any other intensive parameter. For a closed system at a given $T$ and $P$, it is

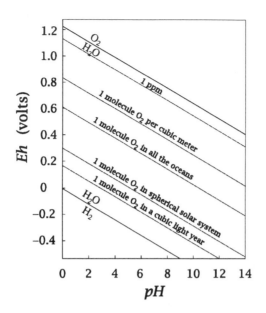

Figure 9.9: An $Eh$-$pH$ diagram with contours of the oxygen concentrations that would result in the redox conditions shown. Obviously, redox conditions over most of the diagram represent solutions that contain no oxygen molecules at all.

completely determined by the bulk composition, as are all intensive parameters, and changes in redox state are accomplished by changing bulk composition.

The measurement of redox conditions by means of a cell voltage, where one electrode has a fixed reference potential and the other is expected to react reversibly with natural systems, is attended by a number of problems. The platinum electrode works well only under certain conditions: It is difficult to get the electrode into reducing environments without allowing some oxidation, and the method is restricted to ambient conditions except in research laboratories. We put up with these problems because there is little choice.

Oxygen fugacity, on the other hand, although a much simpler concept, can be directly measured only at high temperatures, a fact that might seem to rule out its use at Earth-surface conditions, but in fact it does not. Oxygen fugacity can be used (as opposed to *measured*) under any redox conditions, including systems that contain no oxygen molecules whatsoever, as illustrated in Figures 9.9 and 9.10 (they may contain oxygen in combined form, such as $H_2O$, of course).    Oxygen fugacity is an index of redox conditions, not always an approximation to the partial pressure of $O_2(g)$ in a system. It can be calculated because *any* redox reaction can be written to include oxygen, whatever is actually happening in the system. In a similar way, it has been pointed out at times that $pe$ is a somewhat fictitious quantity, in that there is good evidence

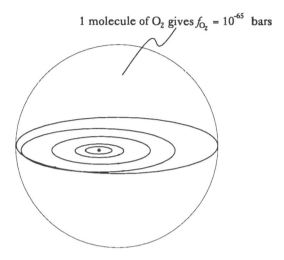

Sphere Containing Pluto's Orbit

Figure 9.10: The meaning of $f_{O_2} = 10^{-65}$ bars.

that there are no hydrated electrons in aqueous solutions. In Chapter 1 we emphasized that we would be describing a *model* of chemical reactions, not real reactions. Parameters such as oxygen fugacity illustrate this point. In a sense, we have oxygen and hydrated electrons in our models, though perhaps not in our systems.

*Eh* has been important in both the measuring and reporting of redox conditions, but an argument can be made that it should be used only in the measurement and not the reporting of redox conditions, that is, that however measurements are *made*, results should be *reported* as $f_{O_2}$ or $\log f_{O_2}$ values. There would be two advantages to this. First, the use of *Eh* entails the use of a relatively complex set of conventions, which are quite difficult to remember unless used continuously. Second, and more importantly, *Eh* is less useful than $f_{O_2}$ because it is so commonly linked in reactions with *pH* (giving the "Nernst slope"). This means that *a value for Eh without an accompanying value for pH is usually meaningless.* This is illustrated by any *Eh–pH* diagram, in which you can see that an *Eh* of 0.0 volts, for example, indicates much more reducing conditions at *pH* 2 than it does at *pH* 10. The conversion from an *Fh–pH* point to a $\log f_{O_2}$ value is very simple [equation (9.32)]. The very small values of $f_{O_2}$ generally obtained at low temperatures should not be a hindrance to its use, which would greatly simplify the reporting and interpretation of any redox conditions.

# PROBLEMS

1. Uranium can occur in $U^{4+}$, $U^{5+}$, and $U^{6+}$ oxidation states. Minerals that contain uranium in its 4+ state, such as uraninite ($UO_2$) and coffinite ($USiO_4$) are very insoluble, so U is normally transported in solutions in which most of it is in the 6+ or possibly 5+ state, and precipitation occurs when U is reduced to the 4+ state. Thus redox reactions are important in the formation of uranium ore deposits.

Extra Data:

| Species | $\Delta_f G°$ (kJ mol$^{-1}$) |
|---|---|
| $UO_2CO_3°(aq)$ | $-1567.2$ |
| $UO_2(CO_3)_2^{2-}(aq)$ | $-2164.1$ |

(a) Calculate the standard potential for the following half-cells. These are required for the Nernst equations in part (b).

$$UO_2^{2+}(aq) + 2e = UO_2(s)$$

$$UO_2CO_3°(aq) + 2H^+ + 2e = UO_2(s) + CO_2(g) + H_2O(l)$$

$$UO_2(CO_3)_2^{2-}(aq) + 4H^+ + 2e = UO_2(s) + 2CO_2(g) + 2H_2O(l)$$

(b) Calculate the pH for equal activities of $UO_2^{2+}(aq)$ and $UO_2CO_3°(aq)$, and for $UO_2CO_3°(aq)$ and $UO_2(CO_3)_2^{2-}(aq)$, which are boundaries 4 and 5. The relevant reactions are

$$UO_2CO_3°(aq) + 2H^+ = UO_2^{2+}(aq) + CO_2(g) + H_2O(l)$$

$$UO_2(CO_3)_2^{2-}(aq) + 2H^+ = UO_2CO_3°(aq) + CO_2(g) + H_2O(l)$$

(c) Write the Nernst equations for boundaries 1, 2, and 3, and derive an equation in Eh and pH for each. All aqueous species have an activity of $10^{-6}$ (concentration $10^{-6} m$), the fugacity of $CO_2(g)$ is constant at 0.1 bar, and water and uraninite have an activity of 1.0.

(d) Construct an Eh-pH diagram for uranium-water-$CO_2$ using these equations. Figure 9.11 shows the arrangement of the fields, but not the correct position of the boundaries, as it was drawn using other data.

(e) Plot $f_{O_2}$ contours of $10^{-20}$, $10^{-40}$, and $10^{-60}$ bar across the diagram.

2. Calculate and draw the Eh-pH diagram for the iron-water system, using aqueous ion activities of $10^{-6}$ (concentrations of $10^{-6} m$). See if you can write the relevant half-cell reactions simply by looking at what lies on opposite sides of the boundaries (e.g., $Fe^{2+}$-hematite) and adding the water, hydrogen ions, and electrons needed to balance. It helps to know that in hematite ($Fe_2O_3$) both iron atoms are ferric ($Fe^{3+}$), while in magnetite

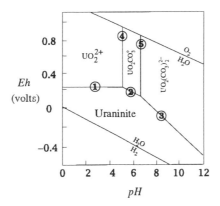

Figure 9.11: *Eh-pH* relations for uraninite and aqueous carbonate species. Note that the diagram shows that uraninite is stable only under quite reducing conditions.

($Fe_3O_4$) one Fe is ferrous ($Fe^{2+}$) and two are ferric. These are the reactions you should write:

$$Fe^{3+} + e = Fe^{2+}$$

$$2\,Fe^{3+} + 3\,H_2O = Fe_2O_3(\textit{hematite}) + 6\,H^+$$

$$Fe_2O_3(\textit{hematite}) + 6\,H^+ + 2\,e = 2\,Fe^{2+} + 3\,H_2O$$

$$3\,Fe_2O_3(\textit{hematite}) + 2\,H^+ + 2\,e = 2\,Fe_3O_4(\textit{magnetite}) + H_2O$$

$$Fe_3O_4(\textit{magnetite}) + 8\,H^+ + 2\,e = 3\,Fe^{2+} + 4\,H_2O$$

(a) Calculate and draw $Fe^{3+}$-hematite, $Fe^{2+}$-hematite, and $Fe^{2+}$-magnetite boundaries for aqueous ion activities of $10^{-4}$ and $10^{-8}$. Make them lighter than the $10^{-6}$ boundaries.

(b) The diagram shows the expected *Eh-pH* relations between ferrous iron and hematite. What usually precipitates when $Fe^{2+}$ oxidizes, however, is not hematite but ferric hydroxide, $Fe(OH)_3(s)$.

  i. Write a balanced reaction between hematite and ferric hydroxide, and show that ferric hydroxide is metastable with respect to hematite.

  ii. Calculate and draw the $Fe^{2+}$-$Fe(OH)_3$ boundary for $a_{Fe^{2+}} = 10^{-6}$. The half-cell reaction is

$$Fe(OH)_3(s) + 3\,H^+ + e = Fe^{2+} + 3\,H_2O$$

(c) On the same diagram, draw the $H_2S$-$SO_4^{2-}$ boundary.

$$SO_4^{2-} + 10\,H^+ + 8\,e = H_2S(aq) + 4\,H_2O$$

(d) A groundwater sample smelling of rotten eggs ($H_2S$) contains 6000 ppm $SO_4^{2-}$ at an $Eh$ of $-0.1\,V$ and a $pH$ of 4.0. Could the $SO_4^{2-}$ and $H_2S$ be in equilibrium, or is sulfate being reduced, or is $H_2S$ being oxidized? Bear in mind that acid water at atmospheric pressure can contain no more than about 0.1 molal $H_2S$. A simple statement of what is happening is not good enough; you have to show why your answer is right.

3. An aqueous solution has a $pH$ of 6.0 and is in equilibrium with hematite [$Fe_2O_3(s)$] and magnetite [$Fe_3O_4(s)$]. What is the $Eh$ and oxygen fugacity of the solution?

4. What is the oxygen fugacity of the same solution at 300°C?

5. Iodine in aqueous solutions exists mostly as iodate ion ($IO_3^-$) in relatively oxidized solutions and as iodide ion ($I^-$) in more reduced solutions. What is the valence of iodine in each of these ions? Calculate the standard electrode potential ($\mathcal{E}°$) for the iodate–iodide redox couple, plus the $Eh$ of a solution having a $pH$ of 6.0 and equal activities of the two iodine species.

6. Calculate the oxygen fugacity of the solution in the previous question. What ionic species of iodine would you expect in a solution in equilibrium with magnetite?

7. Sketch the iodate–iodide boundary on an $Eh$-$pH$ diagram. Sketch a log activity vs. $Eh$ diagram through this boundary to show the nature of the boundary.

8. An electrochemical cell like that in Figure 9.2, but using zinc instead of iron, is called a Daniell cell. This cell was actually widely used as a source of electrical power in the nineteenth century. Do a calculation to show why zinc was favored over iron in making this cell.

9. For some reason students often conclude that an $Eh > 0$ means oxidizing conditions, and $Eh < 0$ means reducing conditions. Show that this cannot be true.

# 10

# MINERALS AND SOLUTIONS

## 10.1 REAL PROBLEMS

We have now completed our survey of the thermodynamic principles required to model natural systems. It only remains to gain practice in formulating problems involving natural systems in thermodynamic terms. Quite often, that is the hardest part. Once the problem is set up in terms of relevant reactions and components, the equations can be solved by anyone who has absorbed the previous chapters. However, choosing the appropriate components and setting up the relevant balanced reactions only comes from experience. In this chapter we explore a few situations that have been investigated by thermodynamic methods.

## 10.2 IS THE SEA SATURATED WITH CALCIUM CARBONATE?

If you have ever been to Florida or the Bahamas, you may be aware that there are vast areas adjacent to the coasts where the sea bottom at shallow levels is a white mud, which turns out to be made of almost pure aragonite. Carbonate muds extend well out to the deep sea as well; in fact, a fairly large proportion of the sea bottom is composed of calcium carbonate. There are also countless calcitic atolls and reefs throughout the tropical zones of the world. Given this amount of contact between the sea and calcium carbonate, both calcite and aragonite, plus the fact that there are vigorous oceanic currents stirring things up constantly, plus the fact that things have not changed drastically for millions of years, you would think that there would be little doubt that the system consisting of the oceans plus their bottom sediments must have reached equilibrium by now. If these were the only factors involved, perhaps they would have, but the situation is quite a bit more complicated. Why would anyone want to know? Reactions involving carbonate in the oceans are fundamental to an understanding of the global $CO_2$ cycle, which in turn is linked to global warming and other things we would like to understand.

### 10.2.1   How Do You Tell If a Solution Is Saturated?

To explore this problem further, we must first find out how to determine whether a solution is saturated, undersaturated, or supersaturated with a given mineral or compound. One answer would be to just observe the solution in contact with the mineral. If the mineral dissolves, the solution is undersaturated. If the mineral grows in size, the solution is supersaturated. If nothing happens, the solution is saturated—it is at equilibrium. This method and variations of it are used, but it is very difficult for a number of reasons. We would like to be able to predict the state of saturation for a sample of water without performing difficult experiments on it. We would like simply to determine the chemical composition of the solution and calculate theoretically the state of saturation.

In other words, we want a thermodynamic answer. Having just spent several chapters developing a method for determining which way a reaction will go, we should be able to put it to use here. The reaction could be written

$$\text{solid mineral} = \text{dissolved mineral} \qquad (10.1)$$

If this reaction goes to the right, the solution is undersaturated. If it goes to the left, the solution is supersaturated. All we need to do is to determine the molar Gibbs free energy of the dissolved mineral and compare it to the molar Gibbs free energy of the pure mineral, and the question is answered.

### 10.2.2   Solubility Products

But dissolution reactions that result in uncharged solutes such as $H_4SiO_4(aq)$ and $H_2CO_3(aq)$ are unusual. Most solutes are ionized to some extent, that is, they break up into charged particles, called ions. In other words, we write the dissolution reaction not as in (10.1), but as

$$\text{solid mineral} = \text{aqueous ions} \qquad (10.2)$$

For example, calcium carbonate (calcite or aragonite), when it dissolves, breaks up into calcium and carbonate ions (Figure 10.1), written as

$$CaCO_3(s) = Ca^{2+} + CO_3^{2-} \qquad (10.3)$$

The equilibrium constant for reaction (10.3) can be found in our routine way:

$$
\begin{aligned}
\Delta_r G° &= \Delta_f G°_{Ca^{2+}} + \Delta_f G°_{CO_3^{2-}} - \Delta_f G°_{CaCO_3(s)} \\
&= -553.58 + (-527.81) - (-1128.79) \\
&= 47.40 \, kJ
\end{aligned}
$$

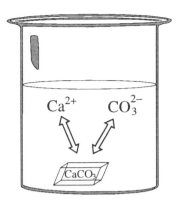

Figure 10.1: When calcite dissolves, the solute consists of electrically charged ions.

for calcite, or

$$= -553.58 + (-527.81) - (-1127.75)$$
$$= 46.36\,kJ$$

for aragonite. This gives

$$\Delta_r G^\circ = -RT \ln K$$
$$47400 = -(2.30259 \times 8.3145 \times 298.15) \log K$$
$$\log K = -8.304$$

for calcite, or $-8.122$ for aragonite.

This equilibrium constant is

$$K = \frac{a_{Ca^{2+}}\, a_{CO_3^{2-}}}{a_{CaCO_3(s)}}$$

Now, if, as in Figure 10.1, we are dealing with the solubility of pure calcium carbonate, its activity according to our rules (§7.4.1) is 1.0, so the equilibrium constant becomes

$$K_{sp} = a_{Ca^{2+}}\, a_{CO_3^{2-}}$$

and is called a *solubility product* constant, or just a solubility product. Whereas in the case of quartz solubility we found the equilibrium constant to be equal to the solubility itself (§8.2.1), here we find the equilibrium constant to be equal to a product of two ion activities. Therefore, "the solubility of calcite" has a somewhat ambiguous meaning. If it refers to the concentration (activity) of calcium in solution, this obviously depends on how much carbonate ion is in solution, and vice versa. It is the *combination* of calcium ion and carbonate ion activities that determines whether calcite is over- or undersaturated in a

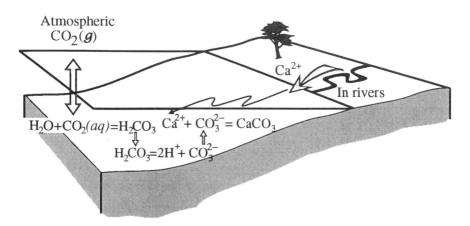

Figure 10.2: Calcite may precipitate in natural bodies of water, but the calcium and the carbonate may come from different sources.

solution. For example, suppose we have determined that the activity of $CO_3^{2-}$ in a solution (say the one in Figure 10.1) is $10^{-5}$. A solution having $a_{Ca^{2+}}\, a_{CO_3^{2-}} = 10^{-8.304}$ will be in equilibrium with calcite, so in this case the equilibrium activity of the calcium ion is $a_{Ca^{2+}} = 10^{-3.304}$  In a solution with $a_{CO_3^{2-}} = 10^{-6}$, the equilibrium value of $a_{Ca^{2+}}$ is $10^{-2.304}$, and so on.

Having now learned about the solubility product, please do not tack the $sp$ subscript on to every equilibrium constant you calculate. There are equilibrium constants for many types of reactions. The solubility product is an equilibrium constant for a reaction having a solid mineral or compound on the left side and its constituent ions on the right.

### 10.2.3   IAP, $K_{sp}$, $\Omega$, and SI

Of course, natural solutions, such as seawater, are not necessarily at equilibrium. In Figure 10.2 we see a river carrying dissolved material, including calcium and carbonate ions, entering the sea. Carbonate ions are already there, because the sea is in contact with the atmosphere, which contains carbon dioxide, and when $CO_2$ dissolves it produces carbonate and bicarbonate ions. Because calcium and carbonate are being added, there may be a tendency for them to increase beyond the equilibrium value, and for calcite to precipitate as a result. The product of the calcium and carbonate ion activities *which are actually present in a solution*, regardless of any theory, is called the *ion activity product* (IAP) for that solution. It follows that when IAP $> K_{sp}$, calcite will precipitate, and when IAP $< K_{sp}$ calcite will dissolve. The IAP/$K_{sp}$ ratio is called $\Omega$, and the logarithm of the ratio is called the *saturation index* (SI), so that when SI $> 0$ calcite precipitates, and when SI $< 0$ calcite dissolves (Table 10.1).

| IAP, $K_{sp}$ | $\Omega$ | SI ($= \log \frac{IAP}{K_{sp}}$) | Result |
|---|---|---|---|
| IAP $< K_{sp}$ | $< 1$ | negative | mineral dissolves |
| IAP $> K_{sp}$ | $> 1$ | positive | mineral precipitates |
| IAP $= K_{sp}$ | 1 | 0 | equilibrium |

Table 10.1: Relations between IAP, $K_{sp}$, and SI.

These relationships should be fairly intuitive, but if you have difficulty convincing yourself, consider equation (8.6).

$$\Delta_r \mu = \Delta_r \mu° + RT \ln Q \tag{8.6}$$

In the case we are considering [calcite and its ions, reaction (10.3)], when $Q = K_{sp}$, $\Delta_r \mu = 0$. Therefore, if $Q = $ IAP $> K_{sp}$, $\Delta_r \mu > 0$, and reaction (10.3) will go to the left (calcite precipitates). And if $Q = $ IAP $< K_{sp}$, $\Delta_r \mu < 0$, reaction (10.3) will go to the right (calcite dissolves).

## 10.2.4 Determining the IAP

Now to answer our question as to whether the sea is saturated with calcite, we need only determine its IAP and compare it with $K_{sp}$ for calcite. Easier said than done. Apart from some specialized electrochemical techniques, it is generally not possible to analyze a solution for the concentration of specific ions such as $Ca^{2+}$ or $CO_3^{2-}$. Analyses are made for the *total* calcium, or the *total* carbonate in the solution. There are many kinds of ions and uncharged molecular species such as ion-pairs in a solution, and so the total calcium concentration would consist of the sum of the concentrations of all the ionic and molecular species containing Ca, and similarly for carbonate. Thus

$$m_{Ca,total} = m_{Ca^{2+}} + m_{CaHCO_3^+} + m_{CaSO_4°} + \cdots$$

$$m_{CO_3,total} = m_{CO_3^{2-}} + m_{HCO_3^-} + m_{H_2CO_3} + \cdots$$

Here, $CaSO_4°$ represents an electrically neutral species resulting from the joining of a $Ca^{2+}$ ion and a $SO_4^{2-}$ ion in solution. There are several more species in each of these summations in real seawater which we won't mention. But however many there are, the concentrations and activities of all of them can be calculated if for every species there is a known equilibrium constant relating it to other species and/or minerals, and if there is a suitable equation for calculating the activity coefficients ($y_H$) of each of the species. What it amounts to is the well-known fact that if in a set of equations you have the same number of equations as you have variables, it is possible to solve for every variable. Given the total concentrations of the various constituents of seawater (not only calcium and carbonate, but however many you are interested in), plus an equilibrium constant for each, plus the knowledge that the total positive charges must equal the total negative charges, it is always possible to achieve this goal. When more than three or four ions are involved, the procedure can be carried out only on a computer, and many programs are now available for doing this. However, to get the feel of it, and to introduce an important type of diagram, let's see how it works for carbonate.

## 10.2.5  The Bjerrum Diagram for Carbonate Species

Let's consider that we have only three carbonate-bearing species, $H_2CO_3$, $HCO_3^-$, and $CO_3^{2-}$. This would result, for example, from just dissolving carbon dioxide gas in water. We have already calculated the equilibrium constant for one of the relevant ionic equilibria (§8.2.1), we repeat the calculation here:

$$H_2CO_3(aq) = HCO_3^- + H^+; \; K = 10^{-6.37} \tag{10.4}$$

Another is

$$HCO_3^- = CO_3^{2-} + H^+; \; K = 10^{-10.33} \tag{10.5}$$

And let's say that our total carbonate is $0.10\,m$, so that

$$m_{CO_3,total} = m_{CO_3^{2-}} + m_{HCO_3^-} + m_{H_2CO_3} = 0.10\,m \tag{10.6}$$

These three equations contain four variables, $m_{CO_3^{2-}}$, $m_{HCO_3^-}$, $m_{H_2CO_3}$, and $m_{H^+}$, so we need another equation, which is the charge balance,

$$m_{H^+} = m_{HCO_3^-} + 2\,m_{CO_3^{2-}} \tag{10.7}$$

An additional four unknowns are the activity coefficients of each of the four species for which we could write four more equations,[1] making a total of eight equations and eight unknowns. However, neglecting these coefficients (assuming each equals 1.0), we can solve the four equations, and without going through the gory details, we get

$$
\begin{aligned}
a_{H^+} &= 10^{-3.685} \\
a_{H_2CO_3} &= 0.0998 \\
a_{HCO_3^-} &= 2.063 \times 10^{-4} \\
a_{CO_3^{2-}} &= 4.677 \times 10^{-11}
\end{aligned}
$$

---

[1] Calculation of activity coefficients is discussed in Appendix C.

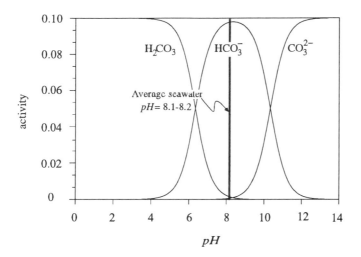

Figure 10.3: Activities of $H_2CO_3(aq)$, and bicarbonate and carbonate ions as a function of $pH$, for a total concentration of $0.10\,m$.

It is much easier to check that these answers are correct than to get the answers. For example we can check that

$$\frac{a_{HCO_3^-}\,a_{H^+}}{a_{H_2CO_3}} = \frac{2.063 \times 10^{-4} \cdot 10^{-3.685}}{0.0998}$$

$$= 10^{-6.37}$$

as it should.

This simple calculation does not illustrate much beyond the solution of four simultaneous equations in four variables. However, one thing it does show is why $H_2CO_3$ is called carbonic acid. You see that when it dissolves in water, the $pH$ becomes fairly acid. This gives carbonated beverages their tart taste.

Much more interesting than this single example, however, would be to see how the activities or concentrations of $H_2CO_3$, $HCO_3^-$, and $CO_3^{2-}$ vary as a function of $pH$. Natural solutions contain many components in addition to $CO_2$ and water, and so the $pH$ can be quite different from the one we have just calculated. To do this, we simply choose specific $pH$ values from 0 to 14 and solve for the activities of the three carbonate species. The result is shown in Figure 10.3. This diagram makes it easy to see which species is dominant (has the largest concentration) at any given $pH$. For example, in seawater, with a $pH$ of about 8.1, carbonate is present almost entirely as the bicarbonate species.

An interesting feature of this diagram is the fact that the intersection of the lines representing $a_{H_2CO_3}$ and $a_{HCO_3^-}$ occurs at a $pH$ of 6.37, which is the $pK$ value of the first ionization constant of $H_2CO_3$,[2] and similarly for the intersection of

---

[2] $pK = -\log K$, as explained in §9.7.3.

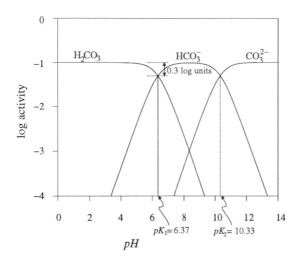

Figure 10.4: Same as Figure 10.3, but with a log activity scale. The $pH$ values of the crossover points are the $pK$ values of carbonic acid ionization.

the $HCO_3^-$ and $CO_3^{2-}$ lines, as shown in Figure 10.4. The reason for this is easy to see when you look at the equilibrium constant expressions, for example

$$\frac{a_{HCO_3^-}\, a_{H^+}}{a_{H_2CO_3}} = 10^{-6.37}$$

At the crossover or intersection point, $a_{HCO_3^-} = a_{H_2CO_3}$, so that

$$a_{H^+} = 10^{-6.37}$$

which means that the intersection of the lines representing $a_{H_2CO_3}$ and $a_{HCO_3^-}$ occurs at a $pH$ of 6.37.

Another interesting fact is that the crossover points occur at an activity of 0.05 (Figure 10.3), because at each of the two crossover points, the activity of the third species (the one not involved in the crossover) is negligibly small, so that the two "crossing species" make up virtually the total activity or concentration, and therefore they both have a value of one-half the total concentration. Because $\log \frac{1}{2} = -0.30$, this means that on the log scale the crossovers occur 0.3 log units below the plateau representing the total concentration, as shown in Figure 10.4. A third interesting fact is that the slopes of the lines representing the activities of the species are either $+1$ or $-1$, just below their intersections (although they may change to $+2$ and $-2$ farther down, as shown in Figure 10.5). The combination of these three properties of these diagrams (called Bjerrum diagrams) makes it very easy to rapidly sketch such a diagram, given some $pK$ values.

Bjerrum diagrams are quite useful in seeing (and remembering) the relationships between species in dissociation reactions. For example, in Figure 10.5 you

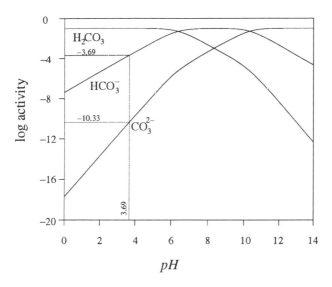

Figure 10.5: Same as Figure 10.4, but with an expanded activity axis. At a $pH$ of 3.69, the $HCO_3^-$ and $CO_3^{2-}$ activities are those calculated in the text.

can see that $a_{CO_3^{2-}}$ goes to some very low values in acid solutions, a fact that explains (when you look at the solubility product) why carbonate minerals have such high solubilities in acid solutions. You also see that the activities of dissolved species never go to zero, at least in the model. In reality, of course, very low activities may mean that that species does not exist in the system.

As another example, Figure 10.6 shows the distribution of phosphate species at the same total concentration of 0.10 $m$. Such diagrams are clearly useful for any solute that can exist in a variety of species, differing only by the number of hydrogen ions (protons) they have.

## 10.2.6 Combining the IAP and the $K_{sp}$

Although we have not discussed solving these simultaneous equations in detail, it should be clear that a solution is always possible as long as the number of equations is the same as the number of unknowns. No one does this by hand anymore, as there are numerous computer programs to do it for us, no matter how many equations there are. The point is that for any chemically analyzed solution, we can calculate the species concentrations and activity coefficients, and therefore the activities of all the species for which we have thermodynamic data (equilibrium constants). This is now done routinely on oceanographic research vessels, and a certain amount of variability is found in the composition of seawater from various locations. However, the composition of *average* seawater is quite well known and is not greatly different from that proposed by

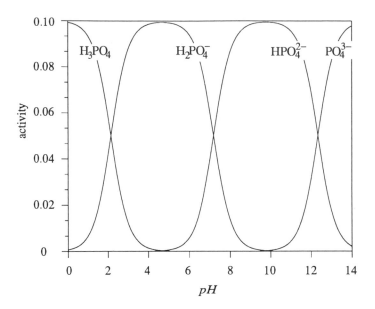

Figure 10.6: The activity distribution diagram for phosphoric acid. The $pK$ values are 2.15, 7.21, and 12.34.

Garrels and Thompson in 1962, from which we get the data in Table 10.2.
From these data, we find

$$
\begin{aligned}
IAP &= a_{Ca^{2+}}\, a_{CO_3^{2-}} \\
&= 0.0025 \times 0.0000054 \\
&= 1.35 \times 10^{-8} \\
&= 10^{-7.870}
\end{aligned}
$$

and we found (§10.2.2) that for calcite $K_{sp} = 10^{-8.304}$, and for aragonite $K_{sp} = 10^{-8.122}$. Therefore, average seawater is slightly supersaturated with calcium carbonate (Table 10.3).

This is a reasonably interesting result, as far as it goes. It means that marine organisms should have no difficulty in precipitating their carbonate shells, and once precipitated, they should not redissolve. This is consistent with the vast amounts of aragonitic mud on the Florida and Bahamas coastlines. These muds are made up almost exclusively of the shells of microscopic marine organisms, which sink to the bottom when the organisms die and do not redissolve.

### 10.2.7  What Part of the Sea is Saturated with $CaCO_3$?

So far, thermodynamics and observations fit together fairly well. However, we know from oceanographic surveys that although vast areas of the sea are un-

| Major ions | Concentration $m$ | Amount occurring as free ions % | Activity coefficient $y_H$ | Activity $a$ |
|---|---|---|---|---|
| $Na^+$ | 0.475 | 99 | 0.76 | 0.357 |
| $Mg^{2+}$ | 0.054 | 87 | 0.36 | 0.017 |
| $Ca^{2+}$ | 0.010 | 91 | 0.28 | 0.0025 |
| $K^+$ | 0.010 | 99 | 0.64 | 0.0063 |
| $Cl^-$ | 0.56 | 100 | 0.64 | 0.36 |
| $SO_4^{2-}$ | 0.028 | 54 | 0.12 | 0.0018 |
| $HCO_3^-$ | 0.0024 | 69 | 0.68 | 0.0011 |
| $CO_3^{2-}$ | 0.0003 | 9 | 0.20 | 0.0000054 |

Table 10.2: Properties of the major ions in near-surface seawater.

| | IAP | $K_{sp}$ | $\Omega$ | SI |
|---|---|---|---|---|
| seawater | $13.5 \times 10^{-9}$ | | | |
| calcite | | $4.97 \times 10^{-9}$ | 2.72 | 0.43 |
| aragonite | | $7.55 \times 10^{-9}$ | 1.79 | 0.25 |

Table 10.3: IAP, $\Omega$ and SI for calcium carbonate in average seawater.

derlain by these carbonate muds, especially in water depths less than 3 to 4 km, the deepest parts of the ocean basins (4 to 6 km depth) have little or no carbonate in the bottom muds. Down to a variable depth, but usually between 3 and 4 km, the bottom muds are close to 100% calcium carbonate. Then within a relatively short increase in depth, the percentage of carbonate in the muds drops off rapidly, becoming zero or close to zero at another variable depth, but usually 4 to 5 km. The depth at which the rapid increase in carbonate dissolution begins is called the Lysocline, and the depth below which there is little or no carbonate is called the Carbonate Compensation Depth (CCD). Carbonate-secreting organisms are active at the surface virtually everywhere, and their carbonate shells are settling down through the water column everywhere, not just in shallow water. But while they coat the ocean floor at shallow depths, they never reach depths greater than 5 km or so—they dissolve completely at these depths. So obviously the oceans are saturated with $CaCO_3$ at and near the surface but undersaturated at great depths.

The exact explanation for this is one of the many continuing problems of chemical oceanography, but from our point of view it illustrates two things.

- The oceans, like most natural systems, are not at chemical, thermal, or mechanical equilibrium, but in spite of this, our equilibrium thermodynamic model is quite useful if we know how to apply it. We have shown how it is useful at the ocean surface; it is also useful at great depth. But obviously it would not be useful to apply it to the oceans as a whole system—they are too far from equilibrium.

- The explanation for the CCD and its variations is complex, involving the kinetics of carbonate dissolution, variations in ocean chemistry, temperature, and pressure, worldwide circulation patterns, and other factors. Equilibrium chemical thermodynamics does not suffice for an understanding of this natural system, but it is invariably the starting point for all other types of investigation. You must have an understanding of the equilibrium state before you can understand the departures from this state.

## 10.3   MINERAL STABILITY DIAGRAMS

The problem of calcite saturation in the sea is only one of a large number of problems in oceanography, geology, soil chemistry, and many other areas of science that involve solid⇌fluid reactions, and a variety of diagrams have been used to illustrate the thermodynamic relationships involved. Humans find two-dimensional diagrams much easier to understand than multidimensional systems of equations.

## 10.3.1 The Reaction of Feldspar With Water

One of the most common minerals on Earth is feldspar, and its reaction with water to form other minerals such as clays is of great interest in several fields. In soils, feldspars (solid solutions of $NaAlSi_3O_8$, $KAlSi_3O_8$, and $CaAl_2Si_2O_8$) react to form clay minerals, helping to control soil acidity. In petroleum reservoirs, the same reaction forming clay minerals can have serious effects on the rock permeability and oil recovery. K-feldspars often have appreciable amounts of Pb substituting for K in their structures, and reaction of K-feldspar with formation fluids is thought to release this Pb to the fluid, which may then go on to form a lead ore-body elsewhere on its travels through the Earth's crust. During metamorphism, when rocks containing feldspars are subjected to high temperatures and pressures deep in the crust, feldspars participate in a variety of reactions with fluids and with other minerals, all of which are of interest to geologists studying the history of the Earth.

The question for us is, how do we apply thermodynamics to these reactions? The first thing we must do is write a reaction that seems interesting. This is one of the most difficult steps, but one that is rarely discussed because there are no rules to guide us. There is a very large number of reactions that could be written involving K-feldspar, depending on what other things are in the system, but only a few of these are useful. Only experience and scientific insight can distinguish between what might turn out to be useful and what will not. The reactions that appear in texts have, of course, proven to be useful.

If you were to write a reaction between K-feldspar and kaolinite with no previous experience or prejudices, but just putting them on opposite sides of an equal sign and adding other compounds to balance, you would likely wind up with something like this:

$$KAlSi_3O_8(s) + \tfrac{3}{2}H_2O(l) = \tfrac{1}{2}Al_2Si_2O_5(OH)_4(s) + 2\,SiO_2 + KOH \qquad (10.8)$$

We have not specified whether $SiO_2$ is quartz or $SiO_2(aq)$—you have your choice. If you choose quartz, and quartz is present in the system, its activity will be 1.0; if you choose $SiO_2(aq)$, its activity will be (a dimensionless number equal to) the molality of silica in solution. Pure KOH is a solid, but it ionizes completely in solution, and you again have a choice of the $(s)$ or the $(aq)$ forms of data. If we choose to use quartz, $KOH(aq)$, and the maximum microcline form of $KAlSi_3O_8$ (or K-feldspar), the equilibrium constant turns out to be $10^{-7\,88}$, and, because all the products and reactants except KOH have $a = 1$, at least if the water is reasonably pure, this turns out to be the molality of KOH in our system. What system? We have not been very specific about what our system consists of, other than that it contains K-feldspar, kaolinite, and water, at 25°C, 1 bar. We might be thinking about a soil with groundwater.

But, you say, groundwater contains organic solutes, $CO_2$, and lots of other things. Why are they not in my system? Well, you can put them in if you wish, in balanced reactions, but you are under no obligation to do so. Your system is a model system, a simplified version of the real thing, and what goes in is under your control. Here we are choosing to look at the predicted KOH concentration. It may be that we have calculated something that is not useful, but that is our fault, not the fault of thermodynamics. Or we may have overlooked some factor which will invalidate our result, but that remains to be seen. We will see examples of such pitfalls shortly.

Knowing the KOH concentration that would equilibrate with K-feldspar and kaolinite is interesting, as far as it goes, but it is not the best way to look at this system. With a little insight, we can see that by subtracting $OH^-$ from each side, and changing from quartz to $SiO_2(aq)$, we get

$$KAlSi_3O_8(s) + \tfrac{1}{2}H_2O(l) + H^+ = \tfrac{1}{2}Al_2Si_2O_5(OH)_4(s) + 2\,SiO_2(aq) + K^+ \quad (10.9)$$

In this reaction, we have three potentially measurable things—the silica concentration $[SiO_2(aq)]$, the $K^+$ concentration, and the $pH$ ($-\log a_{H^+}$). Given any two, we could predict the third, using the thermodynamic model. Or we could construct a three-dimensional diagram using these three parameters as axes. That's a bit too ambitious for us, so we will combine the $a_{H^+}$ and $a_{K^+}$ parameters into a ratio and plot this against $a_{SiO_2}(aq)$. This is now a standard procedure.

The way it is done is worth remembering, because it is used with various kinds of reactions. Write the equilibrium constant, take the logarithm of both sides, put your $y$-axis parameter on the left, the $x$-axis parameter on the right, and combine the rest of the terms. The equilibrium constant is

$$K = \frac{a_{Al_2Si_2O_5(OH)_4}^{\frac{1}{2}}\, a_{SiO_2(aq)}^{2}\, a_{K^+}}{a_{KAlSi_3O_8}\, a_{H_2O(l)}^{\frac{1}{2}}\, a_{H^+}} \quad (10.10)$$

If kaolinite, K-feldspar, and water are reasonably pure, their activities are 1, and

$$\log K = \log \frac{a_{K^+}}{a_{H^+}} + 2\,\log a_{SiO_2(aq)}$$

or

$$\log \frac{a_{K^+}}{a_{H^+}} = -2\,\log a_{SiO_2(aq)} + \log K \quad (10.11)$$

or

$$y = ax + b$$

where $y$ is $\log(a_{K^+}/a_{H^+})$, $x$ is $\log a_{SiO_2(aq)}$, $a = -2$, and $b = \log K$. Thus equation (10.11) is the equation of a straight line having a slope of $-2$, if we plot $\log(a_{K^+}/a_{H^+})$ against $\log a_{SiO_2(aq)}$. To get $K$, we apply our routine method,

$$\begin{aligned}
\Delta_r G^\circ &= \tfrac{1}{2}\Delta_f G^{\circ\,kaolinite}_{Al_2Si_2O_5(OH)_4} + 2\,\Delta_f G^\circ_{SiO_2(aq)} + \Delta_f G^\circ_{K^+} \\
&\quad - \Delta_f G^{\circ\,microcline}_{KAlSi_3O_8} - \tfrac{1}{2}\Delta_f G^\circ_{H_2O(l)} - \Delta_f G^\circ_{H^+}
\end{aligned}$$

Getting numbers from the tables,

$$\Delta_r G^\circ = \tfrac{1}{2}(-3799.7) + 2(-833.411) + (-283.27)$$

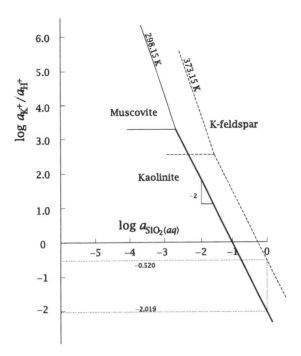

Figure 10.7: A plot of $\log(a_{K^+}/a_{H^+})$ vs. $\log a_{SiO_2(aq)}$ at 298.15 K and 373.15 K.

$$-(-3742.9) - \tfrac{1}{2}(-237.129)$$
$$= \quad 11.523 \, \text{kJ}$$
$$= \quad 11523 \, \text{J}$$

Then

$$\Delta_r G° \quad = \quad -RT \ln K$$
$$11523 \quad = \quad -(8.3145 \times 298.15) \ln K_{298}$$

so

$$\log K_{298} \quad = \quad -11523/(2.30259 \times 8.3145 \times 298.15)$$
$$= \quad -2.019$$

Therefore,

$$\log \frac{a_{K^+}}{a_{H^+}} = -2 \log a_{SiO_2(aq)} - 2.019 \tag{10.12}$$

This is a line on a plot of $\log(a_{K^+}/a_{H^+})$ vs. $\log a_{SiO_2(aq)}$ having a slope of $-2$ and a $y$-intercept (the value of $y$ when $x = 0$) of $-2.019$, as shown in Figure 10.7. The line we have just calculated is the thick line.

We now have a line on a graph. What does it mean? The meaning is implicit in the methods we used to get the line. We put the activities of the minerals and water equal to 1, and we used the equilibrium constant. That means that the line is the locus of solution conditions $(a_{K^+}/a_{H^+})$ and $a_{SiO_2(aq)}$ for which the pure minerals and water are in equilibrium with each other. For any values of $(a_{K^+}/a_{H^+})$ and $a_{SiO_2(aq)}$ that do not lie on the line, our solution cannot be in equilibrium with both minerals, although it might be in equilibrium with one or the other. Applying LeChatelier's principle to reaction (10.9), we see that increasing $a_{SiO_2(aq)}$ or increasing $(a_{K^+}/a_{H^+})$ favors the formation of K-feldspar, so a field of K-feldspar lies to the right and above our line, and a field of kaolinite lies to the left and below.

The next problem is that having chosen a fairly complex system like this, there are more possible reactions than the one we have chosen. K-feldspar can react not only to kaolinite, but also to muscovite, and kaolinite can also react to form muscovite. These reactions are

$$\tfrac{3}{2}KAlSi_3O_8(s) + H^+ = \tfrac{1}{2}KAl_3Si_3O_{10}(OH)_2(s) + 3\,SiO_2(aq) + K^+ \qquad (10.13)$$

and

$$KAl_3Si_3O_{10}(OH)_2(s) + \tfrac{3}{2}H_2O + H^+ = \tfrac{3}{2}Al_2Si_2O_5(OH)_4(s) + K^+ \qquad (10.14)$$

Using the same methods as before, we find that reaction (10.13) has a slope of $-3$ on our graph, and an intercept ($\log K$) of $-4.668$, and reaction (10.14) has a slope of 0 (it is independent of $a_{SiO_2}$) and an intercept of 3.281. It can be shown (using the Phase Rule) that these three lines must intersect at a point, and so an easier way to draw them is to calculate the point of intersection, which is $\log a_{SiO_2(aq)} = -2.650$, $\log(a_{K^+}/a_{H^+}) = 3.281$, and draw lines with slopes 0, $-2$, and $-3$ through this point. We now have a kind of phase diagram, showing which minerals are stable, not as a function of $T$ and $P$, but of the composition of a solution in equilibrium with the minerals.

### Effect of Temperature

There are other mineral phases to be added to our diagram, but first let's look at the effect of temperature. If we want the same diagram for a temperature of 100°C, we must calculate $K$ at this temperature. To do this, we can use (8.14) or (8.17). We'll use both in the following calculations. For reaction (10.9), $\Delta_r H° = 43437\,J$ and $\Delta_r S° = 106.449\,J/K$. Therefore

$$
\begin{aligned}
\Delta_r G^\circ_{373} &= \Delta_r H^\circ_{298} - T\,\Delta_r S^\circ_{298} \\
&= 43437 - 373.15 \times 106.449 \\
&= 3715.556\,J
\end{aligned}
$$

$$
\begin{aligned}
\log K &= -3715.556/(2.30259 \times 8.31451 \times 373.15) \\
&= -0.520 \\
&= \log \frac{a_{K^+}}{a_{H^+}} + 2\log a_{SiO_2(aq)}
\end{aligned}
$$

giving

$$\log \frac{a_{K^+}}{a_{H^+}} = -2 \log a_{SiO_2(aq)} - 0.520$$

as the equation of the K-feldspar–kaolinite boundary at $100°C$.
For reaction (10.13), $\Delta_r H° = 74473$ J and $\log K_{298} = -4.668$, and so

$$\begin{aligned}
\log K_{373} &= \log K_{298} - \frac{\Delta_r H°_{298}}{2.30259\,R}\left(\frac{1}{T} - \frac{1}{298.15}\right) \\
&= -4.668 - \frac{74473}{2.30259 \times 8.31451}\left(\frac{1}{373.15} - \frac{1}{298.15}\right) \\
&= -2.046
\end{aligned}$$

For reaction (10.14), $\Delta_r H° = -18635$ J, $\Delta_r S° = -1.165$ J/K, and so

$$\begin{aligned}
\log K_{373} &= \frac{-\Delta_r H°_{298}}{2.30259\,RT} + \frac{\Delta_r S°_{298}}{2.30259\,R} \\
&= \frac{18635}{2.30259 \times 8.31451 \times 373.15} + \frac{-1.165}{2.30259 \times 8.31451} \\
&= 2.548
\end{aligned}$$

These three lines intersect at $\log(a_{K^+}/a_{H^+}) = 2.55$, $\log a_{SiO_2(aq)} = -1.53$, as shown in Figure 10.7.

### Effect of Choice of Data

This business of looking up data and plotting by hand gets tedious. Surely there are computer programs to do all this for us? There are many, but none of them relieves the user of a fundamental responsibility—you are responsible for choosing your data, and of course for understanding what you are doing. People learning how to use thermodynamics generally assume that the data they find at the back of their text, such as those in Appendix B, are "true" in some absolute sense, and this does no harm while learning the subject. However, when you begin to apply the subject in some area of interest, you sooner or later discover that there are a number of sources of data, that they often disagree, and that the choice of data can affect your results considerably.

One such source of data, incorporated into a program that calculates equilibrium constants over a wide range of $T$ and $P$ using methods mentioned in Appendix C, is SUPCRT92 (Johnson et al., 1992). In Figure 10.8 are shown the mineral boundaries calculated using $K$ values from SUPCRT92 for 373.15 K, in comparison with those from Figure 10.7. This comparison is made to impress on you the fairly large differences that can result from using different sources of data. Most compilations of data these days have been subjected to some sort of quality and consistency analysis, but methods vary. The refinement and improvement of thermodynamic data is a never-ending process, and the choice of thermodynamic data is an integral part of applying thermodynamics to natural systems. The data in SUPCRT92 are especially good for most geochemical appli-

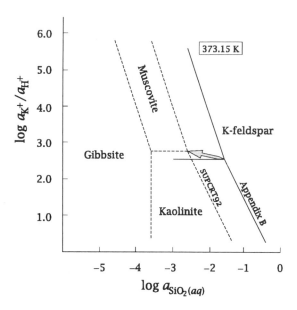

Figure 10.8: A plot of $\log(a_{K^+}/a_{H^+})$ vs. $\log a_{SiO_2(aq)}$ at 373.15 K, showing the effect of changing from the data in Appendix B to data in program SUPCRT92.

| Mineral | SUPCRT92 $\Delta_f G°$ kJ | Appendix B $\Delta_f G°$ kJ | App. B − SUPCRT92 kJ | % Difference |
|---------|------------|-----------|---------|--------------|
| albite | −3708.313 | −3711.5 | −3.2 | 0.08 |
| K-feldspar | −3746.245 | −3742.9 | −3.3 | 0.08 |
| muscovite | −5591.083 | −5608.4 | −17.3 | 0.30 |
| quartz | −856.239 | −856.64 | −0.4 | 0.05 |

Table 10.4: The difference between the data in Appendix B and in SUPCRT92.

cations, because they have been derived using many high pressure–temperature experiments involving minerals that are not always considered by chemists. In other words, they are "tuned" to mineral equilibria. On the other hand, it does not contain many compounds that do not occur in nature which are of great interest to chemists.

The differences between the data in Appendix B and those in SUPCRT92 are relatively small (Table 10.4), yet they result in quite large differences in phase diagrams like Figure 10.8. Let this be a lesson in choosing data—the best are none too good, and no one knows for sure what "the best" data are.

While we are using SUPCRT92, we might as well add two final reactions to our diagram. These relate kaolinite and muscovite to gibbsite:

$$Al_2Si_2O_5(OH)_4(s) + H_2O(l) = 2\,Al(OH)_3(s) + 2\,SiO_2(aq) \qquad (10.15)$$

and

$$KAl_3Si_3O_{10}(OH)_2(s) + 3H_2O(l) + H^+ = 3Al(OH)_3(s) + 3SiO_2(aq) + K^+ \quad (10.16)$$

These two reactions also meet at a point with the muscovite-kaolinite reaction, as shown in Figure 10.8.

If you do not have access to program SUPCRT92, it is of no importance here. We could have illustrated the point that the choice of data is important by using data from any number of other sources.

### The Problem of Metastable Phases

One of the pitfalls in choosing data and drawing diagrams, such as we have done, is that we may not have considered all the possible reactions in our system. That is, there may be phases that are more stable than the ones we have chosen—we may have chosen metastable phases. This is illustrated in our system by the fact that there is another aluminosilicate phase, pyrophyllite, which has more silica in it than does kaolinite and so is stable at higher values of $a_{SiO_2(aq)}$. Considering the reactions

$$KAlSi_3O_8(s) + H^+ = \tfrac{1}{2}Al_2Si_4O_{10}(OH)_2(s) + SiO_2(aq) + K^+ \quad (10.17)$$

and

$$\tfrac{1}{2}Al_2Si_4O_{10}(OH)_2 + \tfrac{1}{2}H_2O(l) = \tfrac{1}{2}Al_2Si_2O_5(OH)_4(s) + SiO_2(aq) \quad (10.18)$$

results in the two boundaries shown in Figure 10.9, which completely enclose the K-feldspar-kaolinite boundary. This means that at 100°C, and according to the SUPCRT92 data, K-feldspar and kaolinite are not stable together in the presence of water. They should react to form pyrophyllite. In nature, of course, they may not. K-feldspar may well react directly to kaolinite, in spite of thermodynamics, but it would be a metastable reaction; that is, a reaction involving metastable phases.

One other thing. The solubility of quartz, calculated by the method in §8.3.1, is $10^{-3.078}$ $m$ at 100°C according to SUPCRT92. This is shown in Figure 10.9 as a vertical line. A solution in equilibrium with quartz at 100°C must lie on this line. Solutions to the right of it are supersaturated, and solutions to the left are undersaturated with quartz. Therefore, if your system contains quartz, as most soils do, then not only is the assemblage K-feldspar + kaolinite metastable with respect to pyrophyllite, but pyrophyllite is itself metastable. According to our diagram, pyrophyllite can only exist in an aqueous solution if that solution is supersaturated with silica. If the silica were to precipitate as quartz, pyrophyllite should break down to form kaolinite, releasing silica to the solution, by reaction (10.18). Again, nature may not do this. It is quite common for natural solutions to be supersaturated with silica, even in the presence of quartz, which is one reason that we use log $a_{SiO_2(aq)}$ as a variable in our diagrams.

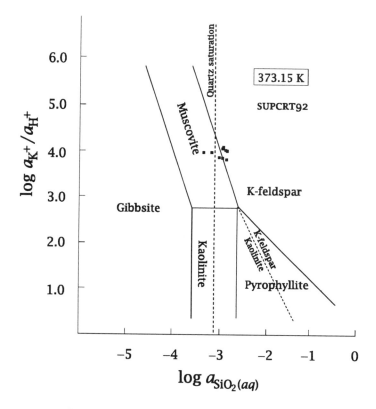

Figure 10.9: A plot of $\log(a_{K^+}/a_{H^+})$ vs. $\log a_{SiO_2(aq)}$ at 373.15 K, using data from SUPCRT92, showing the stability field of pyrophyllite and the quartz saturation line. Each black square represents a brine from the Kettleman North Dome oil field (Merino, 1975).

## Plotting Solution Data

Also shown on Figure 10.9 are a few data points taken at random from Merino (1975). Each one is from a different well in the Kettleman North Dome oil field in California. The numbers for $a_{K^+}$, $a_{H^+}$, and $a_{SiO_2(aq)}$ are obtained from the chemical analyses in Merino (1975) using a computer program (called SOLMINEQ88, see Appendix C) using the principles discussed in §10.2.5. It appears that the solution concentrations are clustering around the K-feldspar-muscovite boundary, near the quartz saturation line, and are likely being controlled by these phases, which are known to be in the sandstones containing the fluids sampled.

## The Albite–K-feldspar Diagram

There are, of course, a great many diagrams we might plot, depending on our interests. To investigate the possibility that the K/Na ratio in the fluids at Kettleman North Dome is controlled by the coexistence of albite and K-feldspar, we could write and plot the reaction

$$NaAlSi_3O_8(s) + K^+ = KAlSi_3O_8(s) + Na^+ \qquad (10.19)$$

Following the procedures outlined above, we find that $\Delta_r G^\circ_{373} - -7052.9\,J$ (Appendix B) or $-14945\,J$ (SUPCRT92). If pure albite and pure K-feldspar are present in the rocks, the equilibrium constant for this reaction becomes

$$K_{373} = \frac{a_{Na^+}}{a_{K^+}} \qquad (10.20)$$

$$= 10^{0.987} \quad \text{Appendix B}$$

$$\text{or} \quad = 10^{2.092} \quad \text{SUPCRT92}$$

and we could plot $a_{Na^+}$ vs. $a_{K^+}$, with a line separating a field of albite from a field of K-feldspar. However, if we want also to have muscovite on the same diagram, it would be convenient to use the $a_{K^+}/a_{H^+}$ parameter from our previous diagrams. So we simply divide both numerator and denominator on the right side of (10.20) by $a_{H^+}$, giving us

$$K_{373} = \frac{(a_{Na^+}/a_{H^+})}{(a_{K^+}/a_{H^+})} \qquad (10.21)$$

$$= 10^{0.987} \text{ (App. B)} \quad \text{or} \quad 10^{2.092} \text{ (SUPCRT92)}$$

To add a boundary between muscovite and albite, we write a reaction between these two minerals involving the same ions and $SiO_2$, which turns out to be

$$KAl_3Si_3O_{10}(OH)_2(s) + 3\,Na^+ + 6\,SiO_2(s) - 3\,NAlSi_3O_8(s) + 2\,H^+ + K^+ \quad (10.22)$$

However, here we choose to use quartz as our $SiO_2$, rather than $SiO_2(aq)$, because the solutions appear to have their silica content controlled by quartz, or near enough, and besides, this diagram has no silica variable. If you prefer, you could calculate an average $a_{SiO_2(aq)}$ of the data points and use this value (or any other value) for our new diagram—the choice is yours. The equilibrium constant for (10.22) at 100°C is $10^{-16.282}$ (Appendix B) or $10^{-14.948}$ (SUPCRT92), and, like (10.20), needs a little manipulation to get it into a form we can use.

$$K_{373} = \frac{a_{K^+}\,a_{H^+}^2}{a_{Na^+}^3}$$

$$= \frac{a_{K^+}\,a_{H^+}^2}{a_{Na^+}^3} \cdot \frac{a_{H^+}}{a_{H^+}}$$

$$= \frac{a_{K^+}/a_{H^+}}{(a_{Na^+}/a_{H^+})^3} \qquad (10.23)$$

so

$$\log K_{373} = \log \frac{a_{K^+}}{a_{H^+}} - 3\log \frac{a_{Na^+}}{a_{H^+}}$$

$$\log \frac{a_{Na^+}}{a_{H^+}} = \tfrac{1}{3}\log \frac{a_{K^+}}{a_{H^+}} + 16.282 \quad \text{or} \quad + 14.948 \qquad (10.24)$$

and (10.24) is the equation of the muscovite–albite boundary on a plot of $a_{Na^+}/a_{H^+}$ vs. $a_{K^+}/a_{H^+}$, shown in Figure 10.10. The black squares in Figure 10.10 represent the same samples as shown in Figure 10.9. They cluster near the muscovite–albite–K-feldspar intersection, which would be expected if the solution compositions are being more or less controlled by the host rocks. However, the data are definitely offset from this intersection into the muscovite field and appear to lie along the metastable extension of the albite–K-feldspar boundary. This may well be an example of fluid compositions being controlled by a metastable assemblage, as was mentioned in connection with pyrophyllite.

## 10.4   SUMMARY

In this chapter we have seen how a knowledge of equilibrium constants and activities can be used to construct diagrams that relate fluid compositions to the minerals in equilibrium with those fluids. Real fluid compositions (activities) can then be plotted on the diagrams to see how well the hypothesis of equilibrium holds, or, assuming equilibrium, to predict what phases are coexisting with fluids, and what will happen if we change the fluid composition. There are a large number of other applications of equilibrium thermodynamics, to be sure, but if you understand equilibrium constants and activities thoroughly, you will have no problem in understanding most other applications.

In the case of the Kettleman North Dome fluids, the mineralogy of the rocks in contact with the fluids is known. You might say, well, if we know what minerals are there, why go to all this trouble to show that the fluid compositions reflect this? Would you not expect them to reflect the mineralogy of their host

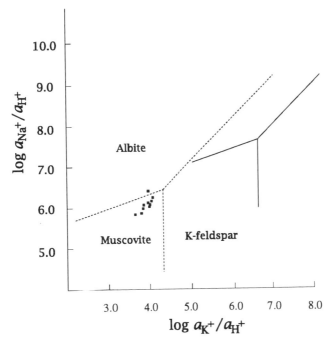

Figure 10.10: A plot of $\log(a_{Na^+}/a_{H^+})$ vs. $\log(a_{K^+}/a_{H^+})$ at 373.15 K for solutions saturated with quartz. Dashed lines from program SUPCRT92, solid lines from Appendix B. A few points from Merino (1975) are shown as black squares.

rocks? For one thing, we didn't know whether the composition of oil field brines or any other kind of natural fluid was controlled by host rocks or not, until this kind of test was done. For all we knew, perhaps disequilibrium reigned supreme. Furthermore, we're still learning.

Why is it important? This question gets at the whole idea of using science to try to understand natural processes. Until you have a quantitative model that can simulate or account for natural data, such as shown in Figures 10.9 and 10.10, you cannot hope to change or control the situation to your benefit. For example, the petroleum geologists at Kettleman North Dome might want to inject something to change the fluid characteristics. Without a chemical model, they would have no way of predicting what would happen. The equilibrium model has its limitations, but it is a good place to start.

## PROBLEMS

1. A sample of seawater has a bicarbonate concentration of 0.09 $m$ ($\approx a_{HCO_3^-}$), a $pH$ of 8.0, and a calcium ion activity ($a_{Ca^{2+}}$) of $10^{-6}$. Will calcite precipitate? What activity of $Mg^{2+}$ would be required to precipitate dolomite?

2. Calculate the calcium ion concentration in a solution having $a_{H_2CO_3(aq)} = 0.1$ at a $pH$ of 5.0, and in equilibrium with (a) calcite; (b) aragonite. What should happen if this same solution was in contact with both minerals? Is this consistent with their values of $\Delta_f G°$?

3. (a) What is an appropriate diagram to use to consider the relationship between water compositions and rocks containing nepheline, wollastonite, and grossularite? What is the slope of the grossularite $\rightarrow$ nepheline + wollastonite boundary on this diagram?

$$2NaAlSiO_4(s) + 2CaSiO_3(s) + Ca^{2+} = SiO_2(aq) + Ca_3Al_2Si_3O_{12}(s) + 2Na^+$$
$$(10.25)$$

   (grossularite = $Ca_3Al_2Si_3O_{12}$; nepheline = $NaAlSiO_4$; wollastonite = $CaSiO_3$).

   (b) What is buffered by coexisting grossularite + nepheline + wollastonite?

4. (a) Calculate equilibrium constants at 25°C for the following reactions, where $NaAlSiO_4$ is nepheline, $NaAlSi_3O_8$ is albite, and $NaAl_3Si_3O_{10}(OH)_2$ is paragonite:

$$NaAlSiO_4 + 2SiO_2(aq) = NaAlSi_3O_8$$
$$\tfrac{1}{2}NaAl_3Si_3O_{10}(OH)_2 + 3H^+ = \tfrac{1}{2}NaAlSiO_4 + SiO_2(aq) + 2H_2O(l) + Al^{·}$$
$$\tfrac{1}{2}NaAl_3Si_3O_{10}(OH)_2 + 3H^+ = \tfrac{1}{2}NaAlSi_3O_8 + 2H_2O(l) + Al^{3+}$$

(b) Use these equilibrium constants to sketch a diagram showing the stability fields of the three minerals as a function of $\log(a_{Al^{3+}}/a_{H^+}^3)$ and $\log a_{SiO_2(aq)}$. The boundaries of the three phase fields intersect at a point. Calculate the value of the two parameters at this point.

(c) After sketching this diagram, you become aware that there is yet another sodium aluminum silicate mineral—analcime ($NaAlSi_2O_6 \cdot H_2O$), which has been neglected. Without knowing its thermodynamic properties, sketch a revised diagram, showing how it might fit in. Remember that the formulae for nepheline, analcime, and albite can be written with the same amounts of Na and Al but increasing amounts of Si. Note too that for the purposes of this question, you can disregard the $H_2O$ in the analcime formula.

(d) Calculate and plot the quartz solubility line and the amorphous silica solubility line.

5. (a) Calculate the equilibrium constants for the following equations:

$$
\begin{aligned}
Al(OH)_3(gibbsite) + 3\,H^+ &= Al^{3+} + 3\,H_2O \\
Al(OH)_3(gibbsite) + 2\,H^+ &= Al(OH)^{2+} + 2\,H_2O \\
Al(OH)_3(gibbsite) + H^+ &= Al(OH)_2^+ + H_2O \\
Al(OH)_3(gibbsite) &= Al(OH)_3^\circ(aq) \\
Al(OH)_3(gibbsite) + H_2O &= Al(OH)_4^- + H^+
\end{aligned}
$$

(b) Derive an equation suitable for plotting each reaction on a graph having *log activity of dissolved Al species* and *pH* as axes.

6. Consider a hot-spring solution in contact with a rock containing microcline, muscovite, kaolinite, quartz, and other minerals (an altered granite) at $100°C$. You have no analysis of this water, but you have a $\log(a_{K^+}/a_{H^+})$ vs. $\log a_{SiO_2(aq)}$ diagram (Figure 10.9). What does the diagram tell you about the composition of the solution,

(a) if quartz does not equilibrate with the solution?

(b) if quartz equilibrates with the solution?

(c) If this hot-spring solution had a *pH* of 5.0 and was observed or believed to be altering the microcline in the granite to kaolinite, what could you say about the activity and the concentration (in ppm) of $K^+$ in the solution? (Note for nongeologists: one of the minerals in all granites is quartz).

7. Gibbsite is a principal mineral in bauxite, which is mined as a source of aluminum. Would you expect to find quartz in bauxite? Would you expect to find kaolinite? Why?

8. Calculate the equilibrium constant of the reaction

$$2\,KAlSi_3O_8(s)+2\,H^+ + H_2O(l) = Al_2Si_2O_7 \cdot 2H_2O(s)+4\,SiO_2(quartz)+2\,K^+$$

How would a change in redox conditions affect this reaction? What would be the potassium ion activity in groundwater in a soil containing microcline, kaolinite, and quartz, at a $pH$ of 5.5? If you measured the $a_{K^+}$ in such a groundwater and found a different result, what reason would you suspect?

9. Calculate and plot the muscovite–gibbsite (10.16) and kaolinite–gibbsite (10.15) boundaries at 373.15 K in Figure 10.9, using the data from Appendix B. You know you have the right answers when these two curves meet the muscovite–kaolinite boundary (10.14) at a single point.

10. If you did Problem 5 at the end of Chapter 5, you will be aware that according to the data in Appendix B, diaspore is more stable than gibbsite in water. Use your knowledge of free energy relationships to *predict* how substituting diaspore for gibbsite in Figure 10.8 or 10.9 will affect the position of the boundaries with muscovite and kaolinite. Calculate the boundaries to confirm your prediction.

11. Show why $\Omega$ (Table 10.3) for calcite must always be greater than $\Omega$ for aragonite.

# 11

# PHASE DIAGRAMS

## 11.1   WHAT IS A PHASE DIAGRAM?

A phase diagram in the general sense is any diagram that shows what phase or phases are stable as a function of some chosen system variable or variables. Therefore, the $Eh$-$pH$, $\log f_{O_2}$-$pH$ and activity-activity diagrams we have been looking at are a kind of phase diagram. However, if you mention the subject of phase diagrams to a petrologist, a metallurgist, or a ceramic scientist, they will immediately think of a particular type of diagram that is of great usefulness in these subjects. In these sciences, the compositions of phases and their relationships during phase changes, particularly solid→liquid and liquid→solid changes, are of particular importance, so diagrams that depict this information as a function of temperature and pressure have come to be the subject of "phase diagrams."

### 11.1.1   Thermodynamics and Phase Diagrams

Though it is true that phase relations can always be described in terms of the thermodynamic principles and equations we have been discussing, and that any phase diagram can in principle be calculated given the appropriate data, the emphasis in this chapter changes from one of calculating what we want to know from numbers in tables of data, to one of simply representing experimentally derived facts in diagrammatic form. The reason for this is that once we get into systems more complex than a single component, and especially when high-temperature melt phases are involved, the calculations (a) are more suited to an advanced course, (b) are often not possible because the data are not available, and (c) even if they are available, are not very accurate, because they are very sensitive to small inaccuracies in the data. Therefore, in this book, although we will show the relationship between functions such as $G$ and our diagrams, this will be in an illustrative rather than a quantitative way.

203

## 11.1.2  Phase Diagrams as Models

Metallurgists and ceramicists quite often deal with simple two- and three-component systems and use phase diagrams to represent their experimental results on the phase relations in these systems. The diagrams therefore truly represent their systems. Petrologists, on the other hand, are interested in the origins of natural rocks, which commonly have ten or more important components (we will give an exact definition of a component shortly). Systems this complex cannot be represented in simple diagrams and, in fact, can hardly even be thought about in a quantitative way. Experiments can and have been done using natural rocks, but the results are complex and may not be generally applicable. Therefore, petrologists use simpler systems such as those having two and three components to better understand the principles involved and to investigate simple *models* of the complex systems in nature.

Phase diagrams represent *equilibrium* relationships. Once these are depicted, simple *processes* such as melting and crystallization can be considered, but because as represented on diagrams these involve continuous successions of equilibrium states, they are *reversible processes* in the sense of §2.6.2. In this sense, then, all branches of science use phase diagrams not only to represent equilibrium relationships, but as models of processes of interest.

## 11.2  THE PHASE RULE

### 11.2.1  Phases, Components, and Degrees of Freedom

Phase relations involve a small number of carefully defined terms.

#### Phases

A *phase* is defined as a homogeneous body of matter having distinct boundaries with adjacent phases, and so is in principle mechanically separable from them. Each mineral in a rock is therefore a single phase, as is a salt solution, or a mixture of gases.

#### Components

Each phase therefore has a definite chemical composition, and the various phases in a system may have the same (polymorphs) or different compositions. The compositions are described in terms of chemical formulae, such as $SiO_2$ or $CaMgSiO_3$. The smallest number of chemical formulae needed to describe the composition of all the phases in a system is called the number of *components* of the system. The *choice* of components is to some extent a matter of convenience, but the *number* of components is not. For example, consider the system $MgO–SiO_2$, which contains the intermediate compounds $MgSiO_3$ (or $MgO \cdot SiO_2$)

and $Mg_2SiO_4$ (or $2\,MgO \cdot SiO_2$). Two formulae are needed to describe the composition of every phase, such as MgO and $SiO_2$. However, the choice could just as well be $MgSiO_3$ and $SiO_2$, or $MgSiO_3$ and $Mg_2SiO_4$, because MgO can be described as $[MgSiO_3 - SiO_2]$, or $[Mg_2SiO_4 - MgSiO_3]$, and so on. There are in fact an indefinite number of possible choices in any system, but the *number* that must be chosen is fixed.

### System Variance or Degrees of Freedom

To understand the Phase Rule and how to use it, you must first understand the concept of *variance* or *degrees of freedom*.

A single homogeneous phase such as an aqueous salt (say NaCl) solution has a large number of properties, such as temperature, density, NaCl molality, refractive index, heat capacity, absorption spectra, vapor pressure, conductivity, partial molar entropy of water, partial molar enthalpy of NaCl, ionization constant, osmotic coefficient, ionic strength, and so on. We know, however, that these properties are not all independent of one another. Most chemists know instinctively that a solution of NaCl in water will have all its properties fixed if temperature, pressure, and salt concentration are fixed. In other words, there are apparently three independent variables for this two-component system, or three variables that must be fixed before all variables are fixed. Furthermore, there seems to be no fundamental reason for singling out temperature, pressure, and salt concentration from the dozens of properties available—it's just more convenient; any three would do. The number of variables (system properties) that must be fixed in order to fix *all* system properties is known as the system variance or degrees of freedom.

Now consider two phases at equilibrium, say solid NaCl and a saturated salt solution. Again, intuition or experience tells us that we no longer have three independent variables, but two, because, for example, we cannot choose the composition of the salt solution once $T$ and $P$ are fixed—it is fixed for us by the solubility of NaCl in water. If we then consider the possibility of having a vapor phase in equilibrium with the salt and the solution, we see that we lose another independent variable because we can no longer choose the pressure on the system independently once the temperature is chosen—it is fixed by the vapor pressure of the system. So it would seem that, in general, we restrict the number of independent variables in a system by increasing the number of phases at equilibrium.

The variance of a chemical system is exactly analogous to the variance of a system of linear equations. For example, for the function

$$x + y + z = 0$$

if we choose $x = 2$, $y = 2$, then $z$ is fixed at $-4$. The equation could be said to have a variance of two, because two variables must be fixed before all variables are fixed. Three variables minus one relationship between them (one equation) leaves two degrees of freedom. If in addition to this function we have another one involving the same variables, such as

$$2x - y + 4z = -19$$

we now have three variables and two functional relationships, and we are free to choose only one of the three variables, the other two then being fixed. For example, if we choose $x = 2$, then there is no further choice—$y = 3$ and $z = -5$. If we choose $x = 3$, then $y = 2.6$ and $z = -5.6$. This situation can be said to be *univariant* or to have one degree of freedom.

And, of course, if we have a third functional relationship, for example,

$$-3x + 2y - 7z = 35$$

then we have no choice: $x$, $y$ and $z$ are fixed at 2, 3, and $-5$, respectively, and the situation is *invariant.*

## 11.2.2   Derivation of the Phase Rule

The reason that the linear equations analogy for phase relationships is so exact is that there is in fact a thermodynamic equation for each phase (see Appendix C), and each of these equations has a number of independent variables equal to the number of components in the system plus two. And this, in turn, is because each component represents a degree of freedom (we can add or subtract each component), and there are two more because we defined our systems at the beginning as being able to exchange energy in only two ways—heat and one kind of work.[1] If the number of components is $c$, then the total number of independent system properties is $c + 2$. If there are $p$ phases in the system, and each phase represents one equation, then there are $p$ equations in $c + 2$ variables, or $c + 2 - p$ degrees of freedom. This is the Phase Rule:

$$f = c - p + 2 \tag{11.1}$$

where $f$ is the number of degrees of freedom.

Degrees of freedom can also be described as the number of intensive variables that can be changed (within limits) without changing the number of phases in a system. This point of view is perhaps more useful to someone looking at a phase diagram; thus divariant, univariant, and invariant systems correspond to areas, lines, and points in a $P$-$T$ projection. I prefer, however, to emphasize the fact that coexisting phases reduce the number of independent variables and that some systems have all their properties determined. This fact is very useful in understanding phase diagrams.

---

[1] If we included other kinds of work in our model, there would be an extra degree of freedom for each.

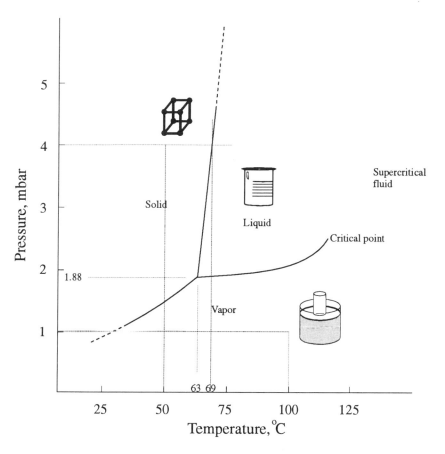

Figure 11.1: Phase diagram for hypothetical compound $\alpha$.

## 11.3 UNARY SYSTEMS

Figure 11.1 shows a typical although hypothetical unary (one component) diagram for compound $\alpha$ ($\alpha$ stands for the formula of some compound, such as NaCl or CaCO$_3$). Although the diagram shows three different *phases* (solid, liquid, and gas), all three have the same composition (whatever the chemical composition of $\alpha$ is), and so the system is unary. This simple diagram contains a surprising amount of information, but you must know how to "read" the diagram. First, note that the diagram contains labeled *areas*, *lines* that separate the areas, and *points*. Every location on the diagram has a pair of $x$-$y$ coordinates, that is, a pressure and a temperature. For example, a pressure of 4 mbar and a temperature of 50°C are the coordinates of a point in the area marked Solid. Under these conditions, the stable form of compound $\alpha$ is observed to be a crystalline solid. If the pressure on $\alpha$ is reduced to 1 mbar, and the temperature increased to 100°C, the stable form of $\alpha$ is gaseous. Similarly, for any combination of pressures and temperatures within the area marked Liquid, the stable form of $\alpha$ is liquid. The phase diagram is in fact a record of these experimental observations about the form of $\alpha$ under various conditions of $P$ and $T$. As mentioned earlier, the vast majority of phase diagrams record the results of experiments—they are not usually the result of theoretical calculations. They are more often a *source* of thermodynamic data than the result of using such data.

Obviously, within these areas, $P$ and $T$ could be changed considerably without changing the nature of the phase, although the *properties* of the phase (its density or heat capacity, say) would certainly change with $P$ and $T$. It appears, then, that for $\alpha$, and for any pure compound, we must choose *two* variables in order to define the state of the compound. Thus to answer the question "what is the density (or heat capacity, refractive index, entropy...) of $\alpha$?" we must first specify two variables—the $P$ and the $T$ we are interested in. One is not enough—at 4 mbar, $\alpha$ can have quite a range of densities, but at 4 mbar, 50°C, its density is fixed and determinable, as are all its other properties. So we say that in each of its three forms—solid, liquid, and gas—$\alpha$ has two *degrees of freedom*—two variables must be specified before *all* are specified. These two variables are in practice usually $T$ and $P$, but in principle any two would do. The Phase Rule summarizes all this discussion by simply saying

$$\begin{aligned} f &= c - p + 2 \\ &= 1 - 1 + 2 \\ &= 2 \end{aligned}$$

With $\alpha$ at 4 mbar and 50°C, consider that we raise the temperature gradually. Nothing much happens, except that the properties of $\alpha$ change continuously, until we reach 69°C, which is the temperature of the boundary between the solid and liquid fields. At this $T$, solid $\alpha$ is observed to begin to melt, and at this $T$, any proportions of solid and liquid $\alpha$ are possible (i.e., almost all solid $\alpha$ with a drop of liquid; or almost all liquid $\alpha$ with a tiny amount of solid; or anything in between). However, if the temperature is held very slightly above the melting $T$, $\alpha$ becomes completely liquid. The solid-liquid boundary line then is a locus of $T$-$P$ conditions that permit the coexistence of solid and liquid $\alpha$. It records the melting temperature of $\alpha$ as a function of pressure.

Note too that because it is a line rather than an area, or because there are two coexisting phases rather than one, we now have only one degree of freedom. In other words, at 4 mbar we now have no choice of temperature. If solid and liquid coexist at equilibrium, the temperature must be 69°C—it is chosen for us. We can still choose whatever $P$ we like (within certain limits), but once we have exercised our one degree of freedom and chosen a pressure, the temperature and all properties of the two phases of $\alpha$ are fixed. Again, we note that the one degree of freedom can be any property, not just $T$ or $P$. We might choose a certain value for the entropy of solid $\alpha$, for example; we would then find that there was only one $T$ and $P$ where solid $\alpha$ with this particular $S_\alpha$ could coexist with liquid $\alpha$. The Phase Rule agrees, saying

$$
\begin{aligned}
f &= c - p + 2 \\
&= 1 - 2 + 2 \\
&= 1
\end{aligned}
$$

Similar comments apply to the boundary between the fields of Liquid and Vapor, which records the *boiling temperatures* of $\alpha$, and the boundary between the Solid and Vapor fields, which records the *sublimation temperatures* of $\alpha$. Where these three boundaries come together at about 63°C, 1.88 mbar (a *triple point*), the three phase fields come together, and solid, liquid and gaseous $\alpha$ can coexist in any proportions at this particular $T$ and $P$. Note that for the coexistence of these three phases, we have lost another degree of freedom. In fact we have no choice at all—if we want three phases to coexist, the $T$ and $P$ must be 63°C and 1.88 mbar. As the number of coexisting phases increases, the number of degrees of freedom decreases. Negative degrees of freedom are not possible, so in a one-component system the Phase Rule predicts that the *maximum* number of phases at equilibrium is three.

$$
\begin{aligned}
f &= c - p + 2 \\
&= 1 - 3 + 2 \\
&= 0
\end{aligned}
$$

## 11.3.1 Free Energy Sections

Despite the fact, mentioned in §11.1.1, that phase diagrams are for the most part experimentally derived, they are controlled by and must conform to fundamental thermodynamic relationships. Understanding phase diagrams is enhanced by examining the relationships between the diagrams and the underlying thermodynamics.

From our study of thermodynamics in previous chapters, we know that the stable state of a system under given conditions is that state having the lowest value of the Gibbs free energy, G (or $G$). If a system does not have the lowest possible value of G, a spontaneous process will take place (according to our model) until this lowest value is achieved. Also, we know that if two phases are in equilibrium in a unary system, the free energy of the component is the same in each phase (§6.2.1; Figure 6.1). Therefore, the phase boundaries in Figure 11.1 are places where $G_\alpha$ is the same in two phases, as shown in Figure 11.2. Note too that we may calculate and plot the free energy (and other properties) of a liquid phase in regions where it is not the stable phase. When we say, for example, that at 4 mbar, 50°C in the solid stability field, $G_\alpha^{solid} < G_\alpha^{liquid}$, we imply that if liquid $\alpha$ could exist at 4 mbar, 50°C, its $G$ would be greater than that of $G_\alpha^{solid}$. We could, in fact, plot the values of $G$ for all possible phases over

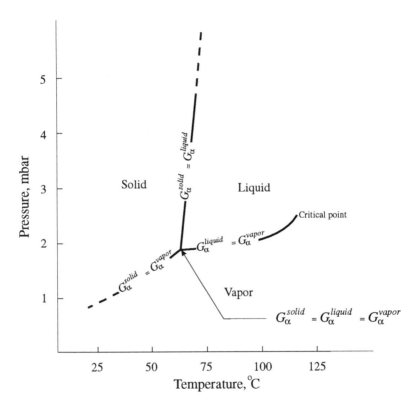

Figure 11.2: Free energy relationships in the phase diagram for compound $\alpha$.

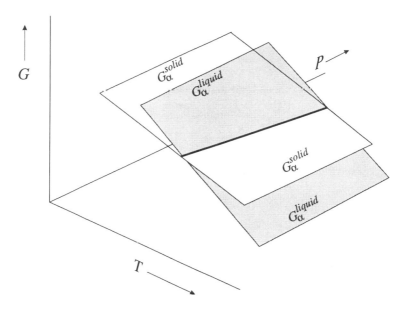

Figure 11.3: $G$-$T$-$P$ diagram for part of Figure 11.1. The heavy line at the intersection of the $G_\alpha^{solid}$ and $G_\alpha^{liquid}$ surfaces is the melting curve.

all parts of the diagram. If we did so and looked at a part near the solid-liquid boundary, we would see something like Figure 11.3.

### $G$-$T$ Sections

Figure 11.4 shows a section through Figure 11.1 at a pressure of 2 mbar. At temperatures below 64°C at 2 mbar pressure, $\alpha$ is solid, and the Gibbs energy of this solid ($G_\alpha^{solid}$) is shown by the line labeled *solid*. Naturally, as we don't *know* the absolute Gibbs energy of any substance, we cannot place any absolute numbers on the $G$-axis. However, we *do* know the slope of this line (the slope is $(\partial G_\alpha / \partial T)_P = -S_\alpha$, and we know $S$ for most compounds), and so we could establish some arbitrary energy divisions on the $G$-axis and plot a line with the correct slope. This line would have a gentle downward curvature because $S$ gradually increases with $T$, but to a first approximation it is a straight line. This line continues to the melting temperature, 64°C, at which point it intersects another line giving the values of $G_\alpha^{liquid}$. This line has a steeper slope, because the entropy of a liquid is always greater than the entropy of a solid of the same composition. At the intersection, $G_\alpha^{solid} = G_\alpha^{liquid}$, as required by phase equilibrium theory (§6.2.1, §6.3).

The $G_\alpha^{liquid}$ line then continues with a gentle downward curvature through the liquid stability region at 2 mbar until it reaches another phase boundary, the boiling curve, at 92°C. Here it intersects the $G_\alpha^{vapor}$ curve, which has a still

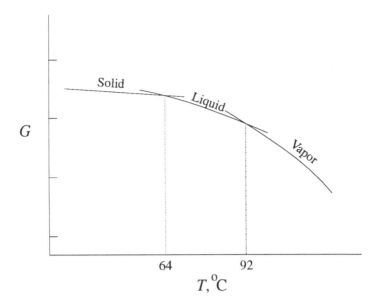

Figure 11.4: $G$–$T$ section through Figure 11.1 at 2 mbar.

steeper slope, because the entropy of gases is always much greater than that of liquids.

Note the similarity of this diagram to Figure 6.5, where we considered $G$–$T$ sections in a quantitative way, to calculate the positions of phase boundaries.

### $G$–$P$ Sections

Figure 11.5 shows a $G$–$P$ section through Figure 11.1 at a temperature of 69°C. At pressures below 1.89 mbar, $\alpha$ is gaseous, and the Gibbs energy of this gas is shown by the line labeled *vapor*. The slope of the line is $(\partial G_\alpha / \partial P)_T = V_\alpha$, and as the molar volume of gases is large, the line has a steep slope. This line intersects another line, giving the values of $G_\alpha^{liquid}$, having a smaller positive slope, because $V_\alpha^{liquid} < V_\alpha^{gas}$. This line continues, again with slight downward curvature because the molar volume of the liquid decreases slightly with increasing pressure, until it reaches the freezing curve at 4 mbar, where it intersects the line giving $G_\alpha^{solid}$. Note the similarity between this diagram and Figure 6.2.

## 11.3.2   Some Important Unary Systems

Substances whose phase relations are interesting for various reasons include carbon (C), iron (Fe), water ($H_2O$), silica ($SiO_2$), aluminum silicate ($Al_2SiO_5$), and calcium carbonate ($CaCO_3$).

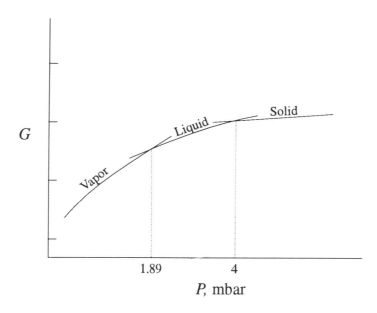

Figure 11.5: $G$-$P$ section through Figure 11.1 at 69°C.

## $H_2O$

The phase diagram for water at relatively low pressures is shown in Figure 11.6. Water is a most unusual substance. It is one of the very few compounds that expands when it freezes, meaning that ice floats. Most substances have solid forms that are denser than their corresponding liquids, and hence will sink during freezing. The fact that ice floats in water is shown in Figure 11.6 by the fact that the liquid–solid boundary (the freezing/melting curve) has a negative slope. In our "typical" unary system (Figure 11.1), this curve has a positive slope. In both cases (Figures 11.1 and 11.6), the denser phase lies at higher pressures, as required by LeChatelier's Principle. The unusual thing is that in the $H_2O$ system, the denser phase is the liquid.

The other term in the slope expression $(dP/dT = \Delta S/\Delta V)$ is $\Delta S$, which is invariably greater in the liquid than in the solid; therefore, the volume change, $\Delta V$, determines whether $dP/dT$ will be positive or negative.

Figure 11.7 shows the same system over a much greater range of pressures. The striking thing about this diagram is the large number of polymorphs of ice, each with its own stability field. These polymorphs give rise to several *triple points*, showing that the solid–liquid–vapor triple point shown in Figure 11.1, which every unary system has, is often not the only one. We came across this phenomenon (a triple point generated by solid polymorphs) previously (§6.4, Figure 6.4). Note the fact that liquid water will freeze (to ice VII) at about 24 kbar at the boiling temperature (100°C). Note too that the negative slope of

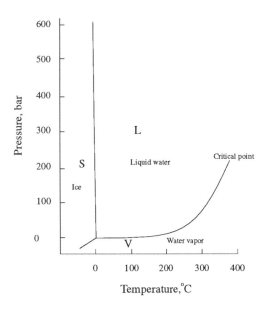

Figure 11.6: Phase diagram for $H_2O$ at relatively low pressures. The solid–liquid boundary is very steep, but in fact has a negative slope.

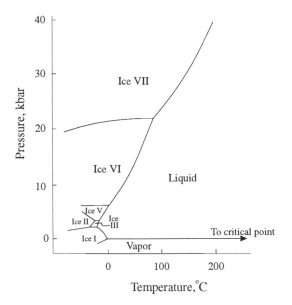

Figure 11.7: Phase diagram for $H_2O$ at high pressures. Ice IV, not shown, is a metastable form of ice in the region of Ice V.

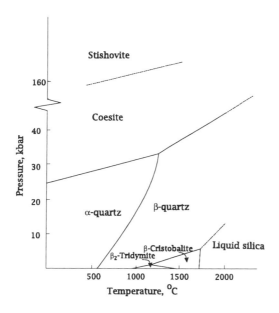

Figure 11.8: Phase diagram for $SiO_2$.

the freezing curve (between Ice I and Liquid) extends to only about 2 kbar.

### $SiO_2$

Silica, one of the most common compounds on Earth, has a number of interesting and complex phase relations, shown in Figure 11.8.

## 11.4 BINARY SYSTEMS

### 11.4.1 Types of Diagrams

When we consider the phase relations in systems having two components instead of one, we add one dimension to our diagrams. That is, in unary diagrams all phases have the same composition, and so we don't need an axis showing compositions—we can use both dimensions available on a sheet of paper for physical parameters, and we choose $T$ and $P$. With two components, we find that phases commonly contain different proportions of these components— they have different compositions. Since this is of great interest, we use one dimension for composition, leaving only one other for either $T$ or $P$. Most commonly temperature variations are of more interest, and so diagrams showing

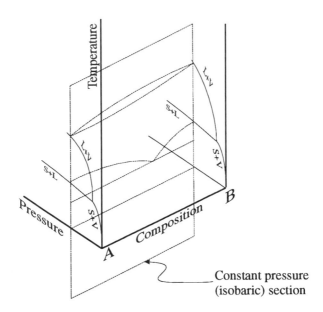

Figure 11.9: The $P$-$T$-$X$ box. Most binary phase diagrams are $T$-$X$ sections through this box. The phase relations will, of course, vary with the pressure chosen for the section. $P$ = 1 bar is the commonest choice.

phase relations on a $T$-$X$ diagram are very common.[2] This relationship is illustrated in Figure 11.9. In this book, we concentrate on $T$-$X$ sections, but it is well to realize that there are other varieties of diagrams. Not only do we have $P$-$T$ sections, but $P$-$T$ or isoplethal (constant composition) sections, as well as various projections.

## 11.4.2   The Melting Relations of Two Components

Suppose that you now understand unary phase relations very well, but have never encountered binary systems, and you are given the following problem. There are two minerals, A and B. We know the melting point of each mineral, $T_{m_A}$ and $T_{m_B}$ at atmospheric pressure. We grind samples of A and B together in various proportions, say 25% A, 75% B; 50% A, 50% B; and 75% A, 25% B, and we perform experiments to determine the melting temperature of these mixtures. Your job is to draw a diagram *predicting* the most likely results. The diagram should show temperature as the vertical axis, and composition as the horizontal axis, and of course the known melting temperatures of the pure minerals A and B should be plotted on the vertical axes at each end of the composition axis.

It seems very likely that your guess would look like Figure 11.10. In other

---

[2]We use the "$X$" in the expressions $T$-$X$ or $P$-$T$-$X$ to mean "composition" generally, whether measured as mole fractions or weight percent, or in some other way.

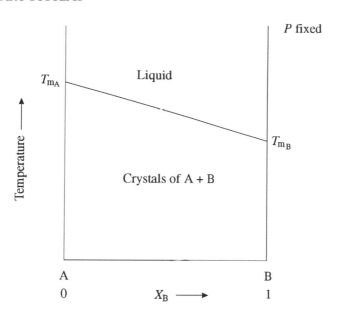

Figure 11.10: An uneducated guess as to the melting temperatures of mixtures of minerals A and B.

words, you would probably suppose that the melting temperature of mixtures of A and B would be some kind of average of the melting temperatures of the pure compounds, much in the way that volumes are averages as shown in Figure 7.2a. But binary systems are not quite that simple. Figure 11.10 is thermodynamically impossible, even if A and B were not separate phases but formed a solid solution, but we will not bother to prove this. Suffice it to say that experiments on hundreds of binary systems have never given results consistent with Figure 11.10.

What *does* happen depends on what compounds A and B actually are. Let's suppose that A is the component $CaMgSi_2O_6$, and B is the component $CaAl_2Si_2O_8$. The stable forms of these components at ordinary temperatures are the minerals diopside ($CaMgSi_2O_6$) and anorthite ($CaAl_2Si_2O_8$), and so we will represent component $CaMgSi_2O_6$ by the symbol Di and component $CaAl_2Si_2O_8$ by the symbol An. Diopside melts at 1392°C, and anorthite melts at 1553°C. We perform the experiments mentioned above, that is, we grind up both samples into fine powders, then mix the powders in various proportions and heat them up in separate experiments and observe what happens at various temperatures. What a surprise—we find that *all mixtures begin to melt at the same temperature!* And when we analyze the composition of the first liquid to form, we find that *the first liquid to form in all mixtures has the same composition!* The temperature is 1274°C (called the *eutectic temperature*), and the composition is 42% An, 58% Di (called the *eutectic composition*). On heating to still higher

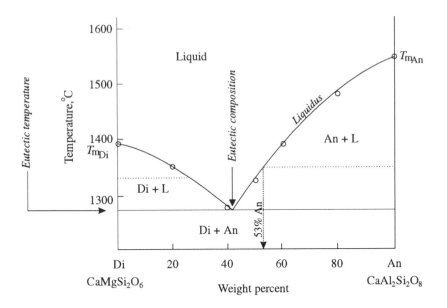

Figure 11.11: The system Di-An at one bar pressure. Two representative tie-lines are shown.

temperatures, another surprise awaits us. For those mixtures having *more* than 42% An, temperatures above 1274°C result in disappearance of all diopside in the mixtures—we are left with only liquid plus anorthite crystals. For those mixtures having *less* than 42% An, temperatures above 1274°C result in the disappearance of all anorthite in the mixtures—only liquid plus diopside crystals are left. So *below* 1274°C, only two phases coexist—crystals of anorthite and diopside. *Above* 1274°C, again only two phases coexist—either liquid and anorthite, or liquid and diopside. Only at 1274°C exactly are three phases observed to coexist at equilibrium—anorthite, diopside, and liquid. And note that in the binary system, we have melting far below the melting temperature of either of the pure components. These relationships are summarized in Figure 11.11. They may seem strange at first, but as we will see, they are one of a rather small set of relationships that satisfy the Phase Rule.

### The Isobaric Phase Rule

But, first, we must mention a slight modification of the regular Phase Rule, equation (11.1). As shown in Figure 11.9, the experiments we are discussing at a fixed pressure of 1 bar can be represented on a plane or section through $P$-$T$-$X$ space. The general Phase Rule (11.1) applies to this $P$-$T$-$X$ space. The fact that we confine ourselves to a fixed $P$ plane within this space means that we have "used" one of our degrees of freedom—we have chosen $P = 1$ bar, and the same would be true for any other constant $P$ section (or constant $T$ section, for that matter). Therefore *on our T-X plane* the Phase Rule is

$$f = c - p + 1 \tag{11.2}$$

This shows that the maximum number of phases that can coexist at equilibrium in a binary system at an arbitrarily chosen pressure (or temperature) is three ($p = 3$ for $c = 2$, $f = 0$), which is consistent with our observations.

### 11.4.3 Reading the Binary Diagram

The main features of the phase relations in Figure 11.11 follow directly from this fact. During the heating of our mixture of Di and An crystals, we have two phases, and

$$
\begin{aligned}
f &= c - p + 1 \\
&= 2 - 2 + 1 \\
&= 1
\end{aligned}
$$

This means that to fix all the properties of both kinds of crystals, we need only choose the temperature (pressure being already fixed at 1 bar). However, when the first drop of liquid forms, $p = 3$ (diopside crystals, anorthite crystals, and liquid), and $f = 0$. Another word for $f = 0$ is *invariant*. When $p = 3$ on an isobaric plane, we have *no* choice as to $T$, $P$, or the compositions of the phases— they are all fixed. This explains why all mixtures begin to melt at the same temperature, and why the liquid formed is always the same composition no matter what the proportions of the two kinds of crystals. No other arrangement would satisfy the Phase Rule.

A line on a phase diagram joining points representing phases that are at equilibrium with each other is called a *tie-line*. Each of the two-phase regions in Figure 11.11 (Di+L; An+L; Di+An) is filled with imaginary tie-lines joining liquid and solid compositions, or two solid compositions, that are at equilibrium. Only two of these tie-lines are shown. Consider the tie-line at 1350°C in the region labeled An+L. One end of the line is on the curved line representing liquid compositions (called the *liquidus*), and the other end is on the vertical line representing 100% An composition. The composition scale across the bottom of the diagram applies at any temperature, so we can get the liquid composition by dropping a perpendicular from the liquidus to the composition scale, showing that the liquid composition at 1350°C in equilibrium with pure anorthite crystals is 53% component An, 47% component Di. The composition of the solid phase is given by the other end of the line, which is at 100% An. In each of these two-phase regions, such as An+L, $f = 1$, which means that once we have chosen the temperature, say 1350°C, all properties of all phases are fixed. Therefore, all proportions of Di and An in this region will have the same liquid and solid compositions. In other words, *any* starting mixture of diopside and anorthite crystals having more than 53% anorthite would, when heated to 1350°C, consist of a liquid of composition 53% An, 47% Di, plus crystals of pure anorthite. Mixtures having between about 20% An and 53% An would be completely liquid at this temperature, and mixtures having 0 to 20% An would consist of pure diopside crystals plus a liquid of composition 20% An, 80% Di.

By imagining tie-lines across the An+L region at successively higher temperatures, we see that the composition of the liquid in equilibrium with anorthite crystals gets progressively richer in component An. Similarly, the tie-lines in the Di+L region show that the liquid gets richer in Di as temperature increases.

Because the temperature of the three-phase tie-line is fixed, it follows that both above and below this temperature there must be regions having only two phases. We already know that below the three-phase line the two phases are Di and An. Above the three-phase line one of the phases must be liquid, because melting has started. Therefore, the liquid can coexist with only one other phase, obviously in this case either Di or An, but not both. As $T$ increases, the proportion of liquid must increase, eventually becoming 100% liquid. This simple analysis is sufficient to explain the main features of the diagram. "Reading" binary diagrams consists largely of distinguishing between one phase regions, which have no tie-lines (e.g., the Liquid region), two-phase regions, which have tie-lines joining two phases at equilibrium, and three-phase tie-lines, which separate two-phase regions, and join three phases at equilibrium.

## 11.4.4   A More General Example

The system Di-An is misleadingly simple in two respects. For one thing, the diagram shows that both diopside and anorthite remain pure while heated in contact with the other component until the melting temperature is reached (1392°C for Di, 1553°C for An). Actually, phases never remain perfectly pure when in contact with other phases—some mutual solution always takes place, although as in the case of Di and An it is sometimes small and does not show on the diagram. A more realistic case is shown in Figure 11.12. The diagram is essentially the same as the Di-An diagram, except that there is a field of $A_{ss}$ and of $B_{ss}$, where subscript ss stands for solid solution.

The other respect in which the Di-An diagram is misleading is that, in fact, it is not strictly speaking a true binary system. This somewhat surprising statement cannot be fully explained without discussing ternary systems. Suffice it to say that just because you choose two components does not necessarily mean that you have a binary system. *To be truly binary, all compositions of all phases must lie on the plane of the diagram.* This must be the case in simple systems such as Cu-Au and with single solution phases such as liquids; but with complex components such as Di and An, although the *bulk composition* must lie on the plane of the diagram, the compositions of coexisting *phases* may each lie "off the plane" of the diagram. Careful work has shown that in the system Di-An, diopside crystals are not pure but contain some Al. This means that, because bulk compositions lie on the Di-An plane, phases coexisting with diopside must be somewhat deficient in Al. To portray this in a diagram, one

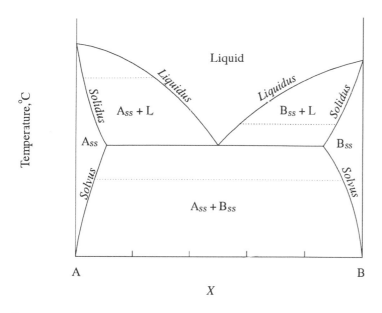

Figure 11.12: A more representative binary system. The difference is that both components show solid solution fields. Three representative two-phase tie-lines are shown.

needs a three-component triangle. Just remember that not all choices of two components are binary systems—some are planes within ternary systems.

**Solid Solutions**

There is no difference in principle between a solid solution and a liquid or gaseous solution. Substances dissolve into one another, like sugar into tea, or like oxygen into nitrogen, because the free energy change of such a process is negative—they are spontaneous processes. Consider the system Di–An at a temperature of 1600°C (Figure 11.11). At this temperature, both pure Di and pure An are liquid phases. If one gram of Di liquid and one gram of An liquid were mixed together, they would dissolve into one another to form a homogeneous liquid solution, represented by a point in the middle of the diagram on the 1600°C isotherm. If pure diopside crystals are mixed with pure anorthite crystals at 1200°C, on the other hand, nothing happens—they do not dissolve into each other.

Components A and B in Figure 11.12, on the other hand, behave differently. Liquid A and liquid B still mix to form a homogeneous solution, but when solid A and solid B are mixed together, they dissolve into one another to a limited extent. Salt will dissolve into water, but not without limit—it will dissolve only until the water becomes saturated. Similarly, solid B will dissolve into solid

A, but not without limit. It dissolves into A until A is saturated with B, and at the same time A dissolves into B until B is saturated with A. The saturation limits of each component in the other is shown by a line called the *solvus*. The existence of a solvus shows that A and B exhibit *limited miscibility* in the solid state. They exhibit *complete miscibility* in the liquid state. "Miscibility" does not really mean "mixability," although they sound similar. "Mixability," if it is a word, just means things can be mixed together—mutual dissolution is not implied. "Miscibility" means the ability to dissolve into something else.

Figure 11.12 shows a eutectic, but the two solid phases in equilibrium with the liquid are not pure A and pure B; A contains some B in solid solution ($A_{ss}$) and B contains some A in solid solution ($B_{ss}$). Similarly, at temperatures above the eutectic, the liquid is not in equilibrium with pure A or pure B, but with $A_{ss}$ and $B_{ss}$. The compositions of the solid solutions in equilibrium with liquid are given by lines called the *solidus*.

## 11.4.5   Freezing Point Depression

Figure 11.11 shows that mixtures of diopside and anorthite become completely liquid at temperatures lower than the melting temperatures of either pure diopside or pure anorthite. This is also shown by the more general system in Figure 11.12 and is, in fact, an extremely common feature of binary systems. It is called *freezing (or melting) point depression* and is, in fact, why we put salt on icy roads in winter. The melting temperature of ice is lowered in the presence of the second component (salt, NaCl), and so the ice melts and the resulting salty water corrodes our cars. But why is the freezing point depressed?

The answer is found in the basic thermodynamic relationships between the phases. Figure 11.4 shows the absolute free energies of the solid, liquid, and vapor phases of our compound $\alpha$ as a function of temperature at a pressure of 2 mbar. Let's call compound $\alpha$ component A (just as we called component $CaMgSi_2O_6$ Di). Figure 11.4 therefore shows $G_A^{solid}$, $G_A^{liquid}$, and $G_A^{vapor}$ as a function of $T$. If we now add a second component B to A, what happens to these free energies? To start with the simplest case, we will suppose that B does not dissolve into solid A or into vapor A, but it does dissolve into liquid A (if you add NaCl to $H_2O$, the salt will not dissolve into ice or into steam, but it will dissolve into liquid water). Therefore, the curves for $G_A^{solid}$ and $G_A^{vapor}$ will not change, because $A^{solid}$ and $A^{vapor}$ are unchanged in the presence of B. But what is the free energy of component A in a liquid containing both A and B?

The answer is shown in Figure 7.4, which shows that when B dissolves into A, the molar free energy of component A in the solution at a given concentration (which we call $\mu_A$), is *lower* than the molar free energy of pure A ($G_A^\circ$). This relationship is quite general and without exception, because otherwise A and B would not form a solution. We will be mentioning this relationship at various points throughout this chapter. The consequence of the fact that $G_A^{liquid}$ is lowered but $G_A^{solid}$ is not is shown in Figure 11.13. The shaded surface in this

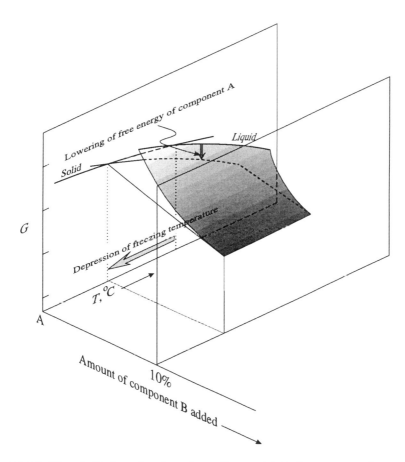

Figure 11.13: The *G–T* plane is taken from Figure 11.4. Component B enters the liquid phase and causes a lowering of $\mu_A^{liquid}$, which in turn causes a depression of the freezing temperature.

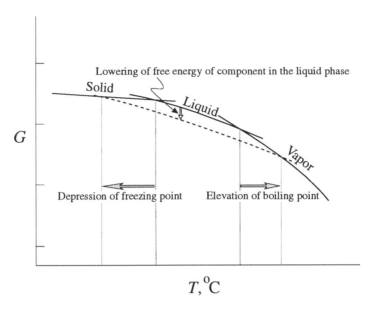

Figure 11.14: $G$-$T$ section, showing lowering of the free energy of A in the liquid phase, causing depression of the freezing point and elevation of the boiling point.

figure represents the free energy curve from Figure 7.4, extended into a range of temperatures. It shows the lowering of the total free energy of the liquid phase as component B is added. At the arbitrary amount of 10% B, a tangent surface to the free energy surface extends back to the 0% B plane, analogous to the tangent at $X_B = 0.4$ in Figure 7.4. The trace of this tangent surface on the $G$-$T$ section for component A gives $\mu_A$, the molar free energy of A in the solution containing 10% B, 90% A. It of course lies below the curve of $G_A^{liquid}$ for pure A. But because the curve for $G_A^{solid}$ has not moved, the *intersection* of the $G_A^{solid}$ and $\mu_A^{liquid}$ curves is moved to a lower temperature. The intersection of the $\mu_A^{liquid}$ and $G_A^{solid}$ curves is the point where these two quantities are equal, and it defines the temperature at which A in the solid state and A in the liquid state are in equilibrium. For pure A, this is the melting or freezing temperature; for the system A-B, it defines a point on the liquidus of A and is the result of freezing point depression.

This relationship is shown in again in Figure 11.14, this time including the vapor curve. If the vapor curve does not move (no B dissolves into vapor A), depression of $G_A^{liquid}$ results in a raising of the boiling temperature as well as a lowering of the freezing temperature. This is also an extremely common effect.

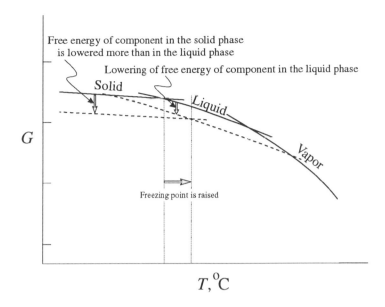

Free energy of component in the solid phase
is lowered more than in the liquid phase

Lowering of free energy of component in the liquid phase

Solid

Liquid

$G$

Vapor

Freezing point is raised

$T, {}^{0}C$

Figure 11.15: $G$-$T$ section, showing a greater lowering of the solid free energy than liquid free energy, causing elevation of the freezing point.

## 11.4.6 Freezing Point Elevation

But suppose our simplifying assumption that no B enters the solid phase is not true? There is no difference in principle between the thermodynamics of solid and liquid solutions, and so if B dissolves into solid A the curve for $G_A^{solid}$ will be lowered for the reasons just discussed. Normally, B is less soluble in solid A than in liquid A, and so the amount of lowering is less for the solid phase, and the freezing point is still lowered. This is shown by systems like that in Figure 11.12, where the liquidus of A slopes downward, even though B is shown as entering both the liquid and the solid phases of A.

However, what of the possibility that the $G_A^{solid}$ curve might be lowered *more* than the $G_A^{liquid}$ curve? This would happen if more B dissolved into solid A than into liquid A and would result in a *freezing point elevation* as shown in Figure 11.15. This explains an important feature of many binary systems.

## 11.4.7 Systems Having Complete Solid Miscibilty

A and B in Figure 11.12 show limited solid miscibility, but some important systems show *complete* miscibility in the solid state, giving rise to a diagram that looks quite different, as shown in Figure 11.16. In a sense, it is simpler than the ones we have looked at so far—in fact it looks rather like Figure 11.10, except that the "melting line" in Figure 11.10 is a *melting loop* in Figure 11.16. But the

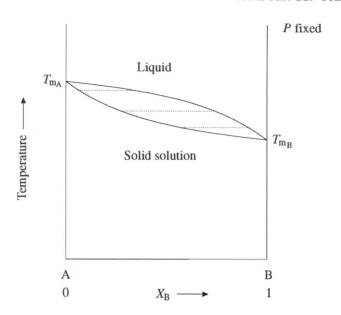

Figure 11.16: A binary system A–B showing complete miscibility in both the solid and liquid states. Three representative tie-lines are shown.

most important difference is that in Figure 11.16 A and B *dissolve completely into one another in the solid state.* This takes some getting used to. We are quite familiar with sugar dissolving into tea, but the idea of placing two solid objects together and observing one disappear into the other is not something in our experience. But this is just another example of something that thermodynamics says *should* happen but in fact does not, because of energy barriers. The thermodynamic model does not consider these barriers, and so does not always work. These solid solutions do exist, however, because they do not form from solids dissolving into one another at low temperatures. They form at high temperatures, sometimes over long periods of time, and then cool down in their mutually dissolved state. Many important alloys and mineral groups are such complete solid solutions, including the feldspars, olivines, and some pyroxenes and amphiboles.

Note that in Figure 11.16, $T_{m_A}$ is lowered by adding B, but $T_{m_B}$ is raised by adding A. This is because more B enters liquid A than solid A, but more A enters solid B than liquid B, and the free energy consequences of this are shown in Figures 11.14 and 11.15, respectively.

The most important mineralogical example of this type of system is the plagioclase feldspar system, shown in Figure 11.17. Plagioclase is a mineral whose composition may vary from virtually pure albite ($NaAlSi_3O_8$), or component Ab, to almost pure anorthite ($CaAl_2Si_2O_8$), or component An, depending on the composition of the liquid from which it crystallizes. The melting behavior of

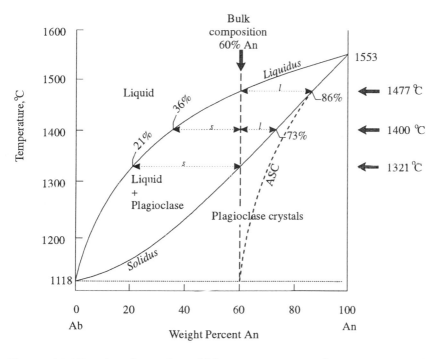

Figure 11.17: The plagioclase feldspar system at 1 bar pressure. The curve labeled ASC is the Average Solid Composition during fractional crystallization of the 60% An bulk composition.

a complete solid solution such as this is a simple melting loop—a combined liquidus and solidus that goes from one pure component over to the other. The melting loop is filled with imaginary horizontal tie-lines, three of which are shown in Figure 11.17. They indicate the compositions of liquids, on the liquidus, and the compositions of plagioclase crystals, on the solidus, which are in equilibrium with each other.

## 11.4.8   Equilibrium vs. Fractional Cooling and Heating

Binary phase diagrams show phase compositions that are at equilibrium—they show what you would obtain if you heated a bulk composition to a certain temperature and waited long enough for equilibrium to be attained. The time required to reach equilibrium after a change in temperature or pressure varies greatly with the system, but equilibrium is *never* achieved instantaneously. Therefore, if we use the diagram to consider what would happen during continuous cooling or heating a given bulk composition, we cannot be considering what would *really* happen in our system during cooling or heating we are considering *model* processes, as usual.

There are any number of models of processes we could devise involving phase changes in binary systems, but two are especially common—complete equilibrium (reversible) processes, and "surface equilibrium" (perfect fractional) processes. We will discuss only cooling processes. Heating processes are the exact reverse of cooling processes in the equilibrium case, but not always the exact reverse in the case of fractional processes.

### Perfect Equilibrium Crystallization

Suppose we had a liquid having a composition of 60% An, 40% Ab at a temperature of about 1600°C (Figure 11.17). On cooling this liquid, nothing much happens (except that the properties of the liquid, such as its density, refractive index, entropy, free energy, etc., etc., change) until it reaches a temperature of 1477°C, the liquidus temperature for this composition. At this point, the bulk composition is still 100% liquid, but the first tiny crystal of plagioclase appears. Its composition, given by the solidus, is 86% An, 14% Ab. As cooling continues, plagioclase crystals continue to form, and previously formed crystals change their composition so that all crystals always have the equilibrium composition, with no compositional gradients. When the temperature reaches 1400°C, the liquid has composition 36% An, and the crystals 73% An. (These compositions are obtained by dropping a perpendicular line from the point of interest to the compositional axis at the bottom of the diagram.) When the composition of the solids reaches the bulk composition of 60% An, the liquid must disappear, and this happens at a temperature of 1321°C. Further cooling results in no further changes in composition of the crystals.

## Perfect Fractional Crystallization

Maintaining perfect equilibrium while cooling is one end of a complete spectrum of possibilities. The other end of the spectrum is that crystals form, but always completely out of equilibrium. This end of the spectrum involves an infinite number of cases and so is rather difficult to discuss in a finite number of words. A subset of these possibilities is the case where crystallization produces crystals in equilibrium with the liquid, as required by the diagram, but after forming, they do not react with the liquid in any way. This is called surface equilibrium (because the liquid is at all times in equilibrium with the surface of the crystals) or fractional crystallization, and is a model process just as much as is equilibrium crystallization. It is also used in connection with liquid–vapor processes (fractional distillation; fractional condensation), as well as isotope fractionation processes.

There are two ways of imagining a process of perfect fractional crystallization.

- As soon as a tiny crystal forms, it is removed from the liquid. This might be by reaching into the liquid with a pair of tweezers and physically removing the crystal, or the crystal might immediately sink to the bottom or float to the top of the liquid, where it becomes covered by other crystals and is removed from contact with the liquid.

- As soon as a tiny crystal forms with a composition given by the solidus, it is covered by a layer of another composition, given by the solidus at a slightly lower temperature. Successive layers are formed, each controlled by the position of the solidus, but after forming, the various layers do not homogenize in the slightest. The result is a compositionally zoned crystal.

Note that crystals are removed from contact with the liquid in both cases. This is the essential element of fractional crystallization.

Considering the same bulk composition, 60% An, the cooling history is the same as before until the first tiny crystal forms at 1477°C, having a composition of 86% An. On further cooling, the liquid composition follows the liquidus, as before, and any *new* crystals that form have compositions given by the solidus at that temperature; but previously formed crystals, being removed from contact with the liquid, do not change their original compositions. The net result is that at any temperature below 1477°C, the *average* composition of all solids formed is more An-rich than would be the case in equilibrium crystallization, that is, more An-rich than the solidus at that temperature. Because of this, at each temperature below 1477°C, there must be a larger proportion of liquid of Ab-rich composition to balance the solid composition, that is, to give the known bulk composition. Therefore, whereas in equilibrium crystallization the last drop of liquid must disappear at 1321°C, in fractional crystallization it does not, and in fact liquids continue to exist right down to pure Ab composition, where the last liquid crystallizes pure albite. This is the important aspect of fractional crystallization from a petrological point of view—that a given bulk composition can generate a much wider range of liquid compositions, and hence a wider range of igneous rocks, than can equilibrium crystallization.

It is possible to calculate the average composition of the solids during fractional crystallization, but we will not do this. Just note that a curve indicating the average composition of all solids generated must begin at 1477°C on the solidus, and it must end at 1118°C at a bulk composition of 60% An, when the last liquid disappears. This curve is shown in Figure 11.17, labeled "ASC." For equilibrium cooling, the "ASC" curve is, of course, the same as the solidus.

## 11.4.9   The Lever Rule and Mass Balances

Phase diagrams contain information not only about phase compositions and their temperatures and pressures, but about the *proportions* of phases for a given bulk composition. This is done using what is called the Lever Rule. Look at the three tie-lines we have just been discussing in Figure 11.17. Consider first the line extending from the liquidus (36% An) to the solidus (73% An), at 1400°C. This line is composed of two parts. One part, labeled $l$, represents the proportion of liquid, and the other part, labeled $s$, represents the proportion of solids. The fraction (by weight) of liquid in the bulk composition is thus $l/(l+s)$, and the fraction of solids is $s/(l+s)$. The easiest way to measure the lengths of $l$ and $s$ is probably by comparing the compositions of the end points of the tie-line at the liquidus and solidus with the bulk composition. Thus the $s$ portion of the tie-line has a length of $60 - 36 = 24\%$, and the $l$ portion of the tie-line has a length of $73 - 60 = 13\%$. The total length of the tie-line is $73 - 36 = 37\%$. Therefore the proportion or fraction of solid in the total bulk composition is $24/37 = 0.65$, and the fraction of liquid is $13/37 = 0.35$, and of course $0.65 + 0.35 = 1.0$. If we had a bulk composition weighing 10 g, then at 1400°C, 1 bar, it would be made up of $0.65 \times 10 = 6.5$ g of crystals (73% An composition), and $0.35 \times 10 = 3.5$ g of liquid (composition 36% An). This Lever Rule can be used in any two-phase region, given the bulk composition, and the lengths of the lines can be measured in % composition as we have done, or in millimeters, or inches, or any other units.

Note that at intersection of the bulk composition line and the liquidus, the $s$ portion of the line reduces to zero, because there is 100% liquid, and at the intersection of the bulk composition line and the solidus, the $l$ portion of the line reduces to zero. This provides a way of remembering which side of the tie-line represents which phase.

**Mass Balances**

Because we know the proportions and compositions of the phases, it is a simple matter to combine these to calculate the bulk composition. But, as we *know* the bulk composition (we needed it to get the phase proportions), the calculation is circular. Nevertheless, it is a useful check on our reading and construction of diagrams. For example, at 1400°C in Figure 11.17, the mass balance is

(solid fraction × solid composition)
+(liquid fraction × liquid composition)   –   bulk composition
$$(0.65 \times 73) + (0.35 \times 36) = 60$$

where 60 is the bulk composition in % An.

## 11.4.10   Binary *G–X* Sections

The fact that the Gibbs energy of solutions is represented by a convex-downward curve (a "festoon") was introduced in §7.2.4 and Figure 7.4. Both solid solutions and liquid solutions are represented by such curves, and understanding of binary diagrams is increased by constructing such curves on *G–X* sections. Each *G*-curve moves upward with decreasing temperature ($\partial G/\partial T = -S$), but the liquid curve moves upward faster than the solid curve, because the entropy of liquids is greater than solids. The stable phase for any bulk composition is always indicated by the lowest *G*, either on a solid or liquid curve or on a tangent joining two such curves.

### A Binary Eutectic System

Consider first the Di–An system in Figure 11.18, an example of a simple binary eutectic system. The *T–X* section is shown at the bottom of the diagram, and *G–X* sections through the system at various temperatures are shown above it. In understanding this diagram, it is important to remember that the *G* of a mixture of crystals that are completely immiscible (show no mutual solid solution) is simply the weighted average of the *G* of the two pure end-members, which in this case appears as a straight line joining $G_{Di}$ and $G_{An}$. This straight line appears on all sections, whether the solids are stable phases or not. When components do not mutually dissolve, the *G* behaves the same way as *V*, and equation (7.5) applies. It appears as the line labeled $X_A G_A^\circ + X_B G_B^\circ$ in Figure 7.4. The line labeled "mixture of Di and An crystals" in Figure 11.18 represents the same situation.

When components *do* mutually dissolve to form a solution, equation (7.4) still works for volumes if the solution is ideal, but even for ideal solutions, it does not work for *G*. The free energy of the solution must be less than the weighted average of the *G* of the two pure end-members for mutual solution to take place, and it is represented by the "festoon," or convex-downward loop. Therefore, the liquid solution formed when liquid Di and liquid are mixed together is represented by such a loop. It is shown in all sections, even when the liquid is not the stable phase.

Understanding these *G–X* sections is helped by realizing that

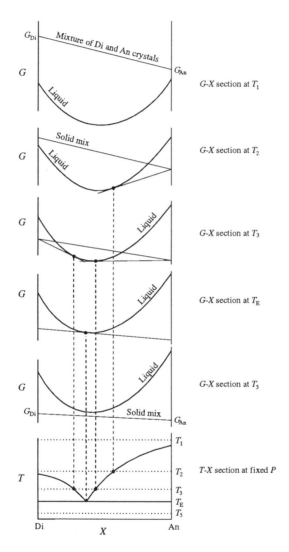

Figure 11.18: $G$-$X$ sections through a binary eutectic diagram.

- The line representing the $G$ of the mixture of solid Di and solid An is shown in every section regardless of whether diopside or anorthite is or is not stable. If they are not stable, the line represents *metastable* free energies.

- The line (festoon) representing the $G$ of the complete liquid solution between Di and An is shown in all sections regardless of whether the liquid is stable or not. If the liquid is not stable, its curve represents *metastable* free energies.

- The stable phase or phases in each section must have the lowest free energy available.

Starting with the $G$-$X$ section at $T_1$ at the top of the diagram, we see that for every bulk composition between Di and An, the liquid free energy is everywhere *lower* than the free energy of a mixture of solid Di and solid An. In other words, the stable phase across the diagram is liquid. On the $T$-$X$ section, note that at $T_1$, we are above the melting temperatures of both components, and in the field of liquid at all compositions.

As we cool from $T_1$ to $T_2$, the free energy of liquids and solids (whether stable or metastable) increases, but that of liquids increases more, so that at $T_2$, the $G$ of liquid An has become greater than that of solid An, but the $G$ of liquid Di remains less than that of solid Di. At some temperature between $T_1$ and $T_2$ we must have passed a point where $G_{An}^{liquid} = G_{An}^{solid}$, that is, the melting temperature of An. At $T_2$, The stable form of pure Di is liquid, but the stable form of pure An is solid. The lowest free energy available to the system as we go from Di toward An is liquid, but just after passing the minimum on the curve, the lowest free energy available is neither liquid nor a mixture of crystals, but a mixture of liquid and An crystals. In this mixture, the free energy of component An in the crystals must be the same as the free energy of component An in the liquid.

Recall from Figure 7.4 that the tangent to a free energy curve of a solution has intercepts giving the chemical potential of each component in the solution. Therefore, a tangent to the liquid curve that has an intercept on the An axis at the free energy of solid An will indicate that liquid composition in which $\mu_{An}^{liquid} = \mu_{An}^{solid}$. That tangent point is, of course, at the composition of the liquidus at that temperature, as shown by the dotted line joining the $G$-$X$ section at $T_2$ with the $T$-$X$ section. As the temperature falls below $T_2$, that tangent, rooted on the An axis at the free energy of solid An ($G_{An}^{solid}$), moves to greater Di compositions, because the liquid loop is moving up with respect to the $G_{An}^{solid}$ point.

At $T_3$, solid diopside is now the stable phase on the Di side of the diagram, and the tangent situation described above holds for both components. With falling temperature, the two tangent points move toward each other, becoming one tangent at $T_E$, the eutectic temperature. Note that at $T_E$ there must be only one tangent because $\mu_{Di}$ and $\mu_{An}$ must be the same in all three phases.[3]

On further cooling, the tangent breaks away from the liquid curve and becomes a straight line below the liquid curve, giving the free energy of a mixture of diopside and anorthite crystals just as in the section at $T_1$. The difference is that now it is completely *below* the liquid curve, and therefore a mixture of crystals is the stable configuration of the system.

## A Melting Loop System

The story is rather similar for $G$-$X$ sections through a melting loop diagram at various temperatures (Figure 11.19). However, instead of dealing with the intersection of a solution curve, or festoon, and a straight line, we have the intersection of two solution curves—one for the liquid solution and one for the solid solution. If this seems confusing, go back to §7.2.4 and recall why a free energy curve for a solution must be convex downward. Then remember that this applies whether the solution is solid, liquid, or gaseous. Finally, remember that in these sections (Figure 11.19) we plot the positions of both solution curves in every section, regardless of whether the solution is the stable phase or not. The point is to determine which parts of which curves give, or combine to give, the lowest free energy available to the system at each composition across the system.

At the top of the diagram, the section is drawn at the melting temperature of anorthite crystals. Therefore, the liquid and solid curves join at the An axis, because $G_{An}^{solid} = G_{An}^{liquid}$. Going toward component Ab, the liquid curve is everywhere below the solid curve, showing that liquid is everywhere the stable phase at this temperature. Similarly, the section at $T_{mAb}$ shows that $G_{Ab}^{solid} = G_{Ab}^{liquid}$, and that at compositions toward component An, the solid curve is below the liquid curve, showing that at this temperature, a solid solution of Ab and An is the stable phase. At intermediate temperatures, the two curves intersect. The liquid curve is lower on one side, and the solid curve is lower on the other side. Intermediate compositions have the lowest possible free energy only by being a mixture of solid and liquid, and because the chemical potentials of both Ab and An must be the same in both phases, the compositions of the two solutions at equilibrium must be given by the only common tangent to the two curves at each temperature.

The sections in Figure 11.19 have been reassembled into a three-dimensional view in Figure 11.20. The only major advantage of this is that you can now see the relative rises of the liquid and solid loops with decreasing temperature, as indicated by the dotted lines on the sides of the box. The line representing liquids has a steeper slope than that representing solids for the same reasons as in Figures 11.4 and 6.5.

---

[3] If you think about this statement, and look at equations (7.7), you will see why we said (§11.4.4) that phases never remain absolutely pure when heated together, at least according to our model.

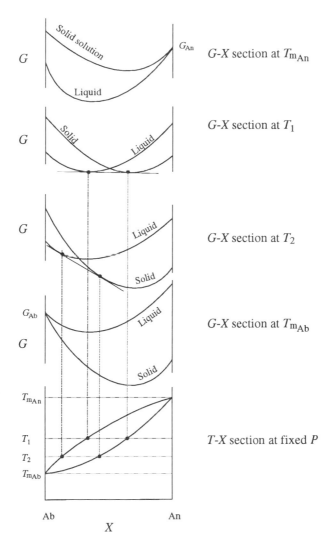

Figure 11.19: $G$–$X$ sections through a melting loop diagram.

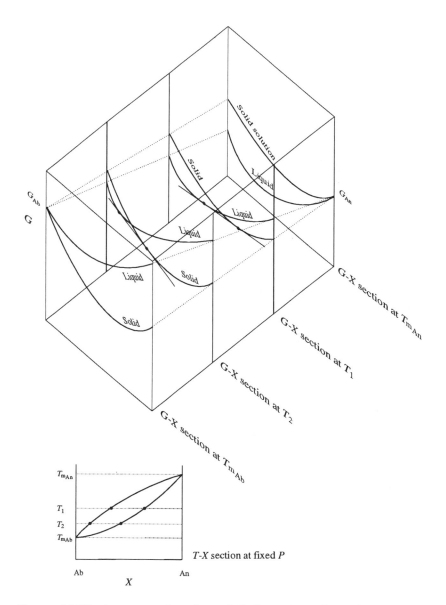

Figure 11.20: A perspective view of $G\text{-}X$ sections through a melting loop diagram.

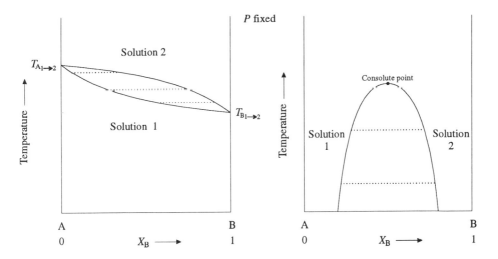

Figure 11.21: The two basic binary diagram elements. In the phase transition loop (left diagram) solution 1 and solution 2 can be solid and solid, solid and liquid, or liquid and vapor, respectively. $T_{A_{1-2}}$ and $T_{B_{1-2}}$ are melting temperatures, boiling temperatures, or polymorphic phase transition temperatures for pure A and B respectively. Three representative tie-lines are shown. In the solvus (right diagram), solution 1 and solution 2 can be two solids or two liquids. Two representative tie-lines are shown.

## 11.4.11 Binary Diagram Elements

Binary phase diagrams can become quite complex, but the complexities are nothing but the elements of simpler diagrams, combined in such a way as to satisfy the Phase Rule. There are essentially only two elements (Figure 11.21), both of which contain two-phase tie-lines, and hence are elements controlling the compositions of coexisting phases.

- The phase transition loop, which separates two different kinds of solutions. This can be a melting loop as shown previously in Figure 11.16, separating a liquid solution from a solid solution, the end points being melting temperatures. But it can also be a boiling loop, separating a gas or vapor from a liquid, the end points being boiling temperatures, and it can also be a polymorphic or solid–solid phase transition loop separating two solid solutions having different structures, the end points being polymorphic transition temperatures. Phase transition loops occur simply because solutions cannot change to other solutions with no change in composition.[4]

---

[4]This can be proven using thermodynamics, but we will not bother. The exception to this rule is a solution having a maximum or minimum in temperature, where it can melt or boil to another solution having the same composition (see Figure 11.23, lower left, upper right.)

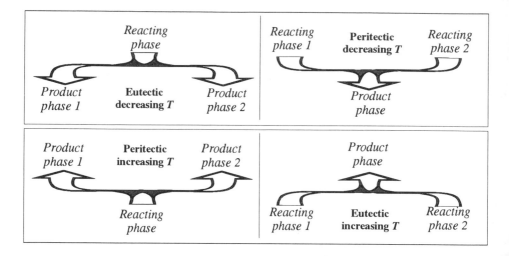

Figure 11.22: The difference between a peritectic and a eutectic.

- The solvus, which separates two solutions of the same kind, such as two solid solutions, or two liquid solutions (Figure 11.21). Increasing the temperature normally increases the solubility of one in the other, and the two phases can become identical (the solvus closes) at an *upper consolute point.* As this name implies, solvi can sometimes close downward (with decreasing temperature), but this is rare. Normally the solvus keeps widening downward.

Figure 11.23 shows some examples of how these two elements combine. For example, a binary minimum melting loop (top of Figure 11.23) can be considered to be produced by combining two simple melting loops. A simple *peritectic* can be considered to be what happens when a solvus intersects a simple melting loop, and a eutectic what happens when a solvus intersects a binary minimum melting loop. You can try to make these intersections with other topologies, but they will generally not obey the Phase Rule. (The difference between a peritectic and a eutectic is illustrated in Figure 11.22. In both, three phases exist together at equilibrium, and so both are represented by a three-phase tie-line. However, the reaction relationships are exactly reversed. What happens at a eutectic during cooling is the same as what happens at a peritectic on heating. Which phases are solids and which are liquids is immaterial.)

Similarly, as conditions change (say, increasing pressure), these configurations can become more complex, but without introducing any new features. For example, at the top of Figure 11.24 is an attempt to portray the effect produced by a solvus moving upward, due to changing conditions. At low pressures, the solvus intersects the melting loop, but does not go through it. If the solvus moves upward in temperature faster than does the melting loop, it must eventually poke its way through the top of the melting loop, as shown. Every such

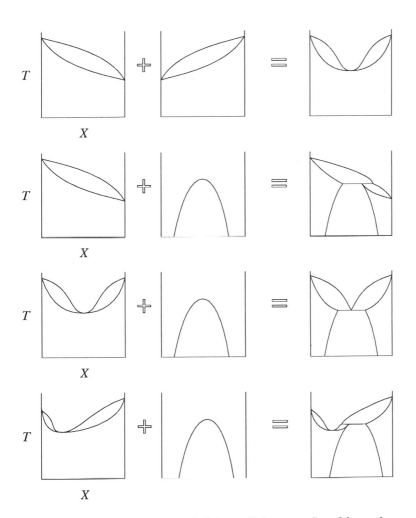

Figure 11.23: Some simple binary "elements," and how they combine.

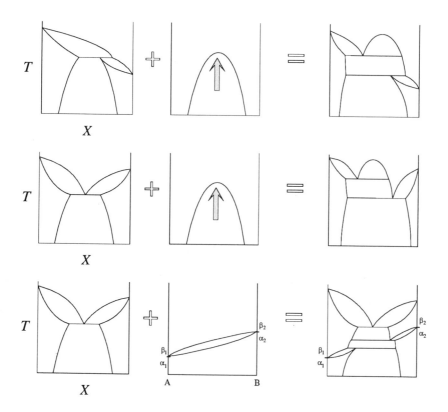

Figure 11.24: What happens when a solvus moves upward through melting loops, and the effect of adding a polymorphic transition.

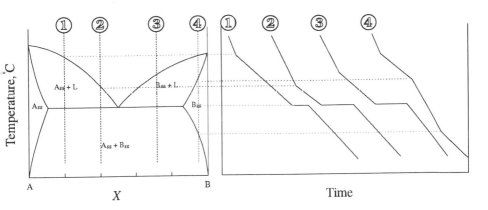

Figure 11.25: Temperature–time curves through a binary eutectic system.

intersection must be accomplished with no more than three coexisting phases, so three-phase tie-lines are produced. Try to imagine the top right diagram on Figure 11.24 as a cross-cutting melting loop and solvus, perhaps with dotted lines completing the individual elements. Then try to satisfy the Phase Rule in some other way; you will find it difficult.

Finally, note that as far as the Phase Rule is concerned, one phase transition is much like another. Thus an $\alpha$–$\beta$ polymorphic transition in a single component behaves just like a melting point or a boiling point when a second component is added. Because one polymorph will in general have a greater capacity for the second component than the other polymorph, a polymorphic phase transition loop is created in the binary system, exactly analogous to boiling and melting loops. These phase transition loops may extend from side to side of the diagram in a completely miscible solid solution, but more likely they will intersect a solvus, as shown at the bottom of Figure 11.24. As before, a three-phase tie-line is created by every such intersection.

It is possible to have a four-phase tie-line in a binary system, but this could only be at a unique temperature and pressure, just like a triple point in a unary system. Binary sections are not usually drawn for such unique conditions. That is, when we chose our pressure for our *T–X* section in §11.4.2 (the Isobaric Phase Rule), it would be extremely unlikely for this choice to be just the pressure needed for four-phase equilibrium, and so three-phase tie-lines are the norm in binary sections.

## 11.4.12 Cooling Curves

Temperature–time curves are often used as a means of experimentally determining the temperatures of phase changes, and thinking about them can add to your understanding of phase relations. In looking at the four different cooling curves in Figure 11.25, you should imagine that you are in a laboratory, con-

242 CHAPTER 11. PHASE DIAGRAMS

ducting an investigation into the system A-B. One way to proceed would be to prepare a number of different bulk compositions (thoroughly mix various proportions of A and B together), then heat each bulk composition to a number of different temperatures, wait long enough to achieve equilibrium, then observe what phases are present, and measure their compositions. Figure 11.25 (left side) would then represent the results you obtained from a large number of such experiments.

Another way to proceed would be to heat each bulk composition to a temperature sufficiently high to produce a homogeneous liquid, and then to cool slowly while observing the temperature. When the liquidus temperature is reached, the latent heat of crystallization is released, resulting in a slower rate of cooling. When the eutectic temperature is reached, cooling will cease completely, while three phases coexist. When the liquid disappears, cooling will resume. By observing the inflection points and plateaus in the temperature-time curves, you may deduce the positions of the liquidus and eutectic for the various compositions. You would still need to do more experiments to determine phase compositions, but the cooling-curve method can often give the general shape of a diagram in a relatively short time. Note that composition number 4 does not pass through the eutectic, and so does not show a temperature plateau.

## 11.4.13   Intermediate Compounds

In all the binary systems we have considered so far, no compounds are formed *between* the compounds A and B. That is, there are no compounds AB, or $A_2B$, or $A_2B_3$, and so on. What happens if these do exist? Consider the binary system A-B that contains the binary compound AB. The simplest possibility is that both A-AB and AB-B are binary systems of the same type, such as simple eutectic systems. Then the two systems are "glued together," as in Figure 11.26.

Another common possibility is that the liquidus for one of the end-member compounds extends completely over the intermediate compound, as in Figure 11.27. When this happens, compound A-B does not melt to a liquid of its own composition—it breaks down at the peritectic temperature to a different compound plus a liquid. This is known as *incongruent melting*. A good example of this is the system $KAlSi_2O_6$-$SiO_2$ (leucite-silica), which contains the intermediate compound $KAlSi_2O_6 \cdot SiO_2$, or $KAlSi_3O_8$, K-feldspar, shown in Figure 11.28. The large liquidus surface extending over the intermediate compound in these diagrams will often "shrink" with increasing pressure, leading ultimately to the "glued together" type of system (Figure 11.26). In other words, AB may melt incongruently at low pressures and congruently at high pressures.

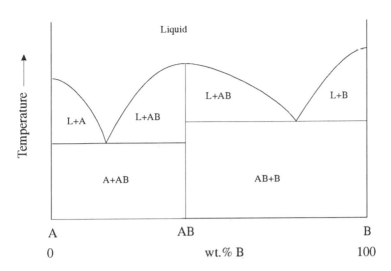

Figure 11.26: Intermediate compound AB divides the binary system A-B into two similar parts. Note that if the composition axis were in mole fraction or mole %, AB would appear midway between A and B, but not if the axis is in weight %.

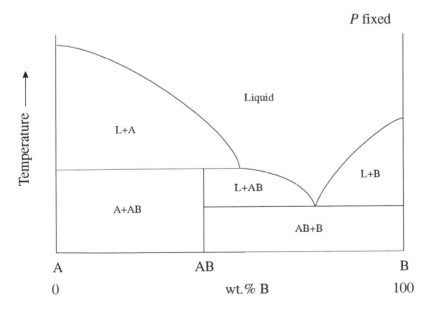

Figure 11.27: Intermediate compound AB displays incongruent melting.

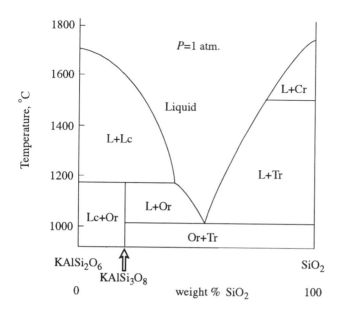

Figure 11.28: The system $KAlSi_2O_6$–$SiO_2$, with intermediate compound $KAlSi_3O_8$. Lc-leucite. Or-orthoclase (K-feldspar). Tr-tridymite. Cr-cristobalite. L-liquid.

## 11.5 TERNARY SYSTEMS

### 11.5.1 Ternary Compositions

With the addition of a third component, we now need two dimensions to display all possible compositions in a system, and so we lose the ability to display composition and temperature simultaneously. We can only display compositions at a chosen $T$ and $P$ on a section, or we can *project* compositions from various conditions onto a single plane, as discussed below. The method of depicting compositions within a triangle is shown in Figure 11.29. Each apex of the triangle represents 100% of one of the components. The proportion of each component in a ternary composition is measured by the distance of a point from the side of the triangle opposite the component in question, as shown. The triangles are usually isometric, but not necessarily. Right angled triangles are also used in some circumstances, and other shapes could be used if desired.

### 11.5.2 Sections and Projections

In discussing binary systems we used only binary *sections*, although we mentioned that various kinds of *projections* could be used. We must now expand on this statement. As shown in Figure 11.9, a section shows what you would

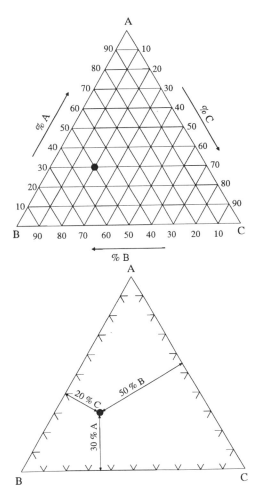

Figure 11.29: Representation of a ternary composition. The dot in the upper triangle represents a composition of 30% A, 50% B, and 20% C, as shown in the lower triangle.

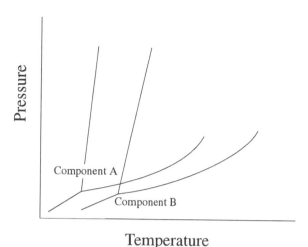

Figure 11.30: A $P$–$T$ projection of Figure 11.9.

see after slicing through a $P$–$T$–$X$ box, as though with a knife. A projection, on the other hand, is what you see by peering through the box at some angle, usually parallel to one of the axes, and seeing the curves on all sides of the box at the same time. For example, Figure 11.30 shows a $P$–$T$ *projection* of Figure 11.9. You see the unary phase diagrams for both components superimposed on one another. In a more complete projection, you would also see various curves projected from *within* the box, as well as the curves on the faces of the box. For example, there is a curve joining the critical points of each pure component, which crosses through the box, showing the critical points of binary compositions, which is not shown.

### Ternary Projections

In looking at ternary systems, we will start with the projection. To best see the meaning of ternary projections, we start with the oblique view of a simple ternary eutectic system, Figure 11.31. In this figure we see that each side of the compositional triangle has a binary isobaric $T$–$X$ section constructed on it, perpendicular to the compositional triangle. We see, too, that the liquidus lines of the binaries are joined into surfaces that extend across the ternary space. Each binary eutectic point becomes a ternary *cotectic line* extending into the ternary, down to a *ternary eutectic* point. Points and lines on these surfaces are projected onto a plane surface, as depicted in Figure 11.31. The projection is what you would see if you looked straight down on the three-dimensional object in Figure 11.31, parallel to the temperature axis. The results of such projection are shown in Figures 11.32 and 11.33.

In Figure 11.32 the cotectic lines and *isothermal contours* on the liquidus

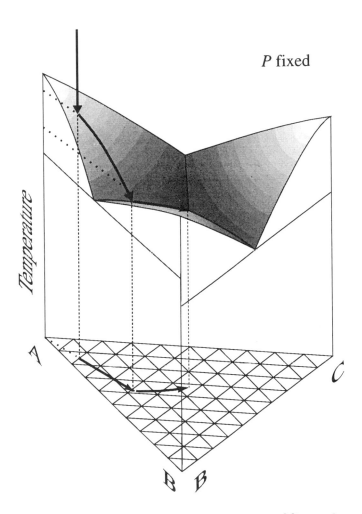

Figure 11.31: A ternary eutectic system ABC in an oblique view. The cooling path of a liquid of composition 80% A, 15% B, 5% C is shown.

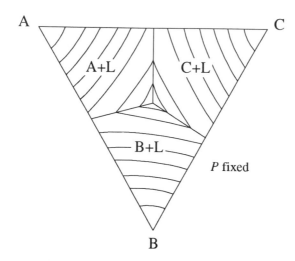

Figure 11.32: A polythermal projection of the liquidus surfaces in Figure 11.31.

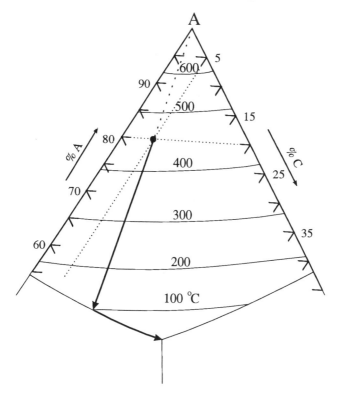

Figure 11.33: Polythermal projection of the liquidus surface of component A in Figure 11.31.

surfaces have been projected onto a plane surface. The triangle is divided by the cotectic lines into three areas, labeled A+L, B+L, and C+L (where A stands for solid A, etc.), because bulk compositions in these areas will consist of these phases at temperatures below those of the liquidus surfaces. This will become clearer by examining what happens to a liquid of composition 80% A, 15% B, 5% C, as shown in Figures 11.31 and 11.33.

Starting at a temperature well above the liquidus surface, as shown in Figure 11.31, the liquid cools vertically downward until it hits the liquidus surface at a temperature of 450°C (Figure 11.33). This liquidus surface is the locus of points indicating the first appearance of crystals of composition A, and so crystals of A start to separate from the liquid. A tie-line joins the liquid composition to the A-axis. Because composition A is being subtracted from the liquid, the liquid composition *must move directly away from* A, as shown in Figure 11.33. The composition of the liquid stays on the liquidus surface, always on the continuation of a straight line through composition A and the point [80% A, 15% B, 5% C]. A continuous series of tie-lines join the liquid composition to the A-axis (two of which are shown in Figure 11.31). This continues until the liquid composition hits the cotectic line (temperature 100°C, Figure 11.33), which joins the liquidus surfaces of A and B. On this line, the liquid is simultaneously in equilibrium with crystals of A and crystals of B, and so B starts to precipitate. On further cooling, both A and B precipitate, and the liquid composition moves down the cotectic line until it hits the ternary eutectic. At this point, C starts to precipitate, and all three solids precipitate until the liquid is used up. At the ternary eutectic, the number of phases is 4 (solid A, solid B, solid C, and L), and

$$f = c - p + 1$$
$$= 3 - 4 + 1$$
$$= 0$$

The ternary eutectic is thus an isobaric invariant point, and no temperature change can take place until the liquid is all used up, at which time the crystals will resume cooling.

### Ternary Sections

Consider a temperature midway between the melting points of the pure components and the three eutectic temperatures in system ABC, as shown in Figure 11.34. An isothermal plane at this temperature will cut through all three liquidus surfaces. Near each apex, the plane lies below each liquidus surface, so it shows an area of solid plus liquid filled with tie-lines. In the center of the diagram, the plane is everywhere above the liquidus surfaces, and so it shows a blank "field of liquid." Sections at successively lower temperatures would show the two-phase fields expanding and coalescing, leaving smaller and smaller liquid fields.

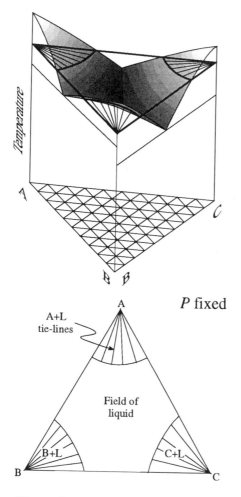

Figure 11.34: An isothermal section through system ABC.

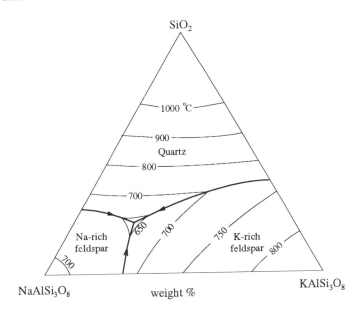

Figure 11.35: The Granite System at 10 kilobars.

## 11.5.3   The Granite System

The simple ternary eutectic system discussed above represents just a beginning to the general subject of ternary phase diagrams. Features such as solid
solutions, peritectics, intermediate compounds, and so on all introduce complications that we will not discuss. However, as an example of the simple ternary
eutectic diagram, let's have a look at the system $SiO_2$–$NaAlSi_3O_8$–$KAlSi_3O_8$–
$H_2O$, which is used as a model for natural granites. Component $SiO_2$ is called q,
component $NaAlSi_3O_8$ is called ab, and component $KAlSi_3O_8$ is called or. This
is because the most common form of $SiO_2$ is quartz (q), pure $NaAlSi_3O_8$ is the
mineral albite (ab), and one of the varieties of $KAlSi_3O_8$ is the mineral ortho-
clase (or). Quartz is abundant in granites, as is plagioclase, made up mostly
of ab, and K-feldspar, made up mostly of or. Granites usually contain several
minerals in addition to quartz, plagioclase, and K-feldspar, but these minerals
(that is, quartz, albite-rich plagioclase and K-feldspar) often account for 80 to
90% of a granite, so the system q-ab-or is quite a useful model in trying to
understand the crystallizing or melting histories of granites in general.

**The Granite System at 10 kbar.**

The Granite System diagram is shown as a polythermal projection in Figure
11.35.
   Compare this diagram with Figures 11.8 and 11.17. In Figure 11.8 you see
that the melting point of solid $SiO_2$ (as $\beta$-cristobalite) is about 1700°C at 1 bar,

whereas in Figure 11.35 it is somewhat less than 1100°C. The melting point of albite at 1 bar in Figure 11.17 is 1118°C, whereas in Figure 11.35 it is just over 700°C. Raising the pressure from 1 bar to 10,000 bar increases the melting points, so that cannot be the explanation. What is going on?

The difference lies in the fact that the Granite System is not the "dry" system q–ab–or, but the "wet" system q–ab–or–$H_2O$. The presence of water at high pressures has the effect of substantially lowering the melting temperatures of the pure minerals. Thus Figure 11.35 is not really a ternary projection, but a quaternary projection. All liquids in the diagram have not ternary compositions, but quaternary compositions; that is, they are all saturated with water, and supercritical water is an extra phase that is not shown in the diagram. The presence of the water brings the liquidus temperatures down well into the range of temperatures found in the Earth's crust and means that water is an important component in the history of real granites. However, as water is an extra phase as well as an extra component, we can treat Figure 11.35 exactly like a ternary eutectic and essentially forget about the water.

According to Figure 11.35, crystallization of any bulk composition within the system will generate a final liquid composition at the ternary eutectic at just under 650°C. Note that the composition of this final liquid (or initial liquid on heating) lies quite close to the albite corner—ternary eutectics do not always occur near the center of the diagram. At other pressures, the position of this eutectic changes. At lower pressures, it moves "northeast," roughly directly away from the albite composition, as shown in Figure 11.36

A complicating factor here is that the system only shows a eutectic at high pressures. Below about 5 kbar, the eutectic changes to a ternary minimum, as indicated by the change from the circle to plus signs in Figure 11.36. The reason for this is shown in the lower two sequences of diagrams in Figure 11.23. In one of these, a binary eutectic is generated when a more-or-less symmetrical melting loop intersects a solvus. This corresponds to the ab–or–$H_2O$ system at high pressures. In the other, the melting loop has a minimum temperature offset to one side, so that even after the solvus intersection takes place, the minimum is preserved. This corresponds to the system ab–or–$H_2O$ at low pressures. This difference between a eutectic and a minimum is preserved in the ternary. However, from the point of view of liquid compositions generated during cooling in ternary systems, there is little difference between a ternary eutectic and a ternary minimum. Both represent the final liquid composition for many bulk compositions in the system.

## 11.5.4  Granite Compositions

This brief explanation is not sufficient for you to understand all the details of this system, but it is sufficient to understand one important result. When the compositions of natural granites are normalized to q–ab–or and plotted in the q–ab–or triangle, a remarkable coincidence of compositions and the ternary minima and eutectics results, as shown in Figure 11.37. As natural granites have

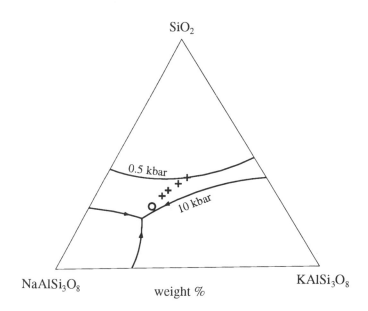

Figure 11.36: The Granite System at pressures from 0.5 to 10 kbar. The plus signs show the position of the ternary minimum at 0.5, 2, 3, and 4 kbar. The circle shows the position of the ternary eutectic at 5 kbar.

undoubtedly crystallized at a variety of pressures, the compositions would be expected to be strung out along the track defined by the ternary minima and eutectics, as they are, *providing that natural granites actually form by crystallizing from silicate liquids*. The demonstration by Tuttle and Bowen (1958) that this was the case provided strong evidence for the magmatic origin of granites. The slight offset of the highest frequency of natural compositions toward the $KAlSi_3O_8$ apex, as well as other aspects of the diagram, have been the subject of much discussion.

## 11.6  SUMMARY

Phase diagrams are a kind of concise representation of the equilibrium relationships between phases as a function of chosen intensive variables, such as temperature, pressure, composition, $pH$, oxidation potential, activity ratio, and so on. They are extremely useful, not only in representing what is known about a system, but in thinking about processes involving phase changes. Most of the diagrams in Chapters 9 and 10 are phase diagrams, although the term is most often used in relation to the $P-T-X$ type of diagram discussed in this chapter.

Although in principle phase diagrams can be calculated from thermodynamic data, in complex systems the relationships are generally determined ex-

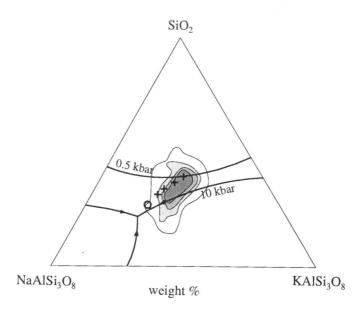

Figure 11.37: The compositions of natural granites superimposed on Figure 11.36. Several hundred individual points are contoured, with the darker colors indicating a higher frequency of points.

perimentally. The diagrams must nevertheless obey rules established by equilibrium thermodynamics. Therefore an understanding of the material in all the previous chapters is a prerequisite to a real understanding of the simple points, lines, and surfaces found in phase diagrams.

## PROBLEMS

1. If the slope of a phase transition of a mineral from phase $\alpha$ to phase $\beta$ is $-21.0$ bar deg$^{-1}$ at a temperature of 600 K, the $\Delta_{\alpha \to \beta} V$ of the transition is $+0.150$ cal bar$^{-1}$, and $\Delta_f H^\circ_{600}$ of phase $\alpha$ is $-17,000$ cal mol$^{-1}$, what is $\Delta_f H^\circ_{600}$ of phase $\beta$? Sketch and label the phase diagram.

2. The slope of the ice-water phase boundary is $-131.7$ bar deg$^{-1}$. Knowing the heat of fusion of ice ($\Delta_r H^\circ_{ice \to water} = 6010$ J mol$^{-1}$) and the molar volume of water at 0°C ($V^\circ_{H_2O(l)} = 18.01826$ cm$^3$ mol$^{-1}$), calculate the molar volume of ice at 0°C.

3. A hypothetical compound $\beta$ has been found to have at least 7 phases, named A, B...G, arranged as shown in Figure 11.38. Point 6 is a critical point. The locations of points 1, 2...7 and some of the thermodynamic properties have been found experimentally to have the following values:

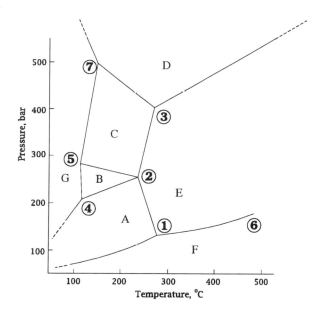

Figure 11.38: Phase diagram for compound $\beta$.

| Point | $T°C$ | $P$ bar |
|-------|-------|---------|
| 1 | 277 | 132 |
| 2 | 236 | 256 |
| 3 | 270 | 404 |
| 6 | 483 | 179 |
| 7 | 157 | |

| Phase | $V°\,\mathrm{cm^3\,mol^{-1}}$ | $S°\,\mathrm{J\,deg^{-1}\,mol^{-1}}$ |
|-------|-------------------------------|--------------------------------------|
| A | 15.0 | 22.0 |
| C | 11.62 | |
| D | 10.0 | 21.4 |
| E | | 22.3 |

(a) There are two parts of this diagram that are thermodynamically impossible. Where are they, and why are they incorrect? Sketch possible correct relationships at each location.

(b) Sketch $G$-$P$ and $V$-$P$ sections at 250°C.

(c) Sketch $G$-$T$ and $H$-$T$ sections at 150 bar.

(d) Calculate $V_E°$.

(e) Using this result, calculate $S_C°$.

(f) Calculate the pressure at point 7. If it doesn't look about right compared to the diagram, you have made a mistake.

(g) What are the upper and lower limits for possible values for $V_B^\circ$?

(h) Identify each phase as solid, liquid, or gas. Which solids float and which sink in liquid $\beta$?

(i) Why are boundaries A-F and E-F curved, while all the others are straight?

4. (a) Describe in detail the equilibrium cooling history of a liquid of composition 6 in Figure 11.39. (b) Describe in detail the perfect fractional cooling history of a liquid of composition 2 in Figure 11.39. This diagram looks frighteningly complex at first, but it contains nothing more than the diagram elements already discussed, and working through such a diagram is not more difficult than a simpler diagram; it just takes longer.

5. You have performed a series of experiments in the lab involving the cooling of various mixtures of compound A and compound B at 1 atm. The results of the experiments are given in the table below, in which $T$ represents the temperature at which you first observed crystals in the cooling mixtures.

| Run no. | Composition (wt.%) |      | $T°C$ |
| :---: | :---: | :---: | :---: |
| 1 | 100% compound A |   | 63 |
| 2 | 100% compound B |   | 88.9 |
| 3 | 90% | " | 85.4 |
| 4 | 75% | " | 75 |
| 5 | 60% | " | 63.9 |
| 6 | 45% | " | 50.5 |
| 7 | 30% | " | 42.9 |
| 8 | 15% | " | 54.4 |

(a) Construct a phase diagram using as axes $T°C$ and wt.% B and label all the phase fields. Assume no appreciable solid solution in the solid phases.

(b) What are the eutectic temperature and composition?

(c) Construct a schematic cooling curve (temperature vs. time) for run no. 4.

(d) Consider 1.5 g of a system containing 35 wt.% compound A and 65 wt.% compound B at a temperature of 60°C. What phases are present at equilibrium, and how much of each phase is there? (Note that this information is contained in the diagram, even though no such experiment has been conducted.)

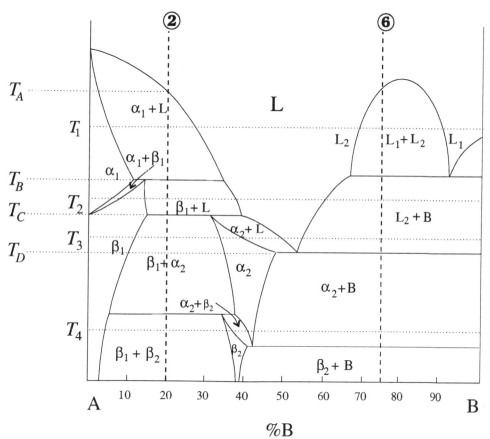

Figure 11.39: $T$-$X$ phase diagram for the system A-B.

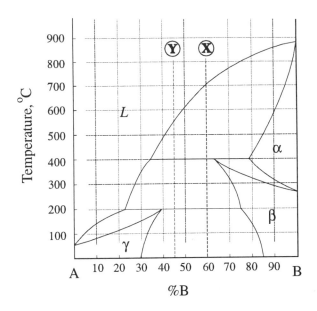

Figure 11.40: System A-B at 1 bar pressure.

6. (a) Describe the equilibrium cooling history of a liquid of composition Y in Figure 11.40. Give the compositions and proportions of the phases before and after the phase transitions. Calculate at least one mass balance.

   (b) Describe the fractional crystallization of a liquid of composition X in Figure 11.40.

   (c) What are the melting points of pure A and pure B? How many polymorphs of component B are there? What is the temperature of transition between them? If the entropy of $\alpha$ is 10 J mol$^{-1}$ K$^{-1}$, what is the entropy of $\beta$? $\Delta_f G^\circ_\alpha = -100$ kJ mol$^{-1}$, $\Delta_f G^\circ_\beta = -101$ kJ mol$^{-1}$.

   (d) Figure 11.40 could be described as being made up of a simple melting loop intersected by a miscibility gap, complicated slightly by the polymorphs of component B. If component B had only one crystal form ($\alpha$), what would the phase diagram look like?

   (e) If liquid immiscibility developed at higher pressures, what might the diagram look like: (a) B having only one crystal form; (b) B having two polymorphs?

7. In §11.4.11 it says that there are two basic binary diagram elements. However, note that if the phase transition loop (Figure 11.21, left) becomes detached from the vertical axes on both sides, and if the solvus (Figure 11.21, right) closes downward as well as upwards (i.e., has both upper and

lower consolute points), the two elements become indistinguishable. So perhaps there is only one "basic element." Then again, these elements are created by tie-lines, so perhaps the "basic elements" are the two-phase and three-phase tie-lines. Which point of view is right? Under what circumstances does the phase transition loop become detached from the temperature axes?

8. A central part of the novel *Cat's Cradle* by Kurt Vonnegut, Jr., involves the existence of a new polymorphic form of ice (ice-IX), which is more stable than ordinary ice (ice-I) at atmospheric pressure and has a melting point of 114.4°F (45.8°C). When ice-IX gets loose it acts as a seed, and all the water on Earth, which is metastable with respect to ice-IX, freezes immediately. This, of course, means the end of life on Earth. Is there any thermodynamic reason why ice-IX could not exist? If it did exist, is 114.4°F a reasonable melting point for it? Answer this with the help of a $G$-$T$ section through system $H_2O$ at 1 bar.

9. Construct the phase diagram for $H_2O$ in the range 0 to 7000 bar (Y-axis), −40°C to 20°C (X-axis), using the data tabulated below. Label each phase boundary and triple point with the coexisting phases and degrees of freedom. Show the metastable extensions of the phase boundaries.

| Point | Phases | $T°C$ | $P$ bar |
|---|---|---|---|
| a | ice-I, ice-III, L | −22 | 2215 |
| b | ice-III, ice-V, L | −17 | 3530 |
| c | ice-I, ice-II, ice-III | −35 | 2170 |
| d | ice-II, ice-III, ice-V | −24 | 3510 |
| e | ice-V, ice-VI, L | 0.2 | 6380 |
| f | ice-VI, ice-VII, L | 81.6 | 22400 |
| | (point f is off the diagram). | | |

(a) The slope of the ice-II—ice-V boundary is negative, and the slope of the ice-I—ice-II boundary is positive (you don't need to know the exact slopes). Sketch a diagram of density vs. pressure (0 to 5000 bar) for $H_2O$ at −40°C and at −10°C, with no numbers on the density axis.

(b) Does ice-V sink or float in water?

(c) Sketch a $G$-$P$ section through the diagram at 10°C from 0 to 7000 bar showing ice-I, water, ice-V, and ice-VI. Show the location on this section of the freezing of water to ice-VI and ice-V as discussed in the quotation from Bridgman, below. Why does water freeze sometimes to ice-V and sometimes to ice-VI, according to Bridgman?

The following extract is from *The Physics of High Pressure*, by P.W. Bridgman (G.W. Bell and Sons, London, 1958, page 423). It may help to dispel any notion that determining phase relations is a simple and straightforward exercise.

Bridgman was a professor at Harvard University and won the Nobel Prize for his research on materials at high pressures.

The conditions under which the different modifications of ice appear are somewhat capricious, and often inconvenient manipulation is necessary to arrive in the part of the phase diagram desired. The behaviour is particularly striking in the neighbourhood of the V-VI-liquid triple point, say, between 0°C and −10°C and between 5000 and 6000 bar. With the ordinary form of apparatus, water in a steel piezometer with pressure transmitted to it by mercury, it is very difficult to produce the modification V. For example, if liquid water at −10°C is compressed across the melting curve, the liquid will persist in the sub-cooled condition for an indefinite time, without freezing to V, the stable form. But if the compression is carried further, to the unstable prolongation of the liquid-VI line, freezing to the form VI will take place almost at once on crossing the line, in spite of the fact that VI is unstable with respect to V. Now in order to make V appear, VI may be cooled 30° or 40°, when it will spontaneously change to V, the stable form. If pressure is now released back to 5000, say, and temperature is raised to the melting line of V, melting takes place at once and so the coordinates of the melting line may be found. Suppose now after the melting is completed the liquid water be kept in the neighbourhood of the melting line for several days, and then the pressure increased again across the melting line at −10°C; it will be found that the instant the melting line of V is crossed the liquid freezes to V. This suggests that there has persisted in the liquid phase some sort of structure, not detected by ordinary large-scale experiments, favourable to the formation of V. It is now known, of course, from X-ray analysis, that structures are possible in the liquid; this experiment suggests a specificity in these structures that might well be the subject of further study.

The formation of the nucleus of V may be favoured by the proper surface conditions. Thus if there is any glass in contact with the liquid, either a fragment of glass wool purposely introduced into the liquid, or by enclosing the water in a glass instead of a steel bulb, freezing to V takes place at once without the slightest hesitation immediately on carrying the virgin liquid across the freezing line.

# 12

# WHERE DO WE GO FROM HERE?

## 12.1  A LOOK AHEAD

The purpose of this chapter is to stand on the firm ground of equilibrium thermodynamics that we have established in the previous chapters and look outward at unconquered ground. Assuming that we have fully assimilated all the material up to here, what's next? What subjects build upon this material? What can we do to better understand how natural systems work? Looking back, we see at least three areas that have been mentioned as important, but that do not fall within the subject area of equilibrium thermodynamics:

- We saw (e.g., §1.4.2) that even if reactions do proceed, we are unable to say *how fast* they will proceed. We know that some are very fast, like sugar dissolving in coffee, and others are slow, like eggs going bad in the refrigerator. But how do we model the rates of chemical reactions?

- We also saw that although we have powerful techniques to say which way reactions *should* proceed spontaneously, and what the equilibrium position is if they do proceed, we are powerless to say whether or not the reaction will proceed at all. This has to do with the "energy barrier" between reactants and products (e.g., Figure 7.6)—how big is it, and under what circumstances can reactions get over the hump?

- And, of course, we would like to understand more about our mysterious quantity, entropy. We have used it as a quantity that tells us which way isolated reactions will proceed, and in combination with enthalpy to tell us which way ordinary reactions (constant $T$, $P$) will proceed, but we took only a fleeting glance at what entropy "really is" (§5.5).

In addition to these topics that arise in the course of any discussion of equilibrium thermodynamics, because they are severe limitations on the subject, there are other directions of inquiry that do not so much repair the deficiencies of equilibrium thermodynamics, but build on its strengths. These are the

261

various methods of process modeling, which combine the thermodynamics we have learned with data on fluid flow, temperature and pressure gradients, surface reactions, and reaction kinetics, and other factors, to build increasingly realistic models of complex natural phenomena involving the movement and chemical reactions of fluids in soils and rocks in the Earth's crust.

The material we will look at in this chapter is just a brief survey, to place the subject matter of the previous chapters in a broader context. Not surprisingly, much of it deals with the atomic and molecular nature of substances, in contrast to most of the material we have covered up to here. Thermodynamics deals with energy balances, and the way things should be. To see *why* things should be that way, we need to look in more detail at the way substances are made and how they interact on a molecular level. Some of this material may well involve mathematics or concepts unfamiliar to the reader. I would like such readers to not worry about that; this chapter is in a sense evangelical. The idea here is just to show that, building on thermodynamics, there are numerous ways of further investigating natural processes available to anyone who really wants to understand how the world works.

Several sections in this chapter draw heavily on Lasaga and Kirkpatrick (1981), a valuable reference on kinetics in natural systems.

## 12.2  KINETICS

### 12.2.1  The Myth of Equilibrium

If you look around at the world we live in at the surface of planet Earth, it seems to be characterized by constant change; virtually nothing is permanent—even the solid rocks of the crust are weathering and being washed into the sea. And organic substances, including us, are among the most evanescent of objects— here today, gone tomorrow. Where are those states of equilibrium that are so important in our thermodynamic model? It is in fact a tribute to the creativity of the human mind that scientists in the last century were able to "see through" the constant flux around them and create what is in essence a model of energy relationships in a world that does not exist—the equilibrium world.

As we have seen in the last few chapters, this mathematical model of energy relationships is tremendously useful in the real world, basically because even though the real world is in a constant state of flux or change, there are many situations in which it approaches fairly closely a state of equilibrium, and even in cases where it does not, it is changing in the direction of an equilibrium state, and our equilibrium models are useful in many ways. But obviously we would like to know more about the state of flux itself. How fast do the changes we see take place, and what controls this rate of change? We enter the world of chemical kinetics.

## 12.2.2 The Progress Variable

Consider a generalized chemical reaction

$$aA + bB = cC + dD \tag{12.1}$$

where A, B, C, and D are chemical formulae, and $a$, $b$, $c$, $d$ are the stoichiometric coefficients. We pointed out in §8.2 that when this reaction reaches equilibrium,

$$c\mu_C + d\mu_D = a\mu_A + b\mu_B$$

and

$$
\begin{aligned}
\Delta_r\mu &= c\mu_C + d\mu_D - a\mu_A - b\mu_B \\
&= 0
\end{aligned} \tag{12.2}
$$

Equation (12.1) can be generalized to

$$\sum_i \nu_i M_i = 0 \tag{12.3}$$

where $M_i$ are the chemical formulae and $\nu_i$ are the stoichiometric coefficients, with the stipulation that $\nu_i$ is positive for products and negative for reactants. Equation (12.2) can then be generalized to

$$\sum_i \nu_i \mu_i = 0 \tag{12.4}$$

Up to now, we have been most interested in reactions that reach equilibrium. Now let's look at what happens before that point is reached, that is, while the reaction is taking place. Let's say that the reaction proceeds from left to right as written. It doesn't matter for the moment whether all the reactants and products are in the same phase (a *homogeneous* reaction) or in different phases (a *heterogeneous* reaction). During the reaction, A and B disappear and C and D appear, but the *proportions* of A:B:C:D that appear and disappear are fixed by the stoichiometric coefficients. If the reaction is

$$A + 2B \rightarrow 3C + 4D \tag{12.5}$$

then for every mole of A that reacts (disappears), 2 moles of B must also disappear, while 3 moles of C and 4 moles of D must appear. This is simply a mass balance, independent of thermodynamics or kinetics, and can be expressed as

$$\frac{dn_A}{\nu_A} = \frac{dn_B}{\nu_B} = \frac{dn_C}{\nu_C} = \frac{dn_D}{\nu_D} = \frac{dn_i}{\nu_i} \tag{12.6}$$

where, as emphasized in Appendix C, the differentials $dn_A$, $dn_B$, and so on, refer to a change in the amount of A, B, and so on, of any convenient magnitude, not necessarily an infinitesimal change. We can next imagine the reaction proceeding in a series of such changes, or reaction increments, from the beginning to the end of the reaction (either when equilibrium is reached, or when one of the reactants is used up), and write

$$\frac{dn_A}{v_A} = \frac{dn_B}{v_B} = \frac{dn_C}{v_C} = \frac{dn_D}{v_D} = d\xi \tag{12.7}$$

from which it appears that

$$\frac{dn_A}{d\xi} = v_A; \quad \frac{dn_B}{d\xi} = v_B; \quad \ldots \frac{dn_i}{d\xi} = v_i \tag{12.8}$$

where our new variable $\xi$ is called the reaction progress variable, and in this case represents an arbitrary number of moles. Equation (12.8) says that in reaction (12.5), $dn_A/d\xi = -1$, $dn_B/d\xi = -2$, $dn_C/d\xi = 3$, and so on, which simply means that for every mole of A that disappears, 2 moles of B also disappear, 3 moles of C appear, and so on.

### 12.2.3  The Reaction Rate

Having defined reaction increments $d\xi$, we can now define the *rate of reaction* as

$$\frac{d\xi}{dt} = \frac{1}{v_A}\frac{dn_A}{dt} = \frac{1}{v_B}\frac{dn_B}{dt} = \ldots = \frac{1}{v_i}\frac{dn_i}{dt} \tag{12.9}$$

where $dt$ is an increment of time, and $d\xi/dt$ is the derivative of $\xi$ with respect to $t$ and is an expression of the amount of progress of the reaction as a function of time, or simply the rate of reaction.

This expression (12.9) is written in terms of the absolute number of moles of A, B, and so on, $(n_A, n_B,\ldots)$, but by considering a fixed volume we could change these to concentration terms. Thus

$$\frac{d\xi}{dt} = \frac{1}{v_A}\frac{dC_A}{dt} = \frac{1}{v_B}\frac{dC_B}{dt} = \ldots = \frac{1}{v_i}\frac{dC_i}{dt} \tag{12.10}$$

where $C$ is some unit of concentration such as $mol\,cm^{-3}$.

So evidently, the rate of reaction can be determined by measuring the concentration of *any* of the reactants or products as a function of time. With one important stipulation.

#### Elementary and Overall Reactions

If you actually measure the rate of change of concentration of products and reactants in many ordinary chemical reactions, you find that the relationship in (12.10) is often not obeyed. This is because the reaction does not actually proceed *as written*, at the molecular level. For example, reaction (12.5), taken literally, indicates that a molecule of A reacts with 2 molecules of B, and at that instant, 3 molecules of C and 4 molecules of D are formed. But this might not be what happens at all, and in view of the improbability of three molecules (A + 2 B) meeting at a single point, it probably is not in this case. The reaction as written may well represent the *overall* result of a series of *elementary* reactions. Thus A and B may in fact react to form a number of *intermediate species* such as X and Y, which then react with each other or with A or B to form C and D. In thermodynamics, the existence of such intermediate species is not important to the study of the overall reaction, as long as equilibrium is attained, but in kinetics, they contribute to the overall rate of reaction and may actually be *rate-controlling*, even though their concentrations may be small.

Of course, it is also possible that intermediate species do form, but they achieve a *steady-state* concentration, that is, they break up just as rapidly as they form. In this case, equation (12.10) would be obeyed, even though it did not represent what actually happens at the molecular level.

From now on in this section, we will use = in overall reactions, and → in elementary reactions (those that actually proceed as written).

## 12.2.4 Rate Laws

A rate law is a statement about how the rate of a reaction depends on the concentrations of the participating species. If one thinks about chemical reactions as something that happens at the molecular level when molecules collide with one another, it makes sense that the number of collisions, and hence the rate of reaction, should depend on how many molecules of each type there are; that is, their concentrations.[1] In most cases, a simple power function of concentrations is found to apply. For reaction (12.1) it is

$$\text{rate of reaction} = \frac{d\xi}{dt} = k \cdot C_A^{n_A} C_B^{n_B} C_C^{n_C} C_D^{n_D} \tag{12.11}$$

The constant of proportionality, $k$, is called the *rate constant*. The exponents $n_A \ldots n_D$ are often integers, but can be fractional or decimal numbers, especially in heterogeneous reactions where adsorption and other surface-related effects can influence reaction rates. They define the *order* of the reaction. If $n_A$ is 2, the reaction is said to be second order in A. The sum of the exponents gives the overall order of the reaction.

The rate laws for chemical reactions are expressions that best fit experimental data, and the order of reaction is the sum of the experimentally determined exponents.

---

[1]In thermodynamics, we must use "corrected" concentrations, or activities. In kinetics it is the actual concentrations that are important.

## Rate Laws for Elementary Reactions

Rate laws are determined by analyzing one or more reactant or product species as a function of time as a reaction proceeds, and then inspecting the results to see what theoretical form best fits the data.

**First Order**   Rate laws for elementary reactions are for the most part what one would expect. For example, a simple molecular (or nuclear) decomposition,

$$A \rightarrow \text{products}$$

proceeds at a rate that depends only on the concentration of A; the more A, the more decomposition per unit time. The rate law is

$$\frac{d\xi}{dt} = -\frac{dC_A}{dt} = k \cdot C_A \tag{12.12}$$

and the reaction is first order. The decay of radioactive elements is an example of such reactions.

If we simplify $C_A$ to $C$ and let $C = C^\circ$ at time $t = 0$, integration of (12.12) gives

$$\int_{C^\circ}^{C} \frac{dC}{C} = -k \int_0^t dt \tag{12.13}$$

$$\ln \frac{C^\circ}{C} = kt \tag{12.14}$$

$$\ln C = \ln C^\circ - kt \tag{12.15}$$

$$C = C^\circ e^{-kt} \tag{12.16}$$

These equations suggest various ways of plotting data to see if they fit a first order rate law. For example, a plot of $\ln(C^\circ/C)$ vs. $t$ will give a straight line with a slope equal to the rate constant for concentration data from a first order reaction.

**Second Order**   The most common type of elementary reaction results from bimolecular collisions:

$$A + B \rightarrow \text{products} \tag{12.17}$$

Here we expect the frequency of reaction to be proportional to the concentrations of the reactants and the concentration of products to have no effect, and so the rate law is

$$\frac{d\xi}{dt} = -\frac{dC_A}{dt} = -\frac{dC_B}{dt} = k \cdot C_A^1 C_B^1 \tag{12.18}$$

and the reaction is second order.

If the initial concentrations of A and B are $C_A^\circ$ and $C_B^\circ$, the stoichiometry of (12.17) requires that

$$C_A^\circ - C_A = C_B^\circ - C_B$$

Solving this for $C_B$ and substituting this result in (12.18) gives

$$-\frac{dC_A}{dt} = k \cdot C_A(C_A - C_A^\circ + C_B^\circ)$$ (12.19)

Integration of (12.19) then gives

$$\ln\left(\frac{C_A^\circ C_B}{C_B^\circ C_A}\right) = (C_B^\circ - C_A^\circ)kt$$ (12.20)

Therefore, a plot of

$$\left(\frac{1}{C_B^\circ - C_A^\circ}\right) \ln\left(\frac{C_A^\circ C_B}{C_B^\circ C_A}\right)$$

vs. time $t$ will result in a straight line with a slope equal to the rate constant for concentrations taken from a second order reaction. Similar equations can be derived for reactions with different stoichiometric coefficients.

There are a number of other rate laws, but this will suffice to give an idea of the procedures involved. However, it should be emphasized that most chemical reactions are "overall" reactions, and that their understanding in terms of their fundamental elementary reactions is a goal not often and not easily achieved.

### 12.2.5 Examples

**Ozone Example**

To take a simple example, the gas reaction

$$2\,O_3(g) = 3\,O_2(g)$$ (12.21)

describes the breakdown of ozone to oxygen. If this were a simple bimolecular collision reaction, the rate law would be

$$-\frac{1}{2}\frac{dC_{O_3}}{dt} = k \cdot C_{O_3}^2$$

However, the rate law has been found experimentally to be

$$-\frac{1}{2}\left(\frac{dC_{O_3}}{dt}\right)_{overall} = \frac{1}{3}\left(\frac{dC_{O_2}}{dt}\right)_{overall}$$
$$= k_{21} \cdot C_{O_3}^2 C_{O_2}^{-1}$$ (12.22)

This is a first order reaction $(2 - 1 = 1)$, but it is obviously not an elementary reaction. This result has been interpreted as due to the formation of nascent oxygen (O) as an intermediate species, with the overall reaction being the sum of the two elementary reactions

$$O_3 \rightleftharpoons O_2 + O$$ (12.23)
$$O + O_3 \rightarrow 2\,O_2$$ (12.24)

where $\rightleftharpoons$ signifies that the reaction goes both ways quite rapidly, leading to a constant or steady-state concentration of nascent oxygen (given a constant supply of ozone). If the rate constant in the forward direction in (12.23) is $k_{23}$ and that in the backward direction is $k_{-23}$, and if these rates are much faster than that of (12.24), then

$$C_O \approx \frac{k_{23}}{k_{-23}} \frac{C_{O_3}}{C_{O_2}}$$

Then the slower second step (12.24) gives the net rate of decomposition of $O_3$,

$$\begin{aligned}
-\left(\frac{dC_{O_3}}{dt}\right)_{24} &= \frac{1}{2}\left(\frac{dC_{O_2}}{dt}\right)_{24} \\
&= k_{24}C_O C_{O_3} \\
&= \frac{k_{23}k_{24}}{k_{-23}} \frac{C_{O_3}^2}{C_{O_2}} \\
&= \text{constant} \cdot C_{O_3}^2 C_{O_2}^{-1}
\end{aligned} \tag{12.25}$$

which has the form of (12.22), and hence is a plausible interpretation of the experimental results. This does not by itself prove that the interpretation is correct; further work would be needed to demonstrate this beyond a reasonable doubt, and this is typical of reaction mechanism interpretations.

The example also illustrates how apparently simple reactions can be actually much more complex than they appear, and finding out the "reaction mechanism" (the elementary reactions that add up to the overall reaction) can be very difficult. You can also see how overall reactions could have a fractional or decimal reaction order, given a number of elementary reactions having integer orders. Finally, although determining the reaction mechanism may be difficult, it is always desirable, because only by knowing how reactions actually happen are we be able to control them effectively. In addition, building up information on elementary reactions eventually leads to the ability to calculate overall reactions. For example, in the ozone reaction (12.21), the overall rate of change of $O_2$ [equation (12.22)], is approximately the same as the rate of change of $O_2$ in (12.24), because (12.23) is in a steady state. Equating $(dC_{O_2}/dt)_{overall}$ and $(dC_{O_2}/dt)_{24}$ results in

$$k_{21} = \frac{2}{3} \frac{k_{23}k_{24}}{k_{-23}} \tag{12.26}$$

Thus elementary rate constants can be combined into overall rate constants, much in the way that equilibrium constants can be combined to give overall equilibrium constants.

### Pyrite Oxidation Example

A much more complex example is provided by the oxidation of pyrite, summarized by Nicholson (1994). Pyrite oxidation in piles of mine waste (crushed rock left over after the valuable minerals have been removed) results in highly acid surface waters with high metal contents and is a serious environmental problem in many areas. Even if we were to decide to stop all mining activity, the problem would be with us for many years to come. The problem seems simple enough—we just need to prevent the sulfide minerals in the piles of rock fragments from oxidizing. However, it has resisted all simple approaches, and some not so simple, and the only hope of finding a remedy appears to lie in achieving a better understanding of what is actually happening. Thermodynamic relationships are a good start and can be used to show the overall energy balances and to help elucidate what reactions are occurring, but we soon find that to make realistic models of oxidizing waste dumps, we need to know what the reaction kinetics are.

There are two main reactions in the oxidation of pyrite at the Earth's surface. One of course involves oxygen,

$$FeS_2 + \tfrac{7}{2}O_2 + H_2O = Fe^{2+} + 2\,SO_4^{2-} + 2\,H^+ \tag{12.27}$$

The other involves ferric iron as oxidant,

$$FeS_2 + 14\,Fe^{3+} + 8\,H_2O = 15\,Fe^{2+} + 2\,SO_4^{2-} + 16\,H^+ \tag{12.28}$$

The ferric iron in this reaction is produced from the oxidation of ferrous iron by oxygen,

$$Fe^{2+} + H^+ + \tfrac{1}{4}O_2 = Fe^{3+} + \tfrac{1}{2}H_2O \tag{12.29}$$

All three reactions, along with countless others, proceed simultaneously in the mine waste environment.

From these equations, which by themselves represent a considerable amount of research,[2] we see a number of factors that might influence the rate of oxidation of pyrite—dissolved oxygen, certainly, but also aqueous ferric iron, $pH$, ferrous iron, and possibly sulfate concentration. In addition, because it is a heterogeneous reaction, the surface area of pyrite exposed to the oxidizing fluid is also an important factor, as is temperature. Because the reaction occurs at the Earth's surface, pressure variations are not likely to be important. And, of course, it is always possible that other important factors remain to be discovered, including ones that do not appear in the reactions. That is, there are *catalysts* and *inhibitors*, substances that speed up or slow down reactions, without appearing in the reactions as normally written. For example, a substance that formed a coating on the surface of the pyrite grains might slow down or *inhibit* the oxidation process. Bacteria often are a major factor in speeding up processes that, without them, would be very slow.

A number of experimental studies have been done on these reactions, with results that are not entirely in agreement for various reasons. This is normal in research—nature does not give up her secrets easily. A recent compilation

---

[2]Out there in the field where natural processes occur, no one tells you what reactions are important—you have to figure them out (cf. §10.3.1). In this case, the importance of oxygen (12.27) was clear, but the discovery of the role of aqueous ferric iron (12.28) was an important step forward (Singer and Stumm, 1968).

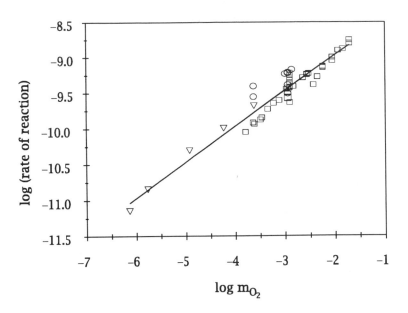

Figure 12.1: The logarithm of the rate of pyrite oxidation (in units of $\mathrm{mol\, m^2\, s^{-1}}$) as a function of dissolved oxygen concentration (in $\mathrm{mol\, kg^{-1}}$). Original data at various *pH* values have been normalized to a *pH* of 2.0 using equation (12.30). The symbols indicate data from various sources referenced in Williamson and Rimstidt (1994). The straight line is an expression of equation (12.30).

of the data shown in Figure 12.1 on pyrite oxidation as a function of oxygen concentration (Williamson and Rimstidt, 1994) gives the relation

$$\text{rate of reaction} = \frac{d\xi}{dt} = 10^{-8.19}\frac{C_{O_2}^{0.50}}{C_{H^+}^{0.11}} \tag{12.30}$$

Thus the rate constant ($k_{27}$) is $10^{-8.19}$, and the order of the reaction is $0.50 - 0.11 = 0.39$. Similar expressions have been derived for the other two reactions [(12.28), (12.29)]. Fractional orders such as this indicate that elementary reactions more complex than simple molecular collisions are important, such as reactions on mineral surfaces, and the details of such processes are a major area of research at the present time.

At the moment it's anybody's guess as to how many elementary reactions are involved in this important oxidation process, and exactly what they are. Even without a complete understanding of the reaction mechanisms, however, models of waste dump oxidation can be formulated, and these will certainly involve not only the rate equations [e.g., (12.30)], but estimates of how fast oxygen can penetrate the pile (which involves understanding fluid flow and diffusion in the pile), the role of iron-oxidizing bacteria, the oxidation mechanisms for sulfides other than pyrite (e.g., pyrrhotite), estimates of the effective surface area of the various sulfides in the pile, how fast the heat generated by oxidation can be dissipated, estimates of the groundwater composition, and so on. There seems to be enough to keep us busy on this problem for some time to come, and this is only one of many important environmental problems.

### 12.2.6 Temperature Dependence of Rate Constants

In §8.5.1 we saw that the temperature dependence of the equilibrium constant $K$ could be expressed as

$$\frac{d \ln K}{d(1/T)} = -\frac{\Delta_r H^\circ}{R}$$

or, alternatively

$$\frac{d \ln K}{dT} = \frac{\Delta_r H^\circ}{RT^2}$$

In 1889, Arrhenius proposed a similar equation for the temperature effect on rate constants,

$$\frac{d \ln k}{d(1/T)} = -\frac{E_a}{R} \tag{12.31}$$

or

$$\frac{d \ln k}{dT} = \frac{E_a}{RT^2} \tag{12.32}$$

where $E_a$ is called the Arrhenius activation energy, or just the activation energy, and turns out to be closely related to the "energy barrier" between products and reactants in chemical reactions (Figure 12.2). The general form of this equation has been shown to be derivable from statistical mechanics.

Experimental data for a great many reactions over a large range of temperatures shows that the Arrhenius equation is usually closely obeyed, showing that $E_a$ is either a constant or a weak function of temperature, and so we can integrate the equation to give

$$\ln k = \ln A - \frac{E_a}{RT} \tag{12.33}$$

or

$$k = Ae^{-E_a/RT} \tag{12.34}$$

where $A$, which enters (12.33) as a constant of integration, is called the *pre-exponential* factor.

The activation energy of an overall reaction is made up of the individual contributions of the elementary reactions making up the overall reaction. The magnitude of the activation energy can vary from virtually zero to hundreds of kJ per mole and, besides controlling the temperature dependence of the rate constant, provides clues as to the nature of the reaction mechanisms, because the energies involved in many types of diffusion, electron exchange, and bond-breaking processes are known.

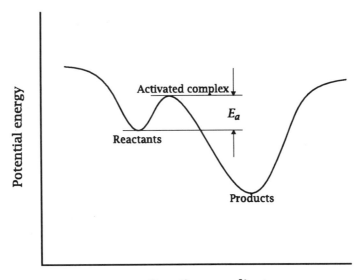

**Reaction coordinate**

Figure 12.2: Schematic representation of the energy difference between reactants and products. An "activated complex" is postulated to exist, in addition to reactants and products. Those reactants that rearrange themselves into this form then are able to proceed to the products form. The "reaction coordinate" (x-axis) is a term used to describe the path taken by reactants changing to products.

Calculation of activation energies for elementary gas phase reactions from relatively simple molecular considerations has been fairly successful, but heterogeneous reactions are still far from such theoretical treatment. Still, it forms the link with our next subject.

## 12.3 TRANSITION STATE THEORY

The rate laws we have examined have to do with reactions that progress at some measurable rate—they are able to get over the *activation energy barrier* which separates reactants from products in all chemical reactions and which in many cases completely prevents reactions from occurring. What do we know about this energy barrier? The most successful approach to an understanding of this subject was suggested independently in 1935 by H. Eyring and by M.G. Evans and M. Polanyi, and is referred to as activated complex theory, or transition state theory.

Consider the bimolecular elementary reaction

$$A + B \rightarrow \text{products}$$

We know that the equilibrium state of chemical reactions is not a static state, but a dynamic state in which the rates of the forward and backward reactions are exactly the same. We sometimes indicate this with the $\rightleftharpoons$ symbol, which symbolizes the forward and backward nature of the equilibrium.

$$A + B \rightleftharpoons \text{products}$$

Activated complex theory treats this process as an equilibrium between the reactants $(A + B)$, and an *activated complex* $AB^{\ddagger}$, with only the complex being able to proceed on to the product state:

$$A + B \rightleftharpoons AB^{\ddagger} \rightarrow \text{products} \qquad (12.35)$$

Rigorous treatment of this activated complex relies on the methods of statistical mechanics, and considerable success has been achieved in understanding this state, particularly in gas phase reactions. However, postulating equilibrium between the reactants and the activated state allows us to use the equations we have developed in previous chapters, as well as allowing some insight.

The goal is to be able to calculate the value of the rate constant for elementary reactions, and hence eventually of overall reactions, from first principles. In terms of an intermediate activated complex, the rate will be a function of the number (concentration) of such complexes and the frequency with which they cross the energy barrier to become reaction products:

$$\text{rate of reaction} = \boxed{\begin{array}{c}\text{concentration of}\\\text{activated complexes}\end{array}} \times \boxed{\begin{array}{c}\text{frequency with which any}\\\text{complex crosses the barrier}\end{array}} \qquad (12.36)$$

The first term on the right of this equation could be evaluated if we could calculate an equilibrium constant for the formation of the activated complex,

$$K^{\ddagger} = \frac{a_{\ddagger}}{a_A a_B}$$

In solutions, these activities will be Henryan (7.16), so that

$$a_i = C_i y_{H_i}$$

where $C_i$ is the concentration of $i$ on some convenient scale. The concentration of the activated complex is then

$$C^{\ddagger} = K^{\ddagger} C_A C_B \frac{y_{H_A} y_{H_B}}{y_{H_{\ddagger}}}$$

The second term on the right side of (12.36) is called the *transmission factor*, $f$, so the rate of reaction for (12.35) is

$$\begin{aligned} \text{rate of reaction} \quad &= \quad f C^{\ddagger} \\ &= \quad f K^{\ddagger} C_A C_B \frac{y_{H_A} y_{H_B}}{y_{H_{\ddagger}}} \end{aligned} \tag{12.37}$$

In normal kinetic terms (12.18), this rate of reaction is also

$$\text{rate of reaction} = k_{35} C_A C_B \tag{12.38}$$

and so combining (12.37) and (12.38), we get

$$k_{35} = f K^{\ddagger} \frac{y_{H_A} y_{H_B}}{y_{H_{\ddagger}}} \tag{12.39}$$

Evaluating the transmission factor involves calculating the proportion of molecules having an energy beyond a certain threshold level, a problem squarely in the domain of statistical mechanics. Borrowing the result from this analysis, we find

$$f = \frac{k_b}{h} T \tag{12.40}$$

where $k_b$ is the Boltzmann constant, $h$ is the Planck constant (Appendix A), and $T$ is the absolute temperature, as usual. Combining (12.39) and (12.40) gives

$$k_{35} = \frac{k_b}{h} T K^{\ddagger} \frac{y_{H_A} y_{H_B}}{y_{H_{\ddagger}}} \tag{12.41}$$

This equation allows calculation of the rate constant, in this case of a bimolecular elementary reaction, in terms of an equilibrium constant for the formation of a fairly hypothetical activated complex. How do we calculate this equilibrium constant?

Following our usual line of thinking with respect to equilibrium constants, we could envisage a standard state free energy of reaction for the formation of the activated complex, such that

$$\Delta G^{\ddagger} = \mu_{\ddagger}^{\circ} - \mu_A^{\circ} - \mu_B^{\circ} \tag{12.42}$$

$$= -RT \ln K^{\ddagger} \tag{12.43}$$

$$= \Delta H^{\ddagger} - T \Delta S^{\ddagger} \tag{12.44}$$

From (12.43)

$$K^{\ddagger} = e^{-\Delta G^{\ddagger}/RT}$$

and so (12.41) could be rewritten

$$k_{35} = \frac{k_b T}{h} \frac{y_{H_A} y_{H_B}}{y_{H_{\ddagger}}} e^{-\Delta G^{\ddagger}/RT} \tag{12.45}$$

$$= \frac{k_b T}{h} \frac{y_{H_A} y_{H_B}}{y_{H_{\ddagger}}} e^{\Delta S^{\ddagger}/R} e^{-\Delta H^{\ddagger}/RT} \tag{12.46}$$

This is called the Eyring equation for the simple bimolecular case. The standard states used in defining the terms in (12.42) depend on the units used at the start in discussing reaction (12.35). In kinetics, these are usually concentrations in $mol\,cm^{-3}$ or $mol\,m^{-3}$.

But we still have the problem of evaluating the properties of the activated state ($K^{\ddagger}$, $\Delta G^{\ddagger}$, etc.). Some insight into this is obtained by comparing our theoretical expression (12.45) or (12.46) with the Arrhenius equation. Differentiating (12.45) with respect to $T$ and assuming that $\Delta S^{\ddagger}$ and $\Delta H^{\ddagger}$ are not functions of $T$ results in

$$\frac{d \ln k_{35}}{dT} = \frac{1}{T} + \frac{\Delta H^{\ddagger}}{RT^2} \tag{12.47}$$

Comparing this with (12.32), we have

$$E_a = RT + \Delta H^{\ddagger} \tag{12.48}$$

and the pre-exponential term $A$ is

$$A = e \frac{k_b T}{h} \frac{y_{H_A} y_{H_B}}{y_{H_{\ddagger}}} e^{\Delta S^{\ddagger}/R} \tag{12.49}$$

These relations[3] provide a link between the measurable quantity $E_a$ and the properties of our hypothetical transition state, so that we can now calculate transition state parameters and hence rate constants from a knowledge of the Arrhenius slope. However, to get the Arrhenius slope we need to have first measured the rate constant; so we seem to have gone in a circle. Actually, what we have done is to analyze the reaction rate concept in terms of an activated state and to express it in terms of the thermodynamic parameters of this state. This opens up the rate concept to exploration by other means, including statistical mechanics, and these other means of investigation are quite independent of particular rate measurements.

---

[3] Slight differences in these equations result from reactions with different stoichiometries.

## 12.3.1   The Relation Between Reaction Rates and $\Delta G$

The introduction of the equilibrium constant $K^{\ddagger}$ has other implications for the relationship between kinetics and thermodynamics. So far, we have considered only the forward rate of reaction (12.35). In the general case, such as (12.1), there are separate rate constants for the forward and backward reactions. Let's say that (12.1) is an elementary reaction, and that the forward rate is called $R_+$, where (with $a = b = c = d = 1$)

$$R_+ = k_+ C_A C_B \tag{12.50}$$

and the backward reaction is called $R_-$, where

$$R_- = k_- C_C C_D \tag{12.51}$$

In terms of the transition state, these rates are

$$k_+ = \frac{k_b T}{h} \frac{y_{H_A} y_{H_B}}{y_{H_{\ddagger}}} e^{-\Delta G_+^{\ddagger}/RT} \tag{12.52}$$

and

$$k_- = \frac{k_b T}{h} \frac{y_{H_C} y_{H_D}}{y_{H_{\ddagger}}} e^{-\Delta G_-^{\ddagger}/RT} \tag{12.53}$$

where

$$\Delta G_+^{\ddagger} = \mu_{\ddagger}^{\circ} - \mu_A^{\circ} - \mu_B^{\circ}$$

$$\Delta G_-^{\ddagger} = \mu_{\ddagger}^{\circ} - \mu_C^{\circ} - \mu_D^{\circ}$$

If the activated complex is the same for both directions, the ratio of the rate constants from (12.52) and (12.53) is

$$\frac{k_+}{k_-} = \frac{y_{H_A} y_{H_B}}{y_{H_C} y_{H_D}} e^{-(\Delta G_+^{\ddagger} - \Delta G_-^{\ddagger})/RT} \tag{12.54}$$

$$= \frac{y_{H_A} y_{H_B}}{y_{H_C} y_{H_D}} e^{-\Delta G^{\circ}/RT} \tag{12.55}$$

where in subtracting $\Delta G_-^{\ddagger}$ from $\Delta G_+^{\ddagger}$ the transition state free energy cancels out, and $\Delta G^{\circ} = \mu_C^{\circ} + \mu_D^{\circ} - \mu_A^{\circ} - \mu_B^{\circ}$, that is, the standard state free energy of reaction (12.1), where standard states are expressed in their kinetic form.

Equation (12.55) can also be expressed as

$$\frac{k_+}{k_-} = \frac{y_{H_A} y_{H_B}}{y_{H_C} y_{H_D}} K_1 \tag{12.56}$$

where $K_1$ is the normal equilibrium constant for (12.1). Furthermore, from (12.50), (12.51), and (12.55),

$$\frac{R_+}{R_-} = \frac{k_+ C_A C_B}{k_- C_C C_D} \tag{12.57}$$

$$= \frac{y_{H_A} y_{H_B}}{y_{H_C} y_{H_D}} \frac{C_A C_B}{C_C C_D} e^{-\Delta G^\circ / RT} \qquad (12.58)$$

$$= \frac{a_A a_B}{a_C a_D} e^{-\Delta G^\circ / RT} \qquad (12.59)$$

$$= \frac{1}{Q_1} e^{-\Delta G^\circ / RT} \qquad (12.60)$$

$$= e^{-\Delta G / RT} \qquad (12.61)$$

where $\Delta G$ is the actual free energy difference between products and reactants. When $\Delta G = 0$, $R_+ = R_-$. That last step, from (12.60) to (12.61), is the important one. Recall that from (8.6),

$$\Delta_r \mu = \Delta_r \mu^\circ + RT \ln Q \qquad (8.6)$$

where in the case of (12.1),

$$Q_1 = \frac{a_C a_D}{a_A a_B}$$

in the general case where the reaction is not necessarily at equilibrium. Equation (8.6) can also be written

$$\begin{aligned}
\ln Q &= (\Delta_r G - \Delta_r G^\circ) / RT \\
Q &= e^{(\Delta_r G - \Delta_r G^\circ)/RT} \\
&= e^{\Delta_r G / RT} e^{-\Delta_r G^\circ / RT}
\end{aligned}$$

and so

$$\frac{1}{Q} e^{-\Delta_r G^\circ / RT} = e^{-\Delta_r G / RT}$$

which is the step from (12.60) to (12.61).

But we're not finished yet. The *net* rate of an elementary reaction is the difference between the forward and reverse rates, or $R_{net} = R_+ - R_-$. Combining this with (12.61), we get

$$R_{net} = R_+ \left( 1 - e^{\Delta G / RT} \right) \qquad (12.62)$$

A property of $1 - e^x$ where $x$ is any real number is that $1 - e^x \approx -x$ when $x$ is small (try it on your calculator!)[4] Therefore, if the reaction is sufficiently close to equilibrium ($\Delta_r G$ very small) so that $|\Delta_r G| \ll RT$, then (12.62) gives

$$R_{net} = -\frac{R_+ \Delta G}{RT} \qquad (12.63)$$

which establishes a *linear* relationship between the net rate of an elementary reaction and the free energy difference of that reaction, as long as the system is

---

[4]This follows from the expansion $e^x = 1 + \frac{x}{1!} + \frac{x^2}{2!} + \frac{x^3}{3!} + \cdots$. This can be used to evaluate $e$ by letting $x = 1$.

not far from equilibrium. In plain language, it says that reactions will proceed faster the farther they are from equilibrium.

Here is a direct link between thermodynamics and kinetics, admittedly an approximation unless close to equilibrium, but nevertheless one that is used considerably in process modeling (§12.5).

### 12.3.2  Comment

The great successes of TST (transition state theory) have so far been with relatively simple gas phase reactions, where the methods of statistical mechanics can be used to calculate the properties of the activated state. If you are interested in natural systems, such as the problems of groundwater contamination, you might wonder what all this has to do with you.

This subject is included for two reasons. The first is to show how the weaknesses of equilibrium thermodynamics have been addressed; the inability to say anything about energy barriers or rates of reaction is a major weakness. Second, the concepts and viewpoint of TST have spread far and wide, and virtually all research on heterogeneous reactions—dissolution, precipitation, adsorption, all of which are vital to an understanding of fluid compositions in the crust of the Earth—makes use of these ideas. The distance between "esoteric" chemical theories and practical environmental management is becoming smaller every day.

## 12.4  STATISTICAL THERMODYNAMICS

### 12.4.1  Thermodynamics and Statistical Mechanics

The thermodynamic concept most difficult to fully understand is without doubt the concept of entropy. In this book we have introduced it in a quite empirical way, avoiding much explanation. We basically said we needed a parameter for our model that would give the direction of spontaneous reactions in isolated systems (Chapter 4), and we said entropy was that parameter without much attempt to justify the claim. It comes complete with a method of measurement (Chapter 5) and a base level (Third Law), and its justification in terms of our empirical or postulational approach is the fact that it works. Over 100 years of experience has shown how useful it is. But what is it?

Equilibrium thermodynamics deals with homogeneous bodies of matter in equilibrium states. Except for applications to things like aqueous solutes, where the exact nature of the molecular species is sometimes considered, thermodynamics deals with compositional variation in terms of fairly abstract components. The mathematics of thermodynamics is the calculus, with its assumption of a continuum in the space occupied by thermodynamic variables, in which reversible processes are possible. As we have mentioned at various points, thermodynamics is a mathematical model of energy relationships, and this model

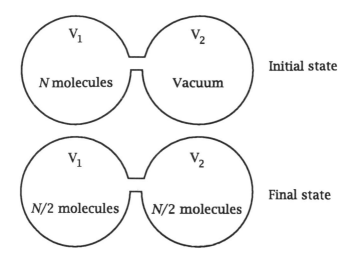

Figure 12.3: The expansion of a gas into a vacuum.

has no way of including individual particles.

But we know beyond any doubt that the universe and everything it contains is made up of particles, which we call atoms and molecules. The universe is not a continuum. Therefore, a model of energy relationships based on a mathematics of particles might be expected to yield deeper insights into the way nature works than one that is not. Such a model is statistical mechanics, which incorporates statistical thermodynamics. This model, initiated by Ludwig Boltzmann in the 1890s, leads to a parameter indistinguishable from entropy in any way, and therefore identified with entropy, but based on completely different reasoning. In this model, entropy is a parameter that is a measure of the probability of a given configuration or system structure, and processes are spontaneous if they lead to a more probable state of the system. This characterization of entropy in terms of the configuration of elementary particles has done a great deal to lift the veil of mystery that surrounds it.

## 12.4.2 Entropy and Probability

The relationship between entropy and probability can be found by analyzing a simple process for which both the entropy change and the probabilities are known. Such a process is the expansion of an ideal gas into a vacuum. Consider the expansion of $N$ molecules of an ideal gas, initially at $P_1$ and $V_1$,[5] into an evacuated chamber of volume $V_2$ (Figure 12.3). The final pressure is $P_2$ and the

[5]Note the use of Roman V for volume here, which denotes total volume, not molar volume (§2.4.1).

final volume is $V_1 + V_2$. For this change, the entropy change is[6]

$$\Delta S = nR \ln \left( \frac{V_1}{V_1 + V_2} \right) \tag{12.64}$$

where $n$ is the number of moles of gas. After expansion the probability of finding any particular molecule in the original volume is $V_1/(V_1 + V_2)$. The probability of finding all $N$ molecules in the original volume $V_1$ at the same time without being constrained to be there is $[V_1/(V_1 + V_2)]^N$ (since there are $N$ molecules and probabilities are multiplicative). Therefore, the probability of the initial state arising spontaneously after expansion is

$$p_{initial} = \left( \frac{V_1}{V_1 + V_2} \right)^N$$

This is an extremely small probability. For example if $V_1/(V_1 + V_2)$ is 0.5 and $N = 10^{23}$ (i.e., $n \approx 1$), it is about $1/(2^{10^{23}})$.

After the expansion, the probability ($p_{final}$) of finding all $N$ molecules in the final volume $V_1 + V_2$ must be 1.0. The ratio of these two probabilities is

$$\frac{p_{final}}{p_{initial}} = \left( \frac{V_1 + V_2}{V_1} \right)^N \tag{12.65}$$

Comparison of (12.64) and (12.65) gives

$$\Delta S = nR \ln \left( \frac{p_{final}}{p_{initial}} \right)^{1/N}$$

or

$$\Delta S = \frac{nR}{N} \ln \left( \frac{p_{final}}{p_{initial}} \right)$$

The number of molecules divided by the number of moles ($N/n$) is Avogadro's number, $N_a$, so that the equation becomes

$$\Delta S = \frac{R}{N_a} \ln \left( \frac{p_{final}}{p_{initial}} \right) \tag{12.66}$$

Thus there is a direct relationship between entropy and probability in this simple case, and by analogy, generally. In the example mentioned above where $V_1/(V_1 + V_2) = 0.5$, (12.66) becomes, for one mole of gas,

$$\begin{aligned} \Delta S &= (R/10^{23}) \ln(2^{10^{23}}) \\ &= R \ln 2 \\ &= 5.76 \, \mathrm{J\,K^{-1}} \end{aligned}$$

($10^{23}$ is an abbreviation for Avogadro's number, $6.022136 \cdot 10^{23}$, Appendix A.)

---

[6]This equation is not derived in this book.

This general relationship between entropy and probability is more generally expressed as Boltzmann's famous equation

$$S = k \ln W \tag{12.67}$$

which is engraved on his tombstone in Vienna. Here $k$ is $R/N_a$, and $W$ is called variously the statistical probability or the statistical weight of a state whose entropy is $S$. Let's see how $W$ is derived in our simple gas expansion case, involving $N = 10^{23}$ molecules. The problem this time is to find the ratio of the probabilities of the initial and final states, where the final state is characterized not, as before, simply by the fact that all molecules are to be found in $V_1 + V_2$, but by the fact that it is the most probable of all possible states after expansion, that is, the state having $N/2$ molecules in each of $V_1$ and $V_2$. If the initial state (Figure 12.3) has entropy $S_{initial}$ and probability $W_{initial}$ and the final state has entropy $S_{final}$ and probability $W_{final}$, then the expansion process will have

$$
\begin{aligned}
\Delta S &= S_{final} - S_{initial} \\
&= R \ln W_{final} - R \ln W_{initial} \\
&= R \ln \frac{W_{final}}{W_{initial}}
\end{aligned}
\tag{12.68}
$$

To calculate $W_{initial}$ and $W_{final}$, we add molecules to $V_1 + V_2$ one at a time, with no control over which volume they go in—the outcome of each addition of a molecule is random. Each molecule has an equal chance of landing in $V_1$ or in $V_2$, so the probability of it landing in $V_1$ is exactly $\frac{1}{2}$. Probabilities are multiplicative, so the probability of any two molecules both landing in $V_1$ is $\frac{1}{2} \times \frac{1}{2}$, and the probability of all $N$ molecules landing in $V_1$ is $(\frac{1}{2})^N$, or $0.5^{-23}$. So

$$W_{initial} = \left(\frac{1}{2}\right)^N$$

Naturally, it is a very small number, reflecting the near impossibility of such an event. Finding all the air in a room to be suddenly all in one corner would have about the same probability.

But there is another way to look at $W$. That is, we could consider, not the probability, which is a number between 0 and 1, but the total number of ways that a state can arise, given the circumstances. For our case, which has two possible outcomes for each molecule added, we can imagine a large number of possible results after adding our $10^{23}$ molecules. One possible result we have already considered—all molecules wind up in $V_1$. Another result would be that $\frac{3}{4}$ of the molecules wind up in $V_1$ and $\frac{1}{4}$ in $V_2$. Another is that they are almost evenly divided, but there are $\frac{N}{2} + 100$ molecules in $V_1$ and $\frac{N}{2} - 100$ molecules in $V_2$. Each of these is a possible outcome, a possible *configuration* of the system, and each final result (except the first) can be arrived at in a large number of ways, called the number of *microstates* in that configuration. If you toss a coin four times, you might get three heads (3H) and one tail (1T) (a configuration of the coin tossing system), but this result can be achieved in four different ways: (HHHT), (HHTH), (HTHH), or (THHH). Each of these four results is a possible microstate of the system, and each has an equal probability of occurring. The general expression for calculating how many ways a given result can be achieved is

$$\frac{N!}{N_1!N_2!} \tag{12.69}$$

where $N$ is the number of events (coin tosses; molecules added), and $N_1$ and $N_2$ are the numbers of the two possible results (head or tails; molecules in $V_1$ or $V_2$). This can be converted into a probability by dividing by the total number of possible results (total number of microstates), which is $2^N$, so the general expression for finding the probability of a particular result in our case is

$$\left(\frac{1}{2}\right)^N \frac{N!}{N_1!N_2!} \tag{12.70}$$

Thus with $N = 4$, $N_1 = 3$, and $N_2 = 1$,

$$\frac{N!}{N_1!N_2!} = \frac{4!}{3!1!}$$
$$= 4$$

and the probability of this result is $4/2^4$ or 0.25. And for the case of all molecules winding up in $V_1$.

$$\frac{N!}{N_1!N_2!} = \frac{10^{23}!}{10^{23}!0!}$$
$$= 1 \quad (\text{recall that } 0! = 1)$$

There is only one way that all molecules can go into $V_1$. If you repeated the process, every toss of a molecule must have the same result as before, to achieve this result. The probability of this result is $(1/2)^{10^{23}}$, as we saw above.

Going back to tossing coins, the number of ways that you could have 2H and 2T after four tosses is

$$\frac{N!}{N_1!N_2!} = \frac{4!}{2!2!}$$
$$= 6$$

and this is the most probable result for precisely the reason that there are more ways that it can be achieved (more microstates available) during the given process. Getting 3H and 1T is not impossible, it's just less likely. It is even less likely to get 4H or 4T, because like all our molecules in $V_1$, there is only one possible way to do it, and many ways to get other results.

The number of ways of achieving what we know to be the equilibrium result ($N/2$ molecules in each volume) is[7]

$$\frac{N!}{N_1!N_2!} = \frac{10^{23}!}{\left(\frac{1}{2}10^{23}\right)!\left(\frac{1}{2}10^{23}\right)!}$$

$$= 6.93147 \times 10^{22}$$

and the probability of this result is

$$\left(\frac{1}{2}\right)^N \frac{N!}{N_1!N_2!} = \left(\frac{1}{2}\right)^{10^{23}} \frac{10^{23}!}{\left(\frac{1}{2}10^{23}\right)!\left(\frac{1}{2}10^{23}\right)!}$$

$$= 0.3466$$

If we are calculating a $\Delta S$, it obviously doesn't matter whether we consider $W$ to be a probability, (12.70), or the number of ways a result can be achieved (called the statistical weight, or number of equally probable microstates), (12.69), because the $(1/2)^N$ factor will cancel out. Both descriptions will be found in textbooks, but it is more common to consider $W$ to be the statistical weight.

Actually, what we should do to find the equilibrium final state of the gas expansion using this statistical approach is to *demonstrate* that equal numbers of molecules in the two volumes is the most probable result, or that there are more ways of achieving this result than any other result. This can certainly be done, but we will save time by considering only one other state.

If as many as $10^{13}$ extra molecules are found in $V_1$ and the same number are missing from $V_2$, the number of molecules in the chambers would be $0.5000000001 \times 10^{23}$ and $0.4999999999 \times 10^{23}$. But even this virtually unmeasurable event is extremely unlikely to happen. If this state is called state *final'*,

$$W_{final'} = \frac{N!}{\left(\frac{1}{2}N + 10^{13}\right)!\left(\frac{1}{2}N - 10^{13}\right)!}$$

and the logarithm of the ratio of our $W_{final}$ to $W_{final'}$ is

$$\ln \frac{W_{final}}{W_{final'}} = \ln \frac{\left(\frac{1}{2}N + 10^{13}\right)!\left(\frac{1}{2}N - 10^{13}\right)!}{\left(\frac{1}{2}N\right)!\left(\frac{1}{2}N\right)!}$$

$$\approx 10^{868}$$

or

$$\frac{W_{final}}{W_{final'}} \approx e^{10^{868}}$$

---

[7]The evaluation of factorials of large numbers is accomplished using Stirling's approximation, which is

$$\ln N! \approx N \ln N - N$$

Similar numbers are obtained for other alternative final states. This means that the equilibrium configuration predominates by such an enormous margin at large values of $N$ that other configurations are extremely unlikely to occur. You need not be concerned that one day all the air in the room will rush into a corner, leaving you in a vacuum.

### Entropy and Disorder

Considering $W$ as the number of equally probable microstates of a system, we see that conditions that are highly ordered tend to have low statistical weights, and conditions that are disordered or chaotic tend to have high statistical weights. The gas in Figure 12.3 is more ordered when all in $V_1$ ($W = 1$) than when expanded to $V_1 + V_2$. Or the number of microstates available to the atoms in a crystal lattice will be much smaller for a given total energy than if the atoms were free to move as a gas in the same volume as the crystal. These observations suggest that $W$ can be interpreted as a measure of the disorder of a system. Therefore, entropy is also a measure of disorder, and $\Delta S_{U,V} \geq 0$ means that isolated systems must always evolve so as to increase their disorder. Given the isolation of the system from outside influences, this result is fairly intuitive, in the sense that, molecular motions being fortuitous or random, it seems likely that any initial structure will tend to naturally dissipate. This interpretation has done much to help us understand the Second Law.

Of course, this idea of disorder is not always as easy to see as in our expanding gas case, or in comparing solids to gases. For example, you could put a hot block of copper and a cold block of copper together in an isolated system. Eventually the two blocks would reach equilibrium at some intermediate temperature. To see exactly how the two blocks at the same temperature are in a more disordered state than the two blocks at different temperatures, you have to consider the energy levels available to the atoms in the solids, which requires the introduction of quantum mechanics. When this is done, the entropy maximum principle is seen to be exactly the same.

## 12.4.3  The Boltzmann Distribution

Actually, we don't need to know much quantum mechanics to explore the idea of energy levels available to atoms in a system; we need only postulate that there are such things as allowed energy levels available to atoms and to molecules—that in a system with high energy (say at a high temperature) more of the atoms will have higher energy levels than in the same system with lower energy. Suppose then that we postulate the existence of $j$ distinct energy levels, called $\epsilon_1, \epsilon_2, \ldots, \epsilon_j$. We would like to know the connection between these energy levels and $W$, and of course entropy. The question that Boltzmann posed, and answered, is this: How will the $N$ particles of the most probable molecular configuration distribute themselves throughout a set of allowed energy levels? The result is the Boltzmann distribution—the keystone of statistical mechanics.

The mathematical solution finds the number of atoms in each energy level, $N_1, N_2, \ldots, N_j$, which maximizes the probability $W$ of a system, subject only to the constraints that the total number of particles and the total energy ($E$) must remain constant,

$$\sum_{i=1}^{j} N_i = N \tag{12.71}$$

$$\sum_{i=1}^{j} N_i \epsilon_i = E \tag{12.72}$$

This is a mathematical problem beyond the scope of this book, but the result is

$$\frac{N_i}{N} = \frac{e^{-\epsilon_i/k_b T}}{\sum_i e^{-\epsilon_i/k_b T}} \tag{12.73}$$

This is the Boltzmann distribution. It tells us that the fraction of molecules ($N_i/N$) in an energy level $\epsilon_i$ increases exponentially with temperature and decreases exponentially with the energy of that level.

**The Partition Function**

The denominator in equation (12.73) is called the partition function $z$:

$$z = \sum_{i=1}^{j} e^{-\epsilon_i/k_b T} \tag{12.74}$$

This is one of the most important parameters in statistical mechanics since it is directly related to the thermodynamic properties of a system. The summation in (12.74) is made over all energy states so $z$ is a function of the partitioning of all particles among all energies for the equilibrium configuration. Table 12.1 shows how 1000 atoms would be distributed over 10 energy levels equally spaced $k_b T$ apart. Note the concentration of atoms in the lower energy levels. As $T$ increases, the lower levels become less populated and the upper levels more populated.

Because the partition function is related to the number of particles occupying energy levels above the ground state, it can be used to calculate the *average internal energy*, $\bar{\epsilon}$, of a particle. From equations (12.71) and (12.72) the average energy is

$$
\begin{aligned}
\bar{\epsilon} &= \frac{\sum N_i \epsilon_i}{\sum N_i} \\
&= \sum_i N_i \epsilon_i / N \tag{12.75}
\end{aligned}
$$

Substituting (12.74) into (12.73) gives the more common form of the Boltzmann distribution:

$$N_i/N = e^{-\epsilon_i/kT}/z \tag{12.76}$$

Table 12.1: The Boltzmann Distribution

| Level No. | $\epsilon/kT$ | $e^{-\epsilon/kT}$ | $N_i$ for $N = 1000$ |
|-----------|-----|--------|------|
| 0 | 0 | 1.0 | 632 |
| 1 | 1 | 0.3679 | 233 |
| 2 | 2 | 0.1353 | 86 |
| 3 | 3 | 0.0498 | 31 |
| 4 | 4 | 0.0183 | 12 |
| 5 | 5 | 0.0067 | 4 |
| 6 | 6 | 0.0025 | 2 |
| 7 | 7 | 0.0009 | 1 |
| 8 | 8 | 0.0003 | 0 |
| 9 | 9 | 0.0001 | 0 |
| 10 | 10 | 0.0000 | 0 |
| Partition Function | | 1.58195 | |

Then combining (12.75) and (12.76)

$$\bar{\epsilon} = \frac{1}{z} \sum_i \epsilon_i e^{-\epsilon_i/kT}$$

and because

$$\frac{d}{d(1/kT)} \left( e^{-\epsilon_i/kT} \right) = -\epsilon_i \cdot e^{-\epsilon_i/kT}$$

and $d(1/(kT)) = -dT/(kT^2)$, it turns out that

$$\bar{\epsilon} = kT^2 \left( \frac{\partial \ln z}{\partial T} \right)_V \tag{12.77}$$

The constant volume restriction is imposed because the dependence of $\epsilon_i$ on volume is not taken into account in this formulation.

**The Partition Function and Thermodynamic Properties**

The partition function we have so far is independent of the number of particles $N$ being considered. It is the partition function per particle. The partition function for a system as a whole depends on the type of system, but the simplest relationship is

$$Z = e^{-E_i/kT}$$

where

$$Z = z^N$$

where $Z$ is the partition function for the entire system, and the allowed molar energy levels of the system are $E_i$. Then

$$E = kT^2 \left(\frac{\partial \ln Z}{\partial T}\right)_V \tag{12.78}$$

where $E$ is the total molar energy of the system.

In Chapter 3 we considered that the total energy of a system is given by relativity theory as $E_r = mc^2$ and that the internal energy of thermodynamics is some unspecified smaller quantity such that $E_r = U + \text{constant}$. In statistical mechanics the energy of a particle or an energy level is generally taken as the difference in energy between the level considered and a "ground state" of zero energy, quite frequently the state of zero particle vibration, rotation, and translation, that is, $0\,K$ on the absolute temperature scale. Thus systems have zero energy at $0\,K$. Thus the E of statistical mechanics is also related to U by $E = U + \text{constant}$, but the constant is different. Therefore, the $E$ term in (12.78) is actually a $\Delta E$ term, and that equation can also be written

$$\Delta E = E - E_0 = kT^2 \left(\frac{\partial \ln Z}{\partial T}\right)_V$$

The *heat capacity at constant volume* is defined as $C_V = (\partial U/\partial T)_V$, or, because $U = E + \text{constant}$ (Chapter 3), $C_V = (\partial E/\partial T)_V$. Differentiating (12.78) gives

$$C_V = \frac{k}{T^2} \left(\frac{\partial^2 \ln Z}{\partial (1/T)^2}\right) \tag{12.79}$$

Thus the internal energy and heat capacity are simply related to the change in the partition function with temperature. For certain simple systems such as gases at low temperatures, the partition function can be estimated theoretically. For most systems of geological interest such as minerals and concentrated salt solutions, additional experimental information is required. This might take the form of spectroscopic data on electronic or molecular vibrational frequencies, or direct measurement of some of the nonideal thermodynamic properties themselves.

**Entropy and the Partition Function**  Substituting equation (12.69) for $W$ into (12.67) and applying Stirling's approximation for the factorials of large numbers gives[8]

$$S = k(N(\ln N) - N) - k\sum_i (N_i(\ln N_i) - N_i)$$

Substituting version (12.76) of the Boltzmann distribution for $N$ and (12.74) for the partition function and simplifying, we obtain

$$S = k(N(\ln N) - N) - kN(\ln N - \ln z - 1 - \bar{\epsilon}/kT)$$

---

[8]In equation (12.69) the index $i$ goes only from 1 to 2. In general, of course, there can be more than two categories $i$.

or

$$S = N\bar{\epsilon}/T + kN \ln z$$

Letting $N$ be Avogadro's number, the entropy per mole is

$$
\begin{aligned}
S &= E/T + k \ln z^{Na} \\
  &= E/T + k \ln Z
\end{aligned}
\tag{12.80}
$$

where $Z = z^{Na}$.

This is the desired link between $S$ and $Z$ and between the molecular world of statistical mechanics and the macroscopic systems of thermodynamics. All other thermodynamic functions can be calculated if we know $Z$ (over a range of temperature and pressure), since $E$ is given by equation (12.78) and $S$ by (12.80). For example, the Helmholtz work function is $A = E - TS$ or

$$
\begin{aligned}
\Delta A &= A - A_0 \\
         &= -kT \ln Z
\end{aligned}
\tag{12.81}
$$

The Gibbs free energy would be simply $\Delta G = \Delta A + P \Delta V$, and so on.

### 12.4.4  Comment

If you are interested in natural systems, it's quite possible you may never need to know about partition functions in detail, despite the fact that they may have been the source of some of your most vital data. However, it is quite certain that you will benefit from the insight into the nature of entropy afforded by the statistical viewpoint. The relationship between entropy and structure, bonding, and disorder in general is part of every scientist's working knowledge, and it could never have come from equilibrium considerations.

As one small example, the magnitude of $\Delta S^{\ddagger}$, the entropy change between reactants and the activated state, is a valuable guide to the nature of that state. If it is close to $\Delta_r S$, the entropy of reaction, the activated state is rather similar to the product state. Actually, most values of $\Delta S^{\ddagger}$ (derived from the pre-exponential $A$) are negative, because the activated complex is more tightly bound or constrained than the reactants. This correlation between entropy and structural thinking permeates physical chemistry. You need to know at least a little statistical mechanics to fully appreciate it.

## 12.5  PROCESS MODELING

### 12.5.1  Quasistatic Reactions

In §2.6 we described the two main types of processes used in the thermodynamic model. These are the *reversible* and *irreversible* reactions. There is also a useful variant of irreversible reactions called the *quasistatic* reaction.[9] Like the reversible reaction, it is a process that takes place only in the thermodynamic model, never in the real world, but it is extremely useful in spite of this.

### Constraints and Metastable States

In our definition of the metastable state (§2.3.1) we said that it was an equilibrium state, but that other states of the same system were available that had lower energy contents under the same conditions. Normally, in this type of discussion, by "energy" we mean the Gibbs free energy, and by "the same conditions" we mean the same $T$ and $P$. But the metastable state is prevented from changing to some lower energy state by some *constraint*; the system is *constrained* from reacting or changing. This constraint is often an activation energy barrier, as in the case of diamond at low $T$ and $P$, but it can also be simply the fact that the reactants are separated, as in the sugar and coffee example (Figure 2.7), or it can be an externally applied voltage, as in the case of electrolytic cells (§9.4.1). In fact, we can actually define the metastable state as a state that has at least one extra constraint, in addition to the two constraints of the chosen $T$ and $P$. This extra constraint always holds the system in a relatively high energy state and can, in fact, be quite a wide variety of things besides the ones we have mentioned.

Sometimes we have complete control over this extra constraint, as when we control the voltage in a potentiometer attached to an electrolytic cell, and sometimes we have no control at all, as in the case of diamond→graphite at low $T$ and $P$. But whether we have control over the constraint in the real world is of no concern to our mathematical model. In the model, we can release this constraint and reapply it at will, and doing this constitutes the quasistatic reaction.

Any two equilibrium states have some value of $G$, the molar Gibbs free energy, which varies as a function of $T$ and $P$. In other words, $G$ is a function of $T$ and $P$ that can be represented by a surface in $G$-$T$-$P$ space for both stable and metastable states of a system. In Figure 12.4 we see two such surfaces, where the *Reactants* surface is a metastable equilibrium state and the *Products* surface is a lower energy stable equilibrium state of the same system. The Reactants and Products can represent any chemical reaction whatsoever. An irreversible reaction between these two states is represented by A′ → A at a fixed $T$ and $P$. At A′, which is metastable because of some unspecified constraint, we release this constraint momentarily and allow the reaction to proceed irreversibly by an amount $\Delta\xi$ (measured in moles), forming some Products. Then we reapply the constraint, and the system settles down into its new metastable state between A′ and A, with both Reactants and a little bit of Products coexisting. Then we release the constraint momentarily again, another $\Delta\xi$ of reaction occurs, and we reapply the constraint again. A continuous succession of such

---

[9]Usage of this term is variable in the thermodynamic literature.

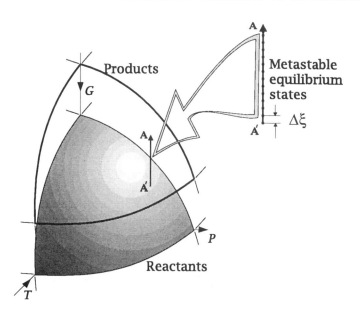

Figure 12.4: A sequence of metastable equilibrium states for the reaction $A' \to A$ at constant $T$, $P$. The progress variable is $\xi$. Compare with Figure C.6, Appendix C.

metastable states in an irreversible reaction constitutes a quasistatic reaction. Note carefully the difference between this and a reversible reaction. The reversible reaction is also a continuous succession of equilibrium states, but they are stable equilibrium states, having only the normal two constraints, such as $T$ and $P$, with no third constraint. In Figure 12.4 a reversible reaction would lie entirely on the Products surface, or on the Reactants surface (see Figure C.6).

## 12.5.2  Using the Progress Variable

Well, that's fine in the abstract, but how do we do it? And why would we want to do it? How we do it is simplicity itself and illustrates once again the difference between reality and our model of reality.

### Aragonite–Calcite Example

Let's consider one of the simplest kinds of reaction, a polymorphic change such as aragonite → calcite. Aragonite on museum shelves actually does not change to calcite at all, but we can do it mathematically with ease. From (12.7) we have

$$\frac{dn_A}{\nu_A} = \frac{dn_B}{\nu_B} = d\xi \tag{12.7}$$

where, if A is aragonite and B is calcite, then $\nu_A = -1$ and $\nu_B = 1$. Thus

$$dn_{aragonite} = -d\xi \tag{12.82}$$
$$dn_{calcite} = d\xi \tag{12.83}$$

where $n_{aragonite}$ is some number of moles of aragonite, and similarly for calcite. If $n°$ is the number of moles of each to start with, then integrating these equations from $n°$ to some new value of $n$ gives

$$\int_{n°}^{n} dn_{aragonite} = -\int d\xi$$
$$n_{aragonite} - n°_{aragonite} = -\Delta\xi \tag{12.84}$$

and similarly

$$n_{calcite} - n°_{calcite} = \Delta\xi \tag{12.85}$$

Equations (12.84) and (12.85) might be rewritten

$$n_{aragonite} = n°_{aragonite} - \Delta\xi \tag{12.86}$$
$$n_{calcite} = n°_{calcite} + \Delta\xi \tag{12.87}$$

which shows that whatever amounts of each mineral we have to start with, this amount is decreased by $\Delta\xi$ moles for aragonite and increased by $\Delta\xi$ moles for calcite, every time we allow the reaction to proceed by $\Delta\xi$. What could be simpler? If we let $n°_{calcite} = 0$ and $n°_{aragonite} = 1$, and we proceed from pure aragonite to pure calcite in four steps of $\Delta\xi = 0.25$ moles, then after one step $n_{aragonite} = 0.75$ moles, $n_{calcite} = 0.25$ moles, and so on, and the result can be diagramed as in Figure 12.5. We could, of course, use as many small steps as we like, changing aragonite into calcite quasistatically, although this never happens in nature.

### Iron Oxidation Example

But we are not restricted to such simple reactions, or to only one reaction. Let's next consider a case where we have two simultaneous reactions,

$$12\,Fe + 8\,O_2(g) = 4\,Fe_3O_4 \tag{12.88}$$
$$4\,Fe_3O_4 + O_2(g) = 6\,Fe_2O_3 \tag{12.89}$$

This is the oxidation of native iron, first to the intermediate stage of magnetite, then to the final product, hematite. In the presence of abundant oxygen at $25°C$, thermodynamics tells us that the stable equilibrium state is hematite plus oxygen; no Fe or $Fe_3O_4$ should remain. But what happens during the reaction? Does magnetite form and then change to hematite as the equations imply, or is magnetite bypassed completely? Well, what *actually* happens is a matter for experimentation—you must bring iron and oxygen together under various

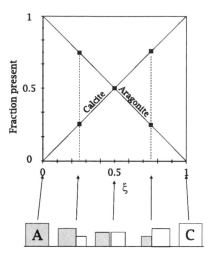

Figure 12.5: The irreversible reaction aragonite (A)→calcite (C) considered as a function of the progress variable $\xi$.

conditions and see what happens. Right now we cannot do that. But what happens in our model of iron oxidation is entirely under our control.

The way we have written the reactions in (12.88) and (12.89), every mole of magnetite that forms eventually gets transformed into hematite. That is, if you add the two reactions together, you get

$$12\,Fe + 9\,O_2(g) = 6\,Fe_2O_3 \tag{12.90}$$

and so magnetite does not appear. Another way to show this is to write the reaction progress equations, analogous to (12.82) and (12.83). This time, however, we have two reactions, and $\Delta\xi$ need not be the same for each. For the moment, we will keep them separate, as $\Delta\xi_{88}$ and $\Delta\xi_{89}$. Thus

$$\left.\begin{array}{l} n_{Fe} = n_{Fe}^\circ - 12\,\Delta\xi_{88} \\ n_{O_2} = n_{O_2}^\circ - 8\,\Delta\xi_{88} - \Delta\xi_{89} \\ n_{Fe_3O_4} = n_{Fe_3O_4}^\circ + 4\,\Delta\xi_{88} - 4\,\Delta\xi_{89} \\ n_{Fe_2O_3} = n_{Fe_2O_3}^\circ + 6\,\Delta\xi_{89} \end{array}\right\} \tag{12.91}$$

showing that if $\Delta\xi_{88} = \Delta\xi_{89}$, $n_{Fe_3O_4}$ stays constant at $n_{Fe_3O_4}^\circ$, or zero if we start with only Fe and oxygen.

But suppose (12.88) proceeds faster than does (12.89), that is, has a greater reaction rate. In (12.9) we defined the reaction rate as $d\xi/dt$, which we could call $r$, so that the rates for (12.88) and (12.89) are

$$r_{88} \quad = \quad \frac{d\xi_{88}}{dt} \tag{12.92}$$

$$r_{89} \quad = \quad \frac{d\xi_{89}}{dt} \tag{12.93}$$

and

$$\frac{r_{88}}{r_{89}} = \frac{d\xi_{88}/dt}{d\xi_{89}/dt}$$

Now if reaction (12.88) actually proceeds twice as fast as (12.99), we say $r_{88} = 2\,r_{89}$, and

$$d\xi_{88} = \frac{d\xi_{89}r_{88}}{r_{89}}$$

$$= 2\,d\xi_{89} \tag{12.94}$$

and after integration,

$$\Delta\xi_{88} = 2\,\Delta\xi_{89} \tag{12.95}$$

Substituting (12.95) into the progress reactions (12.91), and letting the resulting $\Delta\xi_{89}$ be simply $\Delta\xi$, we get

$$\left. \begin{array}{l} n_{Fe} = n_{Fe}^{\circ} - 24\,\Delta\xi \\ n_{O_2} = n_{O_2}^{\circ} - 17\,\Delta\xi \\ n_{Fe_3O_4} = n_{Fe_3O_4}^{\circ} + 4\,\Delta\xi \\ n_{Fe_2O_3} = n_{Fe_2O_3}^{\circ} + 6\,\Delta\xi \end{array} \right\} \tag{12.96}$$

So that this time magnetite does appear and continues to coexist with hematite, at least until the Fe is all used up.

This result should be fairly intuitive. If magnetite appears and disappears at the same rate, the amount present will not change. But if it forms faster than it disappears, it will accumulate, along with hematite. [We could have got the same result by multiplying all the stoichiometric coefficients in (12.88) by two.] To take a specific case, we could let $n_{Fe}^{\circ} = 480$ moles, $n_{O_2}^{\circ} = 510$ moles, $n_{Fe_3O_4}^{\circ} = 0$, and $n_{Fe_2O_3}^{\circ} = 0$. The numbers for Fe and oxygen are chosen so that oxygen is in excess of that required to oxidize all the Fe to hematite, and also so that the Fe will be used up at an even number of $\Delta\xi$ steps. This is not necessary—any starting numbers can be used. The resulting graph of moles vs. $\xi$ is shown in Figure 12.6. Note that after all the Fe is used up at $\xi = 20$ ($480 - 20 \times 24 = 0$), reaction (12.88) is no longer available, and so the progress equations (12.96) change to reflect (12.89) only. Note too that when the magnetite has finally disappeared, we have used 480 moles of Fe and $510 - 150 = 360$ moles of $O_2$ to produce 240 moles of $Fe_2O_3$, or

$$480\,Fe + 360\,O_2(g) = 240\,Fe_2O_3$$

which is consistent with (12.90). The final equilibrium state is independent of how we get there, or the *reaction path*, which depends greatly on kinetics.

### A Solution Example

The iron oxidation example is actually a special case of two simultaneous reactions, where the reactants and products do not change activities as the reaction proceeds. In the more general case of gaseous or aqueous solutions,

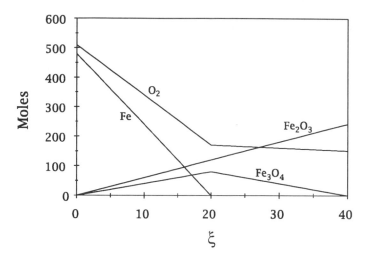

Figure 12.6: Two simultaneous iron oxidation reactions as a function of $\xi$.

two simultaneous reactions involving gases or solutes will be a bit more complex because the concentrations or activities of both reactants and products will change during the reaction, and therefore the rate of each reaction will change continuously (except in the special case of zeroth order reactions). For example,[10] consider the following simultaneous reactions,

$$A \;\rightarrow\; B \tag{12.97}$$

$$B \;\rightarrow\; C \tag{12.98}$$

where A, B, and C can be gaseous compounds or aqueous solutes, and where the rate constants are $k_{97}$ for A $\rightarrow$ B, and $k_{98}$ for B $\rightarrow$ C. The rate equations are [cf. (12.12)]

$$-\frac{dC_A}{dt} \;=\; k_{97}C_A \tag{12.99}$$

$$\frac{dC_B}{dt} \;=\; k_{97}C_A - k_{98}C_B \tag{12.100}$$

$$\frac{dC_C}{dt} \;=\; k_{98}C_B \tag{12.101}$$

If the initial concentrations are $C_A^\circ$, $C_B^\circ = 0$, and $C_C^\circ = 0$, (12.99) gives

$$C_A = C_A^\circ e^{-k_{97}t}$$

Equation (12.100) then becomes

$$\frac{dC_B}{dt} = k_{97}C_A^\circ e^{-k_{97}t} - k_{98}C_B \tag{12.102}$$

---

[10]This example after Noggle, J.H., *Physical Chemistry*, 2nd ed. (Scott, Foresman & Co., 1989), page 550.

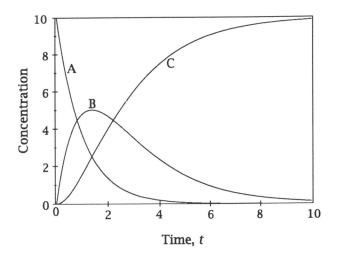

Figure 12.7: Two simultaneous gas reactions as a function of time, $t$.

With a little ingenuity, this is integrated to give

$$C_B = \frac{k_{97} C_A^\circ}{(k_{98} - k_{97})} \left(e^{-k_{97}t} - e^{-k_{98}t}\right) \qquad (12.103)$$

Then $C_C$ is obtained from the mass balance,

$$C_C = C_A^\circ - C_A - C_B \qquad (12.104)$$

The result, for $k_{97} = 1, k_{98} = 0.5$, is shown in Figure 12.7. Note the general similarity to Figure 12.6, in the sense that the intermediate compound (B, $Fe_3O_4$) accumulates, then disappears as equilibrium is approached. The amount of intermediates formed is controlled entirely by kinetics of the reactions.

## 12.5.3 The Reaction of K-feldspar With Water

Finally, having discussed some of the basic concepts, we get to an example that involves some interesting mineralogical results. We have seen that during the course of a reaction, intermediate compounds may appear and disappear, depending on the reaction kinetics. Actually, this may happen in more complex systems even if kinetics are not considered, or more exactly, if the rates of all reactions involved are the same.

The dissolution of K-feldspar in water is now a classic example of the usefulness of the reaction path or process modeling approach, using quasistatic reactions. It was first used by Helgeson and his colleagues in the 1960s when the methods were developed. A useful summary is Helgeson (1979). In this case, the problem is to unravel all the intermediate reactions that might occur

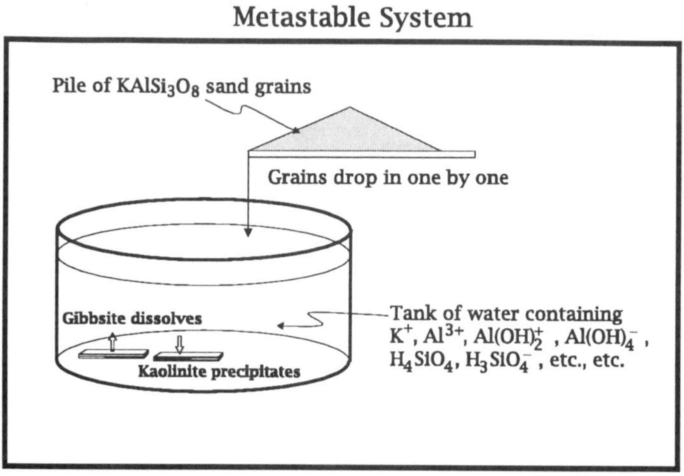

## Metastable System

Figure 12.8: Reaction path model for the reaction of K-feldspar with water. Gibbsite-dissolving, kaolinite-precipitating is represented by path B→C in Figure 12.9.

during the reaction of K-feldspar with water. We have already considered this system in Chapter 10, where we saw that several reactions were possible, involving muscovite, kaolinite, and other minerals. However, in Chapter 10 we did not explicitly consider how the solution might come to have a certain $pH$ or $a_{K^+}/a_{H^+}$ ratio. These were considered to be controlled by outside influences. This time we will start with the metastable system K-feldspar + water (metastable because the two reactants are separated) and follow the reactions that occur when the two react quasistatically. Because we will be considering very small increments of $\xi$, the process can be imagined as dropping tiny grains of K-feldspar into a large tank of water, as in Figure 12.8. After each grain is added, we wait for the water in the tank to reach equilibrium, and then we add another grain, and so on, until K-feldspar is in equilibrium with water.

### The Reaction Path

A surprising number of things happen when the reaction is considered in this way. When the first few grains drop in and dissolve completely, the dissolution must be *congruent* (i.e., the solute produced by K-feldspar dissolution has the same stoichiometry as the feldspar), but of course we must know the nature and thermodynamic properties of the solute species. According to our present knowledge of this system there are quite a few, including $K^+$, $Al^{3+}$, $Al(OH)^{2+}$, $Al(OH)_4^-$, $H_4SiO_4$, and $H_3SiO_4^-$. There are a few others that we can omit without serious error. If you perform a speciation calculation of the kind discussed in §10.2.4 on a solution having K, Al, and Si in the proportions 1:1:3

(as in K-feldspar) at very low concentration (after the first grain has dropped in), you find that the dominant species are $K^+$, $Al(OH)_4^-$, and $H_4SiO_4$. Essentially, $Al(OH)_4^-$ must dominate the Al species to maintain a charge balance with $K^+$. Therefore, the dissolution reaction of K-feldspar can be approximated by

$$KAlSi_3O_8 + 8H_2O \rightarrow K^+ + Al(OH)_4^- + 3H_4SiO_4 \qquad (12.105)$$

and the three species on the right will steadily increase in concentration as more and more grains of K-feldspar dissolve. However, this does not tell the whole story. Also increasing during dissolution of the feldspar are all the other species produced [$Al^{3+}$, $Al(OH)^{2+}$, and $H_3SiO_4^-$], though at lower concentrations. Initially, the concentrations of all these species are so small that the concentrations of $H^+$ and $OH^-$ remain constant at $10^{-7}$. The species all increase from zero, maintaining the overall 1:1:3 stoichiometry, and stop when the solution becomes saturated with K-feldspar. However, *before* that happens, the solution may become saturated with other minerals, which will precipitate as K-feldspar continues to dissolve. This results in *incongruent* dissolution, because the stoichiometry of K:Al:Si in the solution will no longer be the same as in feldspar.

To find out whether the solution has become saturated with another mineral, the solubility products of all minerals in the system considered (i.e., all minerals that contain any combination of the elements in the system) must be compared against the corresponding ion activity product (IAP) (§10.2.3) in the solution after each increment of dissolution $\Delta\xi$. This can be literally hundreds of minerals in large model systems. In the relatively simple K-feldspar case, there are only a few minerals that could possibly form. The first of these is gibbsite. The solubility product for gibbsite is

$$Al(OH)_3(s) = Al^{3+} + 3OH^-; \quad K_{sp} = a_{Al^{3+}} a_{OH^-}^3 \qquad (12.106)$$

Saturation in gibbsite occurs when its activity product exceeds the solubility product for gibbsite.

$$a_{Al^{3+}} a_{OH^-}^3 > K_{sp} \qquad (12.107)$$

When this happens, the $a_{K^+}$, $a_{H^+}$, and $a_{SiO_2(aq)}$ values of the solution result in it plotting at point A in Figure 12.9. Now if K-feldspar continues to dissolve (or, if you prefer, we continue to perform speciation calculations for solutions in which K, Al, and Si continue to increase by $\Delta\xi$ in the ratio 1:1:3), gibbsite will continue to precipitate, and the remaining solution will have compositions that follow the path A→B in Figure 12.9. During this process, the silica content of the solution continues to increase until, at point B, a silica-bearing mineral (kaolinite) becomes stable. The coexistence of gibbsite and kaolinite buffers the activity of silica (see Chapter 8, Problems 7, 14) according to

$$2Al(OH)_3(s) + 2SiO_2(aq) = Al_2Si_2O_5(OH)_4(s) + H_2O$$

Therefore, as K-feldspar continues to dissolve, aqueous $SiO_2$ does not increase, but is used to convert previously precipitated gibbsite into kaolinite. $K^+$ continues to increase, and the net result is the path B→C. This is the part of the

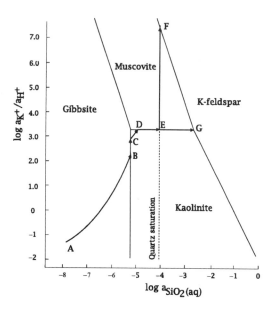

Figure 12.9: The reaction path from Figure 12.8 plotted in $\log a_{K^+}/a_{H^+}$ vs. $\log a_{SiO_2(aq)}$ space at 25°C, 1 atm.

path shown in Figure 12.8. At C, all gibbsite is used up, and the solution composition can resume increasing in silica content, following a path C→D roughly parallel to its original path (A→B), only this time precipitating kaolinite rather than gibbsite. Along C→D, K, Al, and Si are all increasing in solution until, at D, K-mica (muscovite) begins to precipitate, because its solubility product is exceeded. Again, coexistence of minerals buffers a solution parameter, this time $(a_{K^+}/a_{H^+})$, through the relation

$$KAl_3Si_3O_{10}(OH)_2 + \tfrac{3}{2}H_2O + H^+ = \tfrac{3}{2}Al_2Si_2O_5(OH)_4 + K^+$$

Because the ratio $(a_{K^+}/a_{H^+})$ is fixed, but $SiO_2$ continues to increase as K-feldspar dissolves, kaolinite reacts to form muscovite, and the solution follows path D→E, at which point the solution becomes saturated with quartz, and if equilibrium is maintained, quartz will begin to precipitate. With four components ($K_2O$, $Al_2O_3$, $SiO_2$, $H_2O$), a maximum of four phases can coexist at our arbitrarily chosen $T$ and $P$ (25°C, 1 atm) according to the phase rule. With quartz, muscovite, kaolinite, and water, this number has now been reached and cannot be exceeded (K-feldspar doesn't count; it is being used as a source of solutes, and has not yet equilibrated with the solution). Therefore, if we continue to add $K_2O$, $Al_2O_3$, and $SiO_2$ from the K-feldspar to the solution, the solution will stay at point E while kaolinite reacts with the solution to form muscovite, and quartz continues to precipitate. When kaolinite is all used up, additional dissolution of K-feldspar will drive the solution composition along E→F, with the

$SiO_2$ content of the solution buffered by the presence of quartz. At point F, K-feldspar finally becomes stable.

Alternatively, at point E, if quartz does not precipitate, the solution composition could continue from E to G, where K-feldspar would also become stable, but this time in a solution oversaturated with quartz. It would coexist metastably with muscovite and kaolinite, rather than stably with muscovite and quartz. In nature, quartz quite often does not precipitate at low temperatures, and in computer calculations that simulate equilibrium, it can be prevented from "precipitating" by removing it from the list of minerals available to the program.

### The Numerical Model

The description of the reaction path given above is not just imagined; it is the result of reaction path calculations similar in principle to the aragonite→calcite and iron oxidation examples we have already considered. However, the K-feldspar dissolution reaction is obviously a bit more complicated. For one thing, the dissolution reaction considered in the calculations is not (12.105), but the complete reaction involving all known species believed to be significant. This reaction obviously changes from time to time, as minerals appear and disappear. For the path A→B in Figure 12.9, gibbsite is considered to be in equilibrium with the aqueous solution, all dissolved aqueous species are assumed to be equilibrated, and the dissolution reaction is

$$\overline{\nu}_{KAlSi_3O_8}KAlSi_3O_8 + \overline{\nu}_{H_2O}H_2O \rightarrow \overline{\nu}_{Al(OH)_3}Al(OH)_3\,(gibbsite)$$
$$+\overline{\nu}_{K^+}K^+ + \overline{\nu}_{Al^{3+}}Al^{3+} + \overline{\nu}_{Al(OH)^{2+}}Al(OH)^{2+} + \overline{\nu}_{Al(OH)_4^-}Al(OH)_4^-$$
$$+\overline{\nu}_{H_4SiO_4}H_4SiO_4 + \overline{\nu}_{H_3SiO_4^-}H_3SiO_4^- + \overline{\nu}_{H^+}H^+ + \overline{\nu}_{OH^-}OH^- \qquad (12.108)$$

where the quantities $\overline{\nu}_i$ above are the stoichiometric coefficients of each species, but with the overbar added to indicate that they are not constants as before, but variables to be calculated. Although the stoichiometric coefficients in each elementary or individual reaction such as (12.106) are always integers, the overall dissolution reactions such as (12.108) are made up of numerous simultaneous individual reactions, and the coefficients become nonintegral. Calculating them is part of the problem.

A convenient way to start would be to let the system contain 1000 g water and assign $\overline{\nu}_{KAlSi_3O_8} = -1$. Each $\overline{\nu}_i$ then represents the rate of change in the molality of the subscripted species $i$ with a reaction step $d\xi$:

$$\frac{dm_i}{d\xi} = \overline{\nu}_i \qquad (12.109)$$

which is analogous to equation (12.8). To solve for each individual $\overline{\nu}_i$ we need 10 equations in the 10 unknowns $\overline{\nu}_i$, and these are provided by a combination of equilibrium constant expressions and mass balances for K, Al, Si, O, and H. Solving these 10 equations using matrix algebra results in the 10 values of $\overline{\nu}_i$

in (12.108), recalling that $\bar{v}_{KAlSi_3O_8} = -1$. Then using

$$m_i = m_i^\circ + \bar{v}_i \, \Delta\xi \qquad\qquad (12.110)$$

which is equivalent to (12.86) and (12.87) in our aragonite→calcite example, and (12.96) in our iron-oxidation example. New concentrations of all aqueous species are calculated after each reaction increment, as well as new quantities of any solids that are dissolving or precipitating. Inclusion of information on the relative reaction rates, often using the approximation afforded by transition state theory for reactions with unknown rates, is easily done as shown in the iron oxidation example. These calculations, plus calculation of all individual activity coefficients, is repeated over and over (iteratively) until the final equilibrium state is reached.

## 12.5.4   Comment

Clearly there is no theoretical limit to the complexity of the reactions that might be considered in this way. In addition, it is quite possible to *couple* this type of calculation with other types, such as fluid flow, heat flow, pressure changes, diffusion, permeability changes, deformation, and so on, because these other model calculations also are carried out iteratively in a large series of small steps. Thus, for example, after carrying out a $\Delta\xi$ step in a reaction path model, we could then take a small step in a heat flow model, then a small step in a fluid flow model, and then return to the reaction model, and so on. The heat flow calculation would depend on the results of the reaction path, because the reaction would be exo- or endothermic, and the fluid flow calculation would depend on any permeability changes caused by the chemical reactions, as well as on viscosity changes caused by changes in temperature. Each model result is dependent on all the others in a coupled fashion. Naturally, the demands on data, scientific insight, programming talent, and computer resources grows enormously with each additional factor to be modeled. This area is one of considerable activity at the present time.

Why would we want to perform such complex calculations, especially when some of them may be quite unrealistic? For example, we know that quasistatic reactions are unrealistic, yet they are computed everywhere these days. Well, you could easily say that the equilibrium thermodynamic model (Chapters 1–11) is unrealistic too, yet it is quite useful. Reaction path models are useful because they help us shape our ideas about what might be happening in natural processes. No one believes the results of such models in an absolute sense, but they usually reflect some or many aspects of what is really happening in the natural situation being simulated.

Besides, the game is not over. The comparison of model results with natural situations, and the improvement of the models to better simulate nature, is itself an iterative process, which will continue as long as we continue to be interested in understanding the world around us. Most scientists believe that if you can make a mathematical model of a situation, no matter how crude, you will understand that situation at a deeper level than if you just speculate about it. Model building is thus important, but it must be combined with careful observation of nature, whether in the laboratory or in the field, otherwise it will be quite inappropriate and perhaps useless.

## 12.6 SUMMARY

In this chapter we have looked very briefly at a few areas that build on a thermodynamic base and are useful in extending our understanding of how natural systems work. It is perhaps time to take an overview of what we have covered in the book as a whole and to consider what place it has in trying to understand natural systems.

We said at the beginning that our basic goal was to find out which way chemical reactions will proceed. We then went on to develop the concepts of entropy, free energy, equilibrium constants, and so on, which allow us to do this. We said too that although this goal seems too simple to be of any use in understanding volcanic eruptions, global warming, life processes, and other interesting things, we had to break these complex problems down to their more simple component parts, and the simplest parts are often simple chemical reactions that go one way or the other. We also saw that it is quite possible to calculate exact answers that are quite meaningless in the real world (e.g., Chapter 8, Problem 8).

### 12.6.1  What Is Thermodynamics, Once More

We have also emphasized in various places a point of view about thermodynamics which is fairly philosophical, which may also seem of not much use to someone interested in complex natural phenomena. That is the idea that in doing our thermodynamic calculations, we are not really calculating the properties of natural systems, but properties of simplified models of these systems. The models are mathematical, and the properties and processes in the model include some that have no counterpart in the real world. Nevertheless, model results are useful in understanding the real systems, if the models are properly or appropriately constructed. The calculations must satisfy stringent mathematical relationships, often giving them a gloss of certainty to the untrained eye, but actually they are of use only if the model they constitute is appropriate, i.e., if it has some similarity to the problems of interest.

One reason that it is a good idea to make the distinction between our models and reality is that our "explanations" of thermodynamics (and other exact sciences) are often in terms of mathematical planes, surfaces, tangents, and other even more abstract concepts. Students look at these explanations with understanding of a sort, but also with an underlying bafflement as to what these shining, unblemished, perfect, mathematical constructs have to do with anything real. It is best to come to terms with this problem by admitting that thermodynamics is in fact all about these mathematical constructs, and that is why we have emphasized from the beginning the idea that thermodynamics is used in the construction of models of real systems, not in describing real systems. It is not difficult to understand thermodynamics as the mathematics of certain planes, surfaces, tangents, etc. The more difficult problem is to understand why these mathematical constructs have such direct relevance to our universe, but that is best left to the philosophers.

But how do we assure ourselves that our models are appropriate? There is of course no way to be sure. The construction of models useful in understanding nature is the essence of science and relies as much upon creativity and imagination as any painting or musical composition. Unfortunately, the models use the language of mathematics, rather than shapes and colors or musical notes and are hence not understandable to anyone who does not know the language. This is unfortunate, because mathematics and mathematical models have their own kind of beauty, which easily rivals that of the arts. Although we don't know why the universe should be such that mathematics is so useful in describing it, there is no doubt that it is useful. It seems clear that the more mathematical tools you have in your repertoire, the more adept you will be at fitting mathematics to your observations.

## 12.6.2  Conclusion

As mentioned in §12.5.4, the models being developed to simulate complex natural phenomena are becoming very complex. Not only is it difficult to create such models, and to compile enough basic data to enable them to work, but it becomes increasingly difficult to know how well they work. There is a great deal of uncertainty about what nature is actually doing, so that it's often hard to know just how well model results coincide with natural observations. This means that in understanding natural systems, insightful field observations are just as important as model construction. And model results may turn out to be reasonably accurate, based on entirely incorrect ideas. Just because a model "works" does not mean it is right. All this means that there is plenty of scope for thermodynamics and the tools built upon it in the years to come, in our quest to understand our world and how it works.

## PROBLEMS

1. The decomposition of nitrogen pentoxide is $2 N_2O_5 \rightarrow 4 NO_2 + O_2$. Given the following data, calculate the order of reaction and the rate constant.[11] (NB: 1 Torr is 1 mm Hg pressure; 760 mm Hg = 1 atm.)

---

[11]From W.J. Moore, *Physical Chemistry*, 5 th ed. (Prentice-Hall, 1972), page 334.

| Time, $t$ | $P_{N_2O_5}$ | Time, $t$ | $P_{N_2O_5}$ |
|-----------|--------------|-----------|--------------|
| s | Torr | s | Torr |
| 0 | 348.4 | 4200 | 44 |
| 600 | 247 | 4800 | 33 |
| 1200 | 185 | 5400 | 24 |
| 1800 | 140 | 6000 | 18 |
| 2400 | 105 | 7200 | 10 |
| 3000 | 78 | 8400 | 5 |
| 3600 | 58 | 9600 | 3 |

2. Pure $N_2O_5(g)$ at 1 atm starts to dissociate to $NO_2$ and $O_2$ by the above reaction. If the pressure remains constant at 1 atm, what are the equilibrium fugacities of $N_2O_5(g)$, $NO_2(g)$, and $O_2(g)$? You must write three equations in the three unknown quantities and solve them.

3. Calculate the oxygen concentration in water in equilibrium with the atmosphere ($f_{O_2} = 0.21$ bar) under standard conditions (25°C, 1 atm).

4. Calculate the equilibrium constant for reaction (12.27), (12.28), and (12.29) using an aqueous standard state for oxygen [$O_2(aq)$]. What do you conclude about the equilibrium level of molecular oxygen ($O_2$) in groundwaters in contact with pyrite and dissolved ferrous and ferric iron?

5. What is the rate of oxidation of pyrite in water in contact with the atmosphere, if the $pH$ is 2.0? The units of the rate constant are $mol\, m^{-2}\, s^{-1}$. Compare your answer with Figure 12.1.

6. How much heat is generated by each mole of pyrite that oxidizes by (12.27)?

7. Some data from Williamson and Rimstidt (1994) on the oxidation of pyrite by ferric iron in the absence of dissolved oxygen is presented in the following table. The reaction rate has units of $mol\, m^2\, s^{-1}$.

| log (rate of reaction) | $\log m_{Fe^{2+}}$ | $\log m_{Fe^{3+}}$ | $\log m_{H^+}$ | Eh |
|------------------------|--------------------|--------------------|----------------|-----|
| | $mol\, kg^{-1}$ | $mol\, kg^{-1}$ | $mol\, kg^{-1}$ | V |
| −6.67 | −3.83 | −4.94 | −1.97 | 0.838 |
| −6.81 | −4.21 | −5.36 | −1.37 | 0.839 |
| −6.69 | −4.49 | −5.07 | −2.22 | 0.806 |

Calculate the Eh for each of the three data points and compare with that in the table. Assume that all activity coefficients are 1.0. The rate law deduced by the authors from a much larger data set is

$$\log(\text{rate of reaction}) = 10^{-8.58}\frac{m_{Fe^{3+}}^{0.30}}{m_{Fe^{2+}}^{0.47}m_{H^+}^{0.32}}$$

Calculate the rate of reaction from this equation for the three points given above.

8. By measuring the amount of silica in solution as a function of time during the precipitation of cristobalite, Renders et al. (1995) calculated the rate constant for this precipitation.

| $T°C$ | $pH$ | $\log k$ |
|-------|------|----------|
| 150 | 5.54 | −6.665 |
| 150 | 5.54 | −6.690 |
| 200 | 5.40 | −5.891 |
| 200 | 5.40 | −5.914 |
| 213 | 5.39 | −5.798 |
| 250 | 5.38 | −5.234 |
| 300 | 5.47 | −4.854 |
| 300 | 5.47 | −5.006 |

Calculate the activation energy and the value of the pre-exponential factor for this process. Other silica polymorphs have about the same activation energy of precipitation. What would you conclude from this about the transitional state during silica precipitation?

9. Repeat the calculations in Table 12.1, using $2T$ and $3T$ instead of $T$. Plot $N_i$ vs. energy level (1–10), and note how the lower levels become less populated and the upper levels more populated.

10. Repeat the iron oxidation example, assuming that (12.88) occurs three times faster than (12.89).

11. Calculate and plot the gaseous reaction example A → B → C, using $k_{97} = 1$, $k_{98} = 0.2$. Note how the intermediate species B accumulates to a greater extent.

# A

# CONSTANTS AND NUMERICAL VALUES

## The SI (Système International) Units

### Base (fundamental) Units

| Physical quantity | SI unit | Symbol |
|---|---|---|
| length | meter | m |
| mass | kilogram | kg |
| time | second | s |
| electric current | ampere | A |
| temperature | kelvin | K |
| amount of substance | mole | mol |

### Derived SI Units

| Physical quantity | SI unit | Symbol for SI unit | Unit in terms of base units | Unit in terms of other SI units |
|---|---|---|---|---|
| velocity (speed) | | | m/s | |
| acceleration | | | $m/s^2$ | N/kg |
| force | newton | N | $kg\,m/s^2$ | J/m |
| pressure | pascal | Pa | $kg/(m\,s^2)$ | $N/m^2$ |
| energy | joule | J | $kg\,m^2/s^2$ | N m |
| entropy | joule per kelvin | S | $kg\,m^2/(s^2\,K)$ | J/K |
| power | watt | W | $kg\,m^2/s^3$ | J/s |
| momentum | | | $kg\,m/s$ | |
| frequency | hertz | Hz | $s^{-1}$ | |
| electric charge | coulomb | C | A s | V F |
| voltage (emf) | volt | V | $kg\,m^2/(A\,s^3)$ | W/A; C/F |
| electric resistance | ohm | $\Omega$ | $kg\,m^2/(A^2\,s^3)$ | V/A |
| capacitance | farad | F | $A^2\,s^4/(kg\,m^2)$ | C/V |

## Fundamental Physical Constants[a]

| Quantity | Symbol in this text | Value | Units |
|---|---|---|---|
| speed of light in vacuum | $c$ | 299792458 | $m\,s^{-1}$ |
| constant of gravitation | $g$ | 6.67259 | $10^{-11}\,m^3\,kg^{-1}\,s^{-2}$ |
| elementary charge | $e$ | 1.60217733 | $10^{-19}\,C$ |
| Planck constant | $h$ | 6.6260755 | $10^{-34}\,J\,s$ |
| Avogadro constant | $N_a$ | 6.022136 | $10^{23}\,mol^{-1}$ |
| Faraday constant | $\mathscr{F}$ | 96485.309 | $C\,mol^{-1}$ |
| molar gas constant | $R$ | 8.314510 | $J\,mol^{-1}K^{-1}$ |
| Boltzmann constant, $R/N_a$ | $k_b$ | 1.380658 | $10^{-23}\,J\,K^{-1}$ |
| molar volume[b] | $V_m$ | 0.02241410 | $m^3\,mol^{-1}$ |

[a]Cohen, E.R., and Taylor, B.N., 1988, The 1986 CODATA recommended values of the fundamental physical constants: Jour. Phys. Chem. Ref. Data, v. 17, pp. 1795-1803.

[b]The volume per mole of ideal gas at 101325 Pa and 273.15 K

## Miscellaneous Useful Conversions and Older Units

| | |
|---|---|
| $\ln 10$ | 2.302585 |
| $\ln x$ | $\ln 10 \times \log_{10} x$ |
| 1 cal | 4.184 J |
| $R$ | $1.987216\ cal\,K^{-1}\,mol^{-1}$ |
| $\mathscr{F}$ | $96485.309\ J\,V^{-1}\,mol^{-1}$ |
| | $23060.542\ cal\,V^{-1}mol^{-1}$ |
| $RT/\mathscr{F}$ | $0.02569273$ V ($T = 298.15$ K) |
| $2.302585\,RT/\mathscr{F}$ | $0.0591597$ V ($T = 298.15$ K) |
| 1 bar | $10^5$ pascal |
| | 14.504 psi |
| | $0.10\ J\,cm^{-3}$ |
| | $0.0239006\ cal\,cm^{-3}$ |
| 1 atm | 1.01325 bar |
| | 101325 pascal |
| | 14.696 psi |
| 1 $cm^3$ | $0.10\ J\,bar^{-1}$ |
| | $0.0239006\ cal\,bar^{-1}$ |
| 1 Å | 1 angstrom = $10^{-8}$ cm |

# B

# STANDARD STATE THERMODYNAMIC PROPERTIES OF SELECTED MINERALS AND OTHER COMPOUNDS

## Part 1. Inorganic Substances

Data from Wagman et al., The NBS tables of chemical thermodynamic properties. Jour. Physical and Chemical Reference Data, v. 11, Supplement no. 2, 1982; with a few additions from other sources—Al species from Drever (1988); silica species and all volume data from SUPCRT92 (Johnson et al., 1992).

| Formula | Form | Mol. wt. g mol$^{-1}$ | $\Delta_f H°$ kJ mol$^{-1}$ | $\Delta_f G°$ | $S°$ J mol$^{-1}$ K$^{-1}$ | $C_p°$ | $V°$ cm$^3$ mol$^{-1}$ |
|---|---|---|---|---|---|---|---|
| **Aluminum** | | | | | | | |
| Al | s | 26.9815 | 0 | 0 | 28.33 | 24.35 | |
| Al$^{3+}$ | aq | 26.9815 | −531. | −485. | −321.7 | — | −45.3 |
| Al(OH)$^{2+}$ | aq | | −767.0 | −693.7 | — | — | |
| Al(OH)$_2^+$ | aq | | −1010.7 | −901.4 | — | — | |
| Al(OH)$_3°$ ($aq$) | aq | | −1250.4 | −1100.7 | — | — | |
| Al(OH)$_4^-$ | aq | 95.0111 | −1490.0 | −1307.0 | 102.9 | — | 45.60 |
| Al$_2$O$_3$ | $\alpha$, corundum | 101.9612 | −1675.7 | −1582.3 | 50.92 | 79.04 | 25.575 |
| Al$_2$O$_3 \cdot$ H$_2$O | boehmite | 119.9766 | −1980.7 | −1831.7 | 96.86 | 131.25 | 39.07 |
| Al$_2$O$_3 \cdot$ H$_2$O | diaspore | 119.9766 | −1998.91 | −1841.78 | 70.67 | 106.19 | 35.52 |
| Al$_2$O$_3 \cdot$ 3H$_2$O | gibbsite | 156.0074 | −2586.67 | −2310.21 | 136.90 | 183.47 | 63.912 |
| Al$_2$O$_3 \cdot$ 3H$_2$O | bayerite | 156.0074 | −2576.5 | — | — | — | — |
| Al(OH)$_3$ | amorphous | 78.0037 | −1276. | — | — | — | — |
| Al$_2$SiO$_5$ | andalusite | 162.0460 | −2590.27 | −2442.66 | 93.22 | 122.72 | 51.53 |
| Al$_2$SiO$_5$ | kyanite | 162.0460 | −2594.29 | −2443.88 | 83.81 | 121.71 | 44.09 |
| Al$_2$SiO$_5$ | sillimanite | 162.0460 | −2587.76 | −2440.99 | 96.11 | 124.52 | 49.90 |
| Al$_2$Si$_2$O$_7 \cdot$ 2H$_2$O | kaolinite | 258.1616 | −4119.6 | −3799.7 | 205.0 | 246.14 | 99.52 |
| Al$_2$Si$_2$O$_7 \cdot$ 2H$_2$O | halloysite | 258.1616 | −4101.2 | −3780.5 | 203.3 | 246.27 | 99.30 |
| Al$_2$Si$_2$O$_7 \cdot$ 2H$_2$O | dickite | 258.1616 | −4118.3 | −3795.9 | 197.1 | 239.49 | 99.30 |
| Al$_6$Si$_2$O$_{13}$ | mullite | 426.0532 | −6816.2 | −6432.7 | 255. | 326.10 | - |
| Al$_2$Si$_4$O$_{10}$(OH)$_2$ | pyrophyllite | 360.3158 | −5642.04 | −5268.14 | 239.41 | 294.34 | 126.6 |
| **Barium** | | | | | | | |
| Ba | s | 137.3400 | 0 | 0 | 62.8 | 28.07 | |
| Ba$^{2+}$ | aq | 137.3400 | −537.64 | −560.77 | 9.6 | — | −12.9 |
| BaO | s | 153.3394 | −553.5 | −525.1 | 70.42 | 47.78 | |
| BaO$_2$ | s | 169.3388 | −634.3 | — | — | 66.9 | |
| BaF$_2$ | s | 175.3368 | −1207.1 | −1156.8 | 96.36 | 71.21 | |
| BaS | s | 169.4040 | −460. | −456. | 78.2 | 49.37 | |
| BaSO$_4$ | barite | 233.4016 | −1473.2 | −1362.2 | 132.2 | 101.75 | 52.10 |
| BaCO$_3$ | witherite | 197.3494 | −1216.3 | −1137.6 | 112.1 | 85.35 | 45.81 |
| BaSiO$_3$ | s | 213.4242 | −1623.60 | −1540.21 | 109.6 | 90.00 | |
| **Calcium** | | | | | | | |
| Ca | s | 40.0800 | 0 | 0 | 41.42 | 25.31 | |
| Ca$^{2+}$ | aq | 40.0800 | −542.83 | −553.58 | −53.1 | — | −18.4 |
| CaO | s | 56.0794 | −635.09 | −604.03 | 39.75 | 42.80 | |
| Ca(OH)$_2$ | portlandite | 74.0948 | −986.09 | −898.49 | 83.39 | 87.49 | |
| CaF$_2$ | fluorite | 78.0768 | −1219.6 | −1167.3 | 68.87 | 67.03 | 24.542 |
| CaS | s | 72.1440 | −482.4 | −477.4 | 56.5 | 47.40 | |
| CaSO$_4$ | anhydrite | 136.1416 | −1434.11 | −1321.79 | 106.7 | 99.66 | 45.94 |

| Formula | Form | Mol. wt. g mol$^{-1}$ | $\Delta_f H°$ kJ mol$^{-1}$ | $\Delta_f G°$ kJ mol$^{-1}$ | $S°$ J mol$^{-1}$ K$^{-1}$ | $C_p°$ J mol$^{-1}$ K$^{-1}$ | $V°$ cm$^3$ mol$^{-1}$ |
|---|---|---|---|---|---|---|---|
| CaSO$_4$ · 2H$_2$O | gypsum | 172.1724 | −2022.63 | −1797.28 | 194.1 | 186.02 | |
| Ca$_3$(PO$_4$)$_2$ | $\beta$,whitlockite | 310.1828 | −4120.8 | −3884.7 | 236.0 | 227.82 | |
| Ca$_3$(PO$_4$)$_2$ | $\alpha$ | 310.1828 | −4109.9 | −3875.5 | 240.91 | 231.58 | |
| CaCO$_3$ | calcite | 100.0894 | −1206.92 | −1128.79 | 92.9 | 81.88 | 36.934 |
| CaCO$_3$ | aragonite | 100.0894 | −1207.13 | −1127.75 | 88.7 | 81.25 | 34.150 |
| CaSiO$_3$ | wollastonite | 116.1642 | −1634.94 | −1549.66 | 81.92 | 85.27 | 39.93 |
| CaSiO$_3$ | pseudowollastonite | 116.1642 | −1628.4 | −1544.7 | 87.36 | 86.48 | |
| CaAl$_2$SiO$_6$ | Ca-Al pyroxene | 218.1254 | −3298.2 | −3122.0 | 141.4 | 165.7 | |
| CaAl$_2$Si$_2$O$_8$ | anorthite | 278.2102 | −4227.9 | −4002.3 | 199.28 | 211.42 | 100.79 |
| CaTiO$_3$ | perovskite | 135.9782 | −1660.6 | −1575.2 | 93.64 | 97.65 | |
| CaTiSiO$_5$ | sphene | 196.0630 | −2603.3 | −2461.8 | 129.20 | 138.95 | |
| CaMg(CO$_3$)$_2$ | dolomite | 184.4108 | −2326.3 | −2163.4 | 155.18 | 157.53 | 64.365 |
| CaMgSi$_2$O$_6$ | diopside | 216.5604 | −3206.2 | −3032.0 | 142.93 | 166.52 | 66.090 |

**Carbon**

| Formula | Form | Mol. wt. g mol$^{-1}$ | $\Delta_f H°$ kJ mol$^{-1}$ | $\Delta_f G°$ kJ mol$^{-1}$ | $S°$ J mol$^{-1}$ K$^{-1}$ | $C_p°$ J mol$^{-1}$ K$^{-1}$ | $V°$ cm$^3$ mol$^{-1}$ |
|---|---|---|---|---|---|---|---|
| C | graphite | 12.0112 | 0 | 0 | 5.740 | 8.527 | 5.298 |
| C | diamond | 12.0112 | 1.895 | 2.900 | 2.377 | 6.113 | 3.417 |
| CO$_3^{2-}$ | aq | 60.0094 | −677.149 | −527.81 | −56.9 | — | −6.1 |
| HCO$_3^-$ | aq | 61.0174 | −691.99 | −586.77 | 91.2 | — | 24.2 |
| CO | g | 28.0106 | −110.525 | −137.168 | 197.674 | 29.142 | 24465.6 |
| CO$_2$ | g | 44.0100 | −393.509 | −394.359 | 213.74 | 37.11 | 24465.6 |
| CO$_2$ | aq | 44.0100 | −413.80 | −385.98 | 117.6 | — | 32.8 |
| H$_2$CO$_3$ | aq | 62.0254 | −679.339 | −623.109 | 283.65 | — | |
| CH$_4$ | g | 16.0432 | −74.81 | −50.72 | 186.264 | 35.309 | 24465.6 |
| C$_2$H$_6$ | g | 30.0704 | −84.68 | −32.82 | 229.60 | 52.63 | 24465.6 |
| CN | g | 26.0179 | 437.6 | 407.5 | 202.6 | 29.16 | |
| CN$^-$ | aq | 26.0179 | 150.6 | 172.4 | 94.1 | — | |
| HCN | g | 27.0259 | 135.1 | 124.7 | 201.78 | 35.86 | |
| HCN | aq | 27.0259 | 107.1 | 119.7 | 124.7 | — | |

**Chlorine**

| Formula | Form | Mol. wt. g mol$^{-1}$ | $\Delta_f H°$ kJ mol$^{-1}$ | $\Delta_f G°$ kJ mol$^{-1}$ | $S°$ J mol$^{-1}$ K$^{-1}$ | $C_p°$ J mol$^{-1}$ K$^{-1}$ | $V°$ cm$^3$ mol$^{-1}$ |
|---|---|---|---|---|---|---|---|
| Cl$_2$ | g | 70.9060 | 0 | 0 | 233.066 | 33.907 | 24465.6 |
| Cl$^-$ | aq | 35.4530 | −167.159 | −131.228 | 56.5 | −136.4 | 17.3 |
| HCl | aq | 36.4610 | −167.159 | −131.228 | 56.5 | −136.4 | 17.3 |
| HCl | g | 36.4610 | −92.307 | −95.299 | 186.908 | 29.12 | 24465.6 |

**Copper**

| Formula | Form | Mol. wt. g mol$^{-1}$ | $\Delta_f H°$ kJ mol$^{-1}$ | $\Delta_f G°$ kJ mol$^{-1}$ | $S°$ J mol$^{-1}$ K$^{-1}$ | $C_p°$ J mol$^{-1}$ K$^{-1}$ | $V°$ cm$^3$ mol$^{-1}$ |
|---|---|---|---|---|---|---|---|
| Cu | s | 63.5400 | 0 | 0 | 33.15 | 24.435 | |
| Cu$^+$ | aq | 63.5400 | 71.67 | 49.98 | 40.6 | — | |
| Cu$^{2+}$ | aq | 63.5400 | 64.77 | 65.49 | −99.6 | — | |
| CuO | tenorite | 79.5394 | −157.3 | −129.7 | 42.63 | 42.30 | |
| Cu$_2$O | cuprite | 143.0794 | −168.6 | −146.0 | 93.14 | 63.64 | |
| CuS | covellite | 96.6040 | −53.1 | −53.6 | 66.5 | 47.82 | |
| Cu$_2$S | chalcocite | 159.1440 | −79.5 | −86.2 | 120.9 | 76.32 | |

| Formula | Form | Mol. wt. g mol$^{-1}$ | $\Delta_f H°$ | $\Delta_f G°$ kJ mol$^{-1}$ | $S°$ | $C_p°$ J mol$^{-1}$ K$^{-1}$ | $V°$ cm$^3$ mol$^{-1}$ |
|---|---|---|---|---|---|---|---|
| **Fluorine** | | | | | | | |
| $F_2$ | g | 37.9968 | 0 | 0 | 202.78 | 31.30 | |
| HF | g | 20.0064 | −271.1 | −273.2 | 173.779 | 29.133 | |
| HF | aq | 20.0064 | −320.08 | −296.82 | 88.7 | — | |
| $F^-$ | | 18.9984 | −332.63 | −278.79 | −13.8 | −106.7 | — |
| **Hydrogen** | | | | | | | |
| $H_2$ | g | 2.0160 | 0 | 0 | 130.684 | 28.824 | 24465.6 |
| $H^+$ | aq | 1.0080 | 0 | 0 | 0 | 0 | 0 |
| $OH^-$ | aq | 17.0074 | −229.994 | −157.244 | −10.75 | −148.5 | |
| $H_2O$ | l | 18.0154 | −285.830 | −237.129 | 69.91 | 75.291 | 18.068 |
| $H_2O$ | g | 18.0154 | −241.818 | −228.572 | 188.825 | 33.577 | 24465.6 |
| **Iodine** | | | | | | | |
| $I_2$ | s | 253.8088 | 0 | 0 | 116.135 | 54.438 | |
| $I^-$ | aq | 126.9044 | −55.19 | −51.57 | 111.3 | −142.3 | |
| HI | aq | 127.9124 | −55.19 | −51.57 | 111.3 | — | |
| $IO_3^-$ | aq | 174.9026 | −221.3 | −128.0 | 118.4 | — | |
| $IO_4^-$ | aq | 190.9020 | −155.5 | −58.5 | 222.0 | — | |
| **Iron** | | | | | | | |
| Fe | s | 55.8470 | 0 | 0 | 27.28 | 25.10 | |
| $Fe^{2+}$ | aq | 55.8470 | −89.1 | −78.90 | −137.7 | — | |
| $Fe^{3+}$ | aq | 55.8470 | −48.5 | −4.7 | −315.9 | — | |
| $Fe_{0.947}O$ | wüstite | 68.8865 | −266.27 | −245.12 | 57.49 | 48.12 | |
| $Fe_2O_3$ | hematite | 159.6922 | −824.2 | −742.2 | 87.40 | 103.85 | |
| $Fe_3O_4$ | magnetite | 231.5386 | −1118.4 | −1015.4 | 146.4 | 143.43 | |
| FeO(OH) | goethite | 88.8538 | −559.0 | (−487.02) | (60.25) | — | |
| $Fe(OH)_2$ | s | 89.8618 | −569.0 | −486.5 | 88. | — | |
| $Fe(OH)_3$ | s | 106.8692 | −823.0 | −696.5 | 106.7 | — | |
| FeS | troilite | 87.9110 | −100.0 | −100.4 | 60.29 | 50.54 | |
| $FeS_2$ | pyrite | 119.9750 | −178.2 | −166.9 | 52.93 | 62.17 | |
| $FeCO_3$ | siderite | 115.8564 | −740.57 | −666.67 | 92.9 | 82.13 | |
| $Fe_2SiO_4$ | fayalite | 203.7776 | −1479.9 | −1379.0 | 145.2 | 132.88 | |
| **Lead** | | | | | | | |
| Pb | s | 207.1900 | 0 | 0 | 64.81 | 26.44 | |
| $Pb^{2+}$ | aq | 207.1900 | −1.7 | −24.43 | 10.5 | — | |
| PbO | yellow | 223.1894 | −217.32 | −187.89 | 68.70 | 45.77 | |
| PbO | red | 223.1894 | −218.99 | −188.93 | 66.5 | 45.81 | |

| Formula | Form | Mol. wt. g mol$^{-1}$ | $\Delta_f H°$ kJ mol$^{-1}$ | $\Delta_f G°$ | $S°$ J mol$^{-1}$ K$^{-1}$ | $C_p°$ | $V°$ cm$^3$ mol$^{-1}$ |
|---------|------|---------|---------|---------|---------|---------|---------|
| PbF$_2$ | s | 245.1868 | −664.0 | −617.1 | 110.5 | — | |
| PbCl$_2$ | s | 278.0960 | −359.41 | −314.10 | 136.0 | — | |
| PbS | galena | 239.2540 | 100.42 | −98.7 | 91.2 | 49.50 | |
| PbSO$_4$ | anglesite | 303.2516 | −919.94 | −813.14 | 148.57 | 103.207 | |
| PbCO$_3$ | cerussite | 267.1994 | −699.1 | −625.5 | 131.0 | 87.40 | |
| PbSiO$_3$ | s | 283.2742 | −1145.70 | −1062.10 | 109.6 | 90.04 | |
| **Magnesium** | | | | | | | |
| Mg | s | 24.3120 | 0 | 0 | 32.68 | 24.89 | |
| Mg$^{2+}$ | aq | 24.3120 | −466.85 | −454.8 | −138.1 | — | |
| MgO | periclase | 40.3114 | −601.70 | −569.43 | 26.94 | 37.15 | |
| Mg(OH)$_2$ | brucite | 58.3268 | −924.54 | −833.51 | 63.18 | 77.03 | |
| MgF$_2$ | sellaite | 62.3088 | −1123.4 | −1070.2 | 57.24 | 61.59 | |
| MgS | s | 56.3760 | −346.0 | −341.8 | 50.33 | 45.56 | |
| MgCO$_3$ | magnesite | 84.3214 | −1095.8 | −1012.1 | 65.7 | 75.52 | 28.018 |
| MgCO$_3$ · 3H$_2$O | nesquehonite | 138.3676 | — | −1726.1 | — | — | |
| MgSiO$_3$ | enstatite | 100.3962 | −1549.00 | −1462.09 | 67.74 | 81.38 | |
| Mg$_2$SiO$_4$ | forsterite | 140.7076 | −2174.0 | −2055.1 | 95.14 | 118.49 | |
| **Manganese** | | | | | | | |
| Mn | s | 54.9380 | 0 | 0 | 32.01 | 26.32 | |
| Mn$^{2+}$ | aq | 54.9380 | −220.75 | −228.1 | −73.6 | 50. | |
| MnO$_4^-$ | aq | 118.9356 | −541.4 | −447.2 | 191.2 | -82.0 | |
| MnO | manganosite | 70.9374 | −385.22 | −362.90 | 59.71 | 45.44 | |
| Mn$_3$O$_4$ | hausmannite | 228.8116 | −1387.8 | −1283.2 | 155.6 | 139.66 | |
| Mn$_2$O$_3$ | s | 157.8742 | −959.0 | −881.1 | 110.5 | 107.65 | |
| MnO$_2$ | pyrolusite | 86.9368 | −520.03 | −465.14 | 53.05 | 54.14 | |
| Mn(OH)$_2$ | amorphous | 88.9528 | −695.4 | −615.0 | 99.2 | — | |
| MnS | alabandite | 87.0020 | −214.2 | −218.4 | 78.2 | 49.96 | |
| MnCO$_3$ | rhodochrosite | 114.9474 | −894.1 | −816.7 | 85.8 | 81.50 | |
| MnSiO$_3$ | rhodonite | 131.0222 | −1320.9 | −1240.5 | 89.1 | 86.44 | |
| Mn$_2$SiO$_4$ | tephroite | 201.9596 | −1730.5 | −1632.1 | 163.2 | 129.87 | |
| **Mercury** | | | | | | | |
| Hg | l | 200.5900 | 0 | 0 | 76.02 | 27.983 | |
| Hg | g | 200.5900 | 61.317 | 31.820 | 174.96 | 20.786 | |
| Hg$^{2+}$ | aq | 200.5900 | 171.1 | 164.4 | −32.2 | — | |
| Hg$_2^{2+}$ | aq | 401.1800 | 172.4 | 153.52 | 84.5 | — | |
| HgS$_2^{2-}$ | aq | 264.7180 | — | 41.9 | — | — | |
| HgCl$_4^{2-}$ | aq | 342.4020 | −554.0 | −446.8 | 293. | — | |
| Hg$_2$Cl$_2$ | s | 472.0860 | −265.22 | −210.745 | 192.5 | — | |
| HgO | s, red | 216.5894 | −90.83 | −58.539 | 70.29 | 44.06 | |
| HgO | s, yellow | 216.5894 | −90.46 | −58.409 | 71.1 | — | |
| HgS | cinnabar | 232.6540 | −58.2 | −50.6 | 82.4 | 48.41 | |
| HgS | metacinnabar | 232.6540 | −53.6 | −47.7 | 88.3 | — | |

| Formula | Form | Mol. wt. g mol$^{-1}$ | $\Delta_f H°$ kJ mol$^{-1}$ | $\Delta_f G°$ | $S°$ J mol$^{-1}$ K$^{-1}$ | $C_p°$ | $V°$ cm$^3$ mol$^{-1}$ |
|---------|------|---------|--------|--------|------|------|------|
| **Molybdenum** | | | | | | | |
| Mo | s | 95.9400 | 0 | 0 | 28.66 | 24.06 | |
| MoO$_3$ | s | 127.9388 | −745.09 | −667.97 | 77.74 | 74.98 | |
| MoS$_2$ | molybdenite | 160.0680 | −235.1 | −225.9 | 62.59 | 63.55 | |
| **Nickel** | | | | | | | |
| Ni | s | 58.7100 | 0 | 0 | 29.87 | 26.07 | |
| Ni$^{2+}$ | aq | 58.7100 | −54.0 | −45.6 | −128.9 | — | |
| NiO | bunsenite | 74.7094 | −239.7 | −211.7 | 37.99 | 44.31 | |
| NiS | s | 90.7740 | −82.0 | −79.5 | 52.97 | 47.11 | |
| **Nitrogen** | | | | | | | |
| N$_2$ | g | 28.0134 | 0 | 0 | 191.61 | 29.125 | |
| NO | g | 30.0061 | 90.25 | 86.55 | 210.761 | 29.844 | |
| NO$_2$ | g | 46.0055 | 33.18 | 51.31 | 240.06 | 37.20 | |
| N$_2$O | g | 44.0128 | 82.05 | 104.2 | 219.85 | 38.45 | |
| N$_2$O$_4$ | l | 92.0110 | −19.50 | 97.54 | 209.2 | 142.7 | |
| N$_2$O$_4$ | g | 92.0110 | 9.16 | 97.89 | 304.29 | 77.28 | |
| N$_2$O$_5$ | s | 108.0104 | −43.1 | 113.9 | 178.2 | 143.1 | |
| N$_2$O$_5$ | g | 108.0104 | 11.3 | 115.1 | 355.7 | 84.5 | |
| NH$_3$ | g | 17.0307 | −46.11 | −16.45 | 192.45 | 35.06 | |
| NO$_3^-$ | aq | 62.0049 | −205.0 | −108.74 | 146.45 | −86.6 | |
| NH$_4^+$ | aq | 18.0837 | −132.51 | −79.31 | 113.4 | 79.9 | |
| NH$_4$OH | aq | 35.0461 | −366.12 | −263.63 | 181.21 | — | |
| **Oxygen** | | | | | | | |
| O$_2$ | g | 31.9988 | 0 | 0 | 205.138 | 29.355 | |
| O$_2$ | aq | 31.9988 | −11.7 | 16.4 | 110.9 | — | |
| OH$^-$ | aq | 17.0074 | −229.994 | −157.244 | −10.75 | −148.5 | |
| H$_2$O | l | 18.0154 | −285.830 | −237.129 | 69.91 | 75.291 | 18.068 |
| H$_2$O | g | 18.0154 | −241.818 | −228.572 | 188.825 | 33.577 | 24465.6 |
| **Potassium** | | | | | | | |
| K | s | 39.1020 | 0 | 0 | 64.18 | 29.58 | |
| K$^+$ | aq | 39.1020 | −252.38 | −283.27 | 102.5 | 21.8 | 9.0 |
| KCl | sylvite | 74.5550 | −436.747 | −409.14 | 82.59 | 51.30 | |
| KAlSi$_3$O$_8$ | sanidine | 278.3367 | −3959.7 | −3739.9 | 232.88 | 204.51 | |
| KAlSi$_3$O$_8$ | microcline | 278.3367 | −3968.1 | −3742.9 | 214.22 | 202.38 | 108.741 |
| KAlSiO$_4$ | kaliophilite | 158.1671 | −2121.3 | −2005.3 | 133.1 | 119.79 | |
| KAlSi$_2$O$_6$ | leucite | 218.2519 | −3034.2 | −2871.4 | 200.08 | 164.14 | |
| KAl$_3$Si$_3$O$_{10}$OH$_2$ | muscovite | 398.3133 | −5984.4 | −5608.4 | 306.3 | — | 14.087 |

| Formula | Form | Mol. wt. g mol$^{-1}$ | $\Delta_f H°$ kJ mol$^{-1}$ | $\Delta_f G°$ | $S°$ J mol$^{-1}$ K$^{-1}$ | $C_p°$ | $V°$ cm$^3$ mol$^{-1}$ |
|---|---|---|---|---|---|---|---|
| **Silicon** | | | | | | | |
| Si | s | 28.0860 | 0 | 0 | 18.83 | 20.00 | |
| SiO$_2$ | $\alpha$-quartz | 60.0848 | −910.94 | −856.64 | 41.84 | 44.43 | 22.688 |
| SiO$_2$ | $\alpha$-cristobalite | 60.0848 | −909.48 | −855.43 | 42.68 | 44.18 | |
| SiO$_2$ | $\alpha$-tridymite | 60.0848 | −909.06 | −855.26 | 43.5 | 44.60 | 25.740 |
| SiO$_2$ | coesite | 60.0848 | −906.31 | −851.62 | 40.376 | 43.51 | 20.641 |
| SiO$_2$ | amorphous | 60.0848 | −903.49 | −850.70 | 46.9 | 44.4 | |
| SiO$_2$ | aq | 60.0848 | −877.699 | −833.411 | 75.312 | 318.40 | 16.1 |
| H$_4$SiO$_4$ | aq | | −1449.359 | −1307.669 | 215.132 | 468.98 | |
| HSiO$_3^-$ | aq | | −1125.583 | −1013.783 | 41.84 | −137.24 | 9.5 |
| **Silver** | | | | | | | |
| Ag | s | 107.8700 | 0 | 0 | 42.55 | 25.351 | |
| Ag$^+$ | aq | 107.8700 | 105.579 | 77.107 | 72.68 | 21.8 | |
| Ag$_2$O | s | 231.7394 | −31.05 | −11.20 | 121.3 | 65.86 | |
| AgCl | cerargyrite | 143.3230 | −127.068 | −109.789 | 96.2 | 50.79 | |
| Ag$_2$S | acanthite | 247.8040 | −32.59 | −40.67 | 144.01 | 76.53 | |
| Ag$_2$S | argentite | 247.8040 | −29.41 | −39.46 | 150.6 | — | |
| **Sodium** | | | | | | | |
| Na | s | 22.9898 | 0 | 0 | 51.21 | 28.24 | |
| Na$^+$ | aq | 22.9898 | −240.12 | −261.905 | 59.0 | 46.4 | −1.2 |
| NaCl | halite | 58.4428 | −411.153 | −384.138 | 72.13 | 50.50 | 27.015 |
| Na$_2$SiO$_3$ | s | 122.0638 | −1554.90 | −1462.80 | 113.85 | — | |
| NaAlSiO$_4$ | nepheline | 142.0549 | −2092.8 | −1978.1 | 124.3 | — | 54.16 |
| NaAlSi$_3$O$_8$ | low albite | 262.2245 | −3935.1 | −3711.5 | 207.40 | 205.10 | 100.07 |
| NaAlSi$_2$O$_6$ | jadeite | 202.1397 | −3030.9 | −2852.1 | 133.5 | — | 60.40 |
| **Sulfur** | | | | | | | |
| S | orthorhombic | 32.0640 | 0 | 0 | 31.80 | 22.64 | |
| S$^{2-}$ | aq | 32.0640 | 33.1 | 85.8 | −14.6 | — | |
| HS$^-$ | aq | 33.0720 | −17.6 | 12.08 | 62.8 | — | |
| SO$_4^{2-}$ | aq | 96.0616 | −909.27 | −744.53 | 20.1 | −293. | |
| HSO$_4^-$ | aq | 32.0640 | −33.1 | −85.8 | −14.6 | — | |
| S$_2$ | g | 64.1280 | 128.37 | 79.30 | 228.18 | 32.47 | |
| H$_2$S | g | 34.0800 | −20.63 | −33.56 | 205.79 | 34.23 | |
| H$_2$S | aq | 34.0800 | −39.7 | −27.83 | 121. | — | |
| SO$_2$ | g | 64.0628 | −296.830 | −300.194 | 248.22 | 39.87 | |
| SO$_3$ | g | 80.0622 | −395.72 | −371.06 | 256.76 | 50.67 | |

| Formula | Form | Mol. wt. g mol$^{-1}$ | $\Delta_f H°$ kJ mol$^{-1}$ | $\Delta_f G°$ | $S°$ J mol$^{-1}$ K$^{-1}$ | $C_p°$ | $V°$ cm$^3$ mol$^{-1}$ |
|---|---|---|---|---|---|---|---|
| **Titanium** | | | | | | | |
| Ti | s | 47.9000 | 0 | 0 | 30.63 | 25.02 | |
| TiO | s | 63.8994 | −519.7 | −495.0 | 50.0 | 39.96 | |
| TiO$_2$ | anatase | 79.8988 | −939.7 | −884.5 | 49.92 | 55.48 | |
| TiO$_2$ | brookite | 79.8988 | −941.8 | — | — | — | |
| TiO$_2$ | rutile | 79.8988 | −944.7 | −889.5 | 50.33 | 55.02 | |
| **Uranium** | | | | | | | |
| U | s | 238.0290 | 0 | 0 | 50.21 | 27.665 | |
| UO$_2$ | uraninite | 270.0278 | −1084.9 | −1031.7 | 77.03 | 63.60 | |
| UO$_3$ | orthorhombic | 286.0272 | −1223.8 | −1145.9 | 96.11 | 81.67 | |
| U$^{3+}$ | aq | 238.0290 | −489.1 | −475.4 | 192. | — | |
| U$^{4+}$ | aq | 238.0290 | −591.2 | −531.0 | 410. | — | |
| UO$_2^{2+}$ | aq | 270.0278 | −1019.6 | −953.5 | −97.5 | — | |
| **Zinc** | | | | | | | |
| Zn | s | 65.3700 | 0 | 0 | 41.63 | 25.40 | |
| Zn$^{2+}$ | aq | 65.3700 | −155.89 | −147.06 | −112.1 | 46. | |
| ZnO | zincite | 81.3694 | −348.28 | −318.30 | 43.64 | 40.25 | |
| ZnS | wurtzite | 97.4340 | −192.63 | — | — | — | |
| ZnS | sphalerite | 97.4340 | −205.98 | −201.29 | 57.7 | 46.0 | |
| ZnCO$_3$ | smithsonite | 125.3794 | −812.78 | −731.52 | 82.4 | 79.71 | |
| Zn$_2$SiO$_4$ | willemite | 222.8236 | −1636.74 | −1523.16 | 131.4 | 123.34 | |

# Part 2. Organic Substances

**N.B.:** columns for $\Delta_f G°$ and $\Delta_f H°$ are reversed from Part 1, and $\Delta_f G°$ and $\Delta_f H°$ are in J rather than kJ.

Data from Shock and Helgeson, Geochimica et Cosmochimica Acta, v. 54, pp. 915–945, 1990.

| Formula | Form | Name | $\Delta_f G°$ $g\,mol^{-1}$ | $\Delta_f H°$ $J\,mol^{-1}$ | $S°$ $J\,mol^{-1}\,K^{-1}$ | $C_p°$ | $V°$ $cm^3\,mol^{-1}$ |
|---|---|---|---|---|---|---|---|
| **n-Alkanes** | | | | | | | |
| $CH_4$ | aq | methane | −34451 | −87906 | 87.82 | 277.4 | 37.30 |
| $CH_4$ | g | methane | −50720 | −74810 | 186.26 | 35.31 | 24465.6 |
| $C_2H_6$ | aq | ethane | −16259 | −103136 | 112.17 | 369.4 | 51.20 |
| $C_3H_8$ | aq | propane | −8213 | −127570 | 141.00 | 462.8 | 67.00 |
| $C_4H_{10}$ | aq | $n$-butane | 151 | −151586 | 167.44 | 560.2 | 82.80 |
| $C_5H_{12}$ | aq | $n$-pentane | 8912 | −173887 | 198.74 | 640.2 | 98.60 |
| $C_6H_{14}$ | aq | $n$-hexane | 18493 | −198322 | 221.33 | 733.0 | 114.40 |
| $C_7H_{16}$ | aq | $n$-heptane | 27070 | −221543 | 251.04 | 821.7 | 130.20 |
| $C_8H_{18}$ | aq | $n$-octane | 35899 | −248571 | 266.94 | 910.4 | 146.00 |
| **1-Alkenes** | | | | | | | |
| $C_2H_4$ | aq | ethylene | 81379 | 35857 | 120.08 | 261.5 | 45.50 |
| $C_3H_6$ | aq | 1-propene | 74935 | −1213 | 153.55 | 350.2 | 61.30 |
| $C_4H_8$ | aq | 1-butene | 84977 | −23577 | 181.59 | 438.9 | 77.10 |
| $C_5H_{10}$ | aq | 1-pentene | 94014 | −46861 | 209.62 | 527.6 | 92.90 |
| $C_6H_{12}$ | aq | 1-hexene | 101964 | −71233 | 237.65 | 616.3 | 108.70 |
| $C_7H_{14}$ | aq | 1-heptene | 110667 | −94851 | 265.68 | 705.0 | 124.50 |
| $C_8H_{16}$ | aq | 1-octene | 120164 | −117654 | 293.72 | 793.7 | 140.30 |
| **Alkylbenzenes** | | | | | | | |
| $C_6H_6$ | aq | benzene | 133888 | 51170 | 148.53 | 361.1 | 83.50 |
| $C_6H_5CH_3$ | aq | toluene | 126608 | 13724 | 183.68 | 430.1 | 97.71 |
| $C_6H_5C_2H_5$ | aq | ethylbenzene | 135729 | −10460 | 208.36 | 504.2 | 113.80 |

| Formula | Form | Name | $\Delta_f G°$ | $\Delta_f H°$ | $S°$ | $C_p°$ | $V°$ |
|---|---|---|---|---|---|---|---|
| | | g mol$^{-1}$ | J mol$^{-1}$ | | J mol$^{-1}$ K$^{-1}$ | | cm$^3$ mol$^{-1}$ |

**Alcohols**

| | | | | | | | |
|---|---|---|---|---|---|---|---|
| $CH_3OH$ | aq | methanol | −175937 | −246312 | 134.72 | 158.2 | 38.17 |
| $C_2H_5OH$ | aq | ethanol | −181293 | −287232 | 150.21 | 260.2 | 55.08 |
| $C_6H_5OH$ | aq | phenol | −52656 | −153302 | 191.63 | 315.1 | 86.17 |

**Ketones**

| | | | | | | | |
|---|---|---|---|---|---|---|---|
| $C_3H_6O$ | aq | acetone | −161084 | −258236 | 185.77 | 241.4 | 66.92 |

**Carboxylic Acids**

| | | | | | | | |
|---|---|---|---|---|---|---|---|
| HCOOH | aq | formic acid | −372301 | −425429 | 162.76 | 79.5 | 34.69 |
| $CH_3COOH$ | aq | acetic acid | −396476 | −485762 | 178.66 | 169.7 | 52.01 |
| $C_2H_5COOH$ | aq | propanoic acid | −390911 | −512414 | 206.69 | 234.3 | 67.90 |
| $C_3H_7COOH$ | aq | butanoic acid | −381539 | −535343 | 234.72 | 336.8 | 84.61 |
| $C_4H_9COOH$ | aq | pentanoic acid | −373288 | −559359 | 262.76 | 432.2 | 100.50 |
| $C_5H_{11}COOH$ | aq | hexanoic acid | −364343 | −582789 | 292.46 | 523.8 | 116.55 |
| $C_6H_{13}COOH$ | aq | heptanoic acid | −356268 | −607015 | 318.82 | 612.5 | 132.30 |
| $C_7H_{15}COOH$ | aq | octanoic acid | −348946 | −631993 | 346.85 | 701.2 | 148.10 |

**Carboxylate Anions**

| | | | | | | | |
|---|---|---|---|---|---|---|---|
| HCOO$^-$ | aq | formate | −350879 | −425429 | 90.79 | −92.0 | 26.16 |
| $CH_3COO^-$ | aq | acetate | −369322 | −486097 | 86.19 | 25.9 | 40.50 |
| $C_2H_5COO^-$ | aq | propanoate | −363046 | −513084 | 110.88 | 129.3 | 54.95 |
| $C_3H_7COO^-$ | aq | butanoate | −354008 | −535259 | 133.05 | 186.2 | 70.30 |
| $C_4H_9COO^-$ | aq | pentanoate | −345598 | −562371 | 160.25 | 329.7 | 86.31 |
| $C_5H_{11}COO^-$ | aq | hexanoate | −336603 | −585300 | 189.54 | 418.4 | 102.21 |
| $C_6H_{13}COO^-$ | aq | heptanoate | −327984 | −609023 | 217.57 | 469.4 | 118.60 |
| $C_7H_{15}COO^-$ | aq | octanoate | −319407 | −632746 | 242.67 | 558.1 | 134.40 |

| Formula | Form | Name | $\Delta_f G^\circ$ | $\Delta_f H^\circ$ | $S^\circ$ | $C_p^\circ$ | $V^\circ$ |
|---------|------|------|--------------------|--------------------|-----------|-------------|-----------|
|         |      |      g mol$^{-1}$ | J mol$^{-1}$ | | J mol$^{-1}$ K$^{-1}$ | | cm$^3$ mol$^{-1}$ |

**Amino Acids**

| Formula | Form | Name | $\Delta_f G^\circ$ | $\Delta_f H^\circ$ | $S^\circ$ | $C_p^\circ$ | $V^\circ$ |
|---------|------|------|--------------------|--------------------|-----------|-------------|-----------|
| $C_2H_5NO_2$ | aq | glycine | $-370778$ | $-513988$ | 158.32 | 39.3 | 43.25 |
| $C_3H_7NO_2$ | aq | alanine | $-371539$ | $-552832$ | 167.36 | 141.4 | 60.45 |
| $C_5H_{11}NO_2$ | aq | valine | $-356895$ | $-616303$ | 178.24 | 302.1 | 90.79 |
| $C_6H_{13}NO_2$ | aq | leucine | $-343088$ | $-632077$ | 215.48 | 397.9 | 107.57 |
| $C_6H_{13}NO_2$ | aq | isoleucine | $-343925$ | $-631366$ | 220.92 | 383.3 | 105.45 |
| $C_3H_7NO_3$ | aq | serine | $-510866$ | $-714627$ | 194.56 | 117.6 | 60.62 |
| $C_4H_9NO_3$ | aq | threonine | $-502080$ | $-749354$ | 222.59 | 210.0 | 76.86 |
| $C_4H_7NO_4$ | aq | aspartic acid | $-721322$ | $-947132$ | 229.28 | 127.2 | 71.79 |
| $C_5H_9NO_4$ | aq | glutamic acid | $-723832$ | $-970688$ | 294.97 | 177.0 | 89.36 |
| $C_4H_8N_2O_3$ | aq | asparagine | $-538272$ | $-780985$ | 230.96 | 125.1 | 77.18 |
| $C_5H_{10}N_2O_3$ | aq | glutamine | $-529694$ | $-804709$ | 258.99 | 187.0 | 94.36 |
| $C_9H_{11}NO_2$ | aq | phenylalanine | $-207108$ | $-460575$ | 221.33 | 384.1 | 121.92 |
| $C_{11}H_{11}N_2O_2$ | aq | tryptophan | $-112550$ | $-409195$ | 153.13 | 420.1 | 144.00 |
| $C_9H_{11}NO_3$ | aq | tyrosine | $-365263$ | $-658562$ | 190.37 | 299.2 | 123.00 |
| $C_5H_{11}NO_2S$ | aq | methionine | $-502917$ | $-743078$ | 274.89 | 292.9 | 105.30 |

**Peptides**

| Formula | Form | Name | $\Delta_f G^\circ$ | $\Delta_f H^\circ$ | $S^\circ$ | $C_p^\circ$ | $V^\circ$ |
|---------|------|------|--------------------|--------------------|-----------|-------------|-----------|
| $C_4H_8N_2O_3$ | aq | diglycine | $-489612$ | $-734878$ | 226.77 | 158.99 | 319.11 |
| $C_5H_{10}N_2O_3$ | aq | alanylglycine | $-488398$ | $-778684$ | 212.13 | 252.30 | 398.40 |
| $C_8H_{16}N_2O_3$ | aq | leucylglycine | $-462834$ | $-847929$ | 303.76 | 497.06 | 608.10 |
| $C_4H_6N_2O_2$ | aq | diketopiperazine | $-240329$ | $-415471$ | 223.84 | 71.13 | 321.04 |

# C

# MORE DETAILS ON SELECTED TOPICS

## C.1 MATHEMATICS

### C.1.1 Differentials and Derivatives

Figure C.1 illustrates the usual definition of a derivative. For any function $y = y(x)$, the *derivative* of $y$, is a function $y'(x)$ where

$$y'(x) = \lim_{\Delta x \to 0} \left( \frac{y(x + \Delta x) - y(x)}{\Delta x} \right) \tag{C.1}$$

As shown in Figure C.1, the quantity

$$\left( \frac{y(x + \Delta x) - y(x)}{\Delta x} \right)$$

is the slope of a line that intersects the function $y(x)$ at two points, $(x_1, y_1)$ and $(x_0, y_0)$, and $\Delta x = x_1 - x_0$, $\Delta y = y_1 - y_0$. As $\Delta x$ gets smaller, $x_1$ and $y_1$ approach $x_0$ and $y_0$, and in the limit as $\Delta x \to 0$, the line of intersection becomes the tangent to $y(x)$ at $(x_0, y_0)$. The notation $y'(x)$ indicates that the derivative, or the slope of the tangent, is a new function of $x$, quite distinct from the original function $y(x)$.

If we let $\Delta x = dx$ and define $dy$, the *differential of $y$*, as

$$dy = y'(x)\, dx \tag{C.2}$$

then

$$y'(x) = \frac{dy}{dx} \tag{C.3}$$

and the symbol $dy/dx$ is often used to represent the derivative. $dx$ has already been defined as equal to $\Delta x$, that is, any increment of $x$, and $dy$ is a

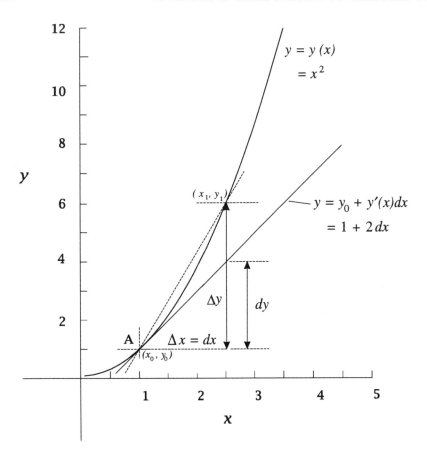

Figure C.1: The meaning of $dy/dx$.

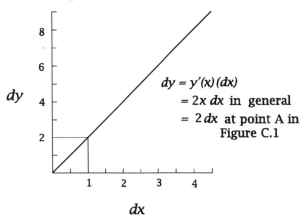

Figure C.2: $dy$ is a homogeneous linear function of $dx$.

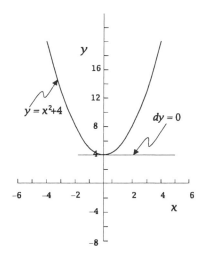

Figure C.3: The differential of $y$ is zero at the minimum in the curve.

linear function of $dx$ as shown in Figure C.2. Obviously neither $dx$ nor $dy$ is necessarily an infinitesimal quantity. It is an unfortunate fact that because $dy/dx$ is equal to the derivative, the concept of which involves allowing $dx$ to become infinitesimally small, many students of thermodynamics get the idea that differentials are infinitesimal quantities, and this is a stumbling block to the intuitive grasp of the many equations involving differentials. During integration, of course, differentials can and do take on infinitesimal values.

Of particular interest in thermodynamics is the extremum value of certain functions, that is, the maximum or minimum point. According to the calculus, this is the point where the derivative passes through zero, or $dy/dx = 0$. In Figure C.3, $dy/dx = 2x$ (or $dy = 2x\,dx$), which equals zero at $x = 0$. In differential form, we say the minimum occurs at $dy = 0$. This means that at the minimum, where the tangent is horizontal, $y$ will not change ($dy = 0$) no matter what the size of $dx$.

## C.1.2   Partial Derivatives and Total Differentials

A function having several variables may be differentiated with respect to one of the variables, keeping all the others at fixed values. Thus the function

$$z = z(x, y)$$

can be differentiated with respect to $x$, keeping $y$ constant, thus evaluating $(\partial z/\partial x)_y$, and it can also be differentiated with respect to $y$, keeping $x$ constant, evaluating $(\partial z/\partial y)_x$. These quantities are termed *partial derivatives*. The new shape of the "$d$" symbol is to remind us of the partial nature of the partial differentiation process, and the subscripts remind us which variables are being held constant. In cases where there is no likelihood of confusion, the subscripts are often omitted. For example, if

$$z = 2x^2 + 4y^3$$

then

$$\partial z / \partial x = (\partial z / \partial x)_y = 4x$$

and

$$\partial z / \partial y = (\partial z / \partial y)_x = 12y^2$$

The *total differential* of $z$, $dz$, is defined as

$$dz = \left(\frac{\partial z}{\partial x}\right)_y dx + \left(\frac{\partial z}{\partial y}\right)_x dy \tag{C.4}$$

For example, if the function $V = V(T, P)$ is

$$V = RT/P \tag{C.5}$$

where $R$ is a constant, then

$$dV = \left(\frac{\partial V}{\partial T}\right)_P dT + \left(\frac{\partial V}{\partial P}\right)_T dP \tag{C.6}$$

$$= (R/P)dT + (-RT/P^2)dP \tag{C.7}$$

Equation (C.4) has a very straightforward geometrical meaning, discussed in connection with the Fundamental Equation, below. Thermodynamics commonly deals with continuous changes in multivariable systems. For this reason, total differentials are frequently used, and it is essential to have a clear idea of their meaning.

## Application to the Fundamental Equation

A particularly important total differential is the Fundamental Equation, equation (4.4). The easiest way to get a clear geometric idea as to what this means is to first realize that by virtue of the First Law, every system at equilibrium, whether as simple as an ideal gas or as complex as a bacterium, has a single fixed energy content for given values of two independent variables, such as $S$ and $V$. Every system can therefore be represented by a surface in $U$-$S$-$V$ space, such as shown in Figure C.4. Note that, in order to make the surface easier to draw, $U$ increases downward in Figure C.4. At every point on the surface, such as point A, there will be a tangent surface, the equation for which is the Fundamental Equation, $dU = T\,dS - P\,dV$. Figure C.5 shows how, starting at point A, increments of $dS$ and $dV$ are combined with the slopes $\partial U / \partial S$ and $\partial U / \partial V$ to produce the total change in $U$ at any other point on the tangent plane. In this case, each of $dU$, $dS$, and $dV$ has any magnitude, however large.

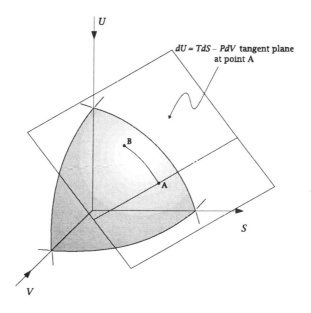

Figure C.4: Every system has a unique $U$-$S$-$V$ surface. The Fundamental Equation (4.4) represents a tangent to this surface when $dS$ and $dV$ are of arbitrary magnitude, and it can be integrated to give the change in $U$, $\Delta U$, between any two points on the surface, such as A and B. The tangent plane is illustrated further in Figure C.5.

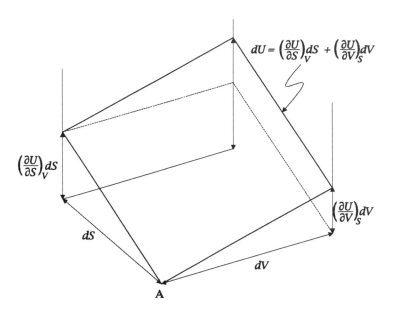

Figure C.5: The tangent surface at point A in Figure C.4, showing how $dU$ is geometrically related to $dS$ and $dV$.

On the other hand, the Fundamental Equation can also be used to calculate values of $\Delta U$ between any two points on the $U$-$S$-$V$ surface itself, such as between points A and B. Because $U$ follows some complex function of $S$ and $V$ between A and B, the Fundamental Equation must be *integrated* between A and B, and this means allowing $dS$ and $dV$ to take on infinitesimal values and performing a summation, symbolized by the $\int$ symbol. This is written

$$\Delta U = \; = \; U_B - U_A$$
$$= \int_A^B T\, dS - \int_A^B P\, dV$$

The *calculation* of this difference follows the reversible path shown on the $U$-$S$-$V$ surface—a continuous succession of equilibrium states. However, the calculated $\Delta U$ is the same no matter how the change from A to B is actually carried out.

The $U$-$S$-$V$ surface referred to above is of course the surface representing stable equilibrium of the system. There are other surfaces possible for the same system having larger values of $U$ for given values of $S$ and $V$— surfaces representing metastable states of the system. Metastable states such as aragonite always have one constraint in addition to the two necessary to define the system, in this case $S$ and $V$. The extra constraint is whatever prevents the system from sliding down the energy ($U$) gradient to the lowest possible value. Sometimes there is a measurable variable associated with this constraint, such

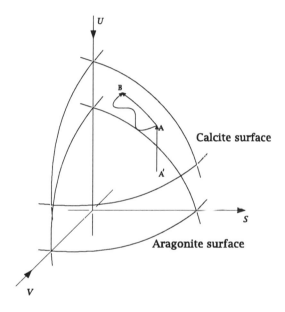

Figure C.6: If the surface in Figure C.4 represents the system calcite, there is another surface having a greater $U$ at every $S$ and $V$, representing aragonite. A′ → A represents an irreversible process. Any line on either surface, such as the two shown (A→ B), represents a reversible process.

as an applied voltage to an electrolytic cell, or a variable amount of atomic disorder in a solid, but usually there is not, as when there is an activation energy barrier. That is, we have no physical control over the activation energy, aside from adding catalysts. But whether we have such control in our real system or not, we have it in our thermodynamic model.

With the normal two constraints, we have two terms on the right side of the Fundamental Equation, derived ultimately from the fact that we chose to consider only two ways of changing the energy of our systems (heat and only one kind of work). A third constraint always constitutes a third way of changing the energy of the system, and therefore should be represented by a third term on the right side of the Fundamental Equation. In the general case, we call the third constraint $\xi$, and write

$$dU = T \, dS - P \, dV + \left( \frac{\partial U}{\partial \xi} \right) d\xi$$

Releasing this constraint allows the system to progress from the metastable state toward the stable state, so $\xi$ is called the *progress variable* (§12.2.2). Because $(\partial U / \partial \xi) \, d\xi$ is always a negative term (because $U_{stable} < U_{metastable}$), the Fundamental Equation for irreversible reactions such as $A' \to A$ becomes $dU < T \, dS - P \, dV$ (or $dG_{T,P} < 0$, because $dG_{T,P} = dU - T \, dS + P \, dV$) if this third term is not included. Changes from one stable equilibrium state to another stable equilibrium state, on the other hand, such as $A \to B$ on the calcite surface, are represented by (integrating) $dU = T \, dS - P \, dV$ (Figure C.6).

### C.1.3 Integration

Integration is the inverse of differentiation. That is, the problem is to find a function when its rate of change is known. It is performed by summing up (functions of) differentials that are chosen to be very small (infinitesimals). This can be done either in the general case, giving indefinite integrals, or between specified limits, giving definite integrals. For example,

$$\int y'(x) \, dx = y(x) + \text{ constant}$$

is the general case, since differentiation of $y(x)$ plus any constant will give $y'(x)$, the derivative of a constant being zero.

$$\int_a^b y'(x) \, dx = y(b) - y(a)$$

is the definite integral between the limits $a$ and $b$ and can be thought of as the area under the curve $y'(x)$ in the $x$-$y$ plane, between the limits $x = a$ and $x = b$. Both methods of integration have been used in applications of thermodynamics in the Earth sciences, but in this text we use the definite integral.

## C.1.4 Molar and Partial Molar Properties

It is possible to subdivide the properties used to describe a thermodynamic system (e.g., $T, P, V, U, \ldots$) into two main classes termed intensive and extensive variables. This distinction is quite important since the two classes of variables are often treated in significantly different fashion. For present purposes, *extensive properties* are defined as those that depend on the mass of the system considered, such as volume and total energy content, indeed all the "total" system properties (Z) mentioned above. On the other hand, *intensive properties* do not depend on the mass of the system, an obvious example being density. For example, the density of two grams of water is the same as that of one gram at the same $P, T$, though the volume is double. Other common intensive variables include temperature, pressure, concentration, viscosity, and all molar ($Z$) and partial molar ($\overline{Z}$, defined below) quantities. [1]

Partial molar quantities are very commonly used to describe solutions, or systems containing more than one component. Mathematically, a partial molar quantity $\overline{Z}_i$ is defined as the partial derivative

$$\overline{Z}_i = \left( \frac{\partial Z}{\partial n_i} \right)_{T,P,\hat{n}_i} \tag{C.8}$$

where $Z$ is an extensive or "total" property of a system that contains constituent $i$ and (usually) other constituents as well, and the partial derivative of Z is taken at constant $T$, $P$, and $\hat{n}_i$, where $n_i$ is the number of moles of constituent $i$, and $\hat{n}_i$ refers to all constituents *other* than the constituent $i$ being considered.

Mathematically, (C.8) is a simple enough definition, but what does it mean physically? Taking volume, V, as an example of property Z, equation (C.8) refers to the change in total volume V of a solution when one mole of component $i$ is added to a quantity of that solution sufficiently large that the concentrations of all other components ($\hat{n}_i$) do not change significantly at constant $T$ and $P$. In other words, $\overline{V}_i$ is the effective volume of one mole of component $i$ in this solution at the $T$, $P$, and concentration of interest. Notice also that while V is obviously an extensive property, $\overline{V}_i$ is intensive since by (C.8) it is defined in terms of volume per mole (and this cannot change with the size of a system). In general, all partial molar quantities such as $\overline{V}_i, \overline{G}_i, \overline{H}_i, \overline{S}_i$, and so on are intensive and derived from their extensive equivalents (V, G, H, S) by (C.8).

## C.1.5 Legendre Transforms

The Legendre Transform allows one to change a function to a different function having as independent variables the partial derivatives of the original function, without losing any information. This description in words is more difficult than the operation itself. To see its usefulness in thermodynamics, one simply needs

---

[1] The word *molar* in the term *partial molar* refers to "per mole" and has no connection with the molarity scale of concentration. An alternative name is *partial molal quantity*, which is synonymous, and has no connection with the molality scale of concentration.

to realize that fundamentally the first and second laws of thermodynamics give us a criterion of system stability in terms of entropy $(S)$, volume $(V)$, and energy $(U)$. In other words, we have some very useful relationships beginning with the function

$$U = U(S, V) \tag{C.9}$$

If you then realize that

$$T = (\partial U / \partial S)_V$$

and

$$P = -(\partial U / \partial V)_S$$

and look at the description we have just given of the Legendre Transform, you will see that it will allow us to define a new function that is just as useful as (C.9) but that uses $T$ and $P$ as independent variables instead of $S$ and $V$. The development of thermodynamics does not depend on the Legendre Transform, but it is elegant and concise. It illustrates quite beautifully the underlying unity and symmetry among the thermodynamic state functions and their independent variables.

The following geometrical approach is modified from Callen (1960). Given the function

$$y = y(x_1, x_2, x_3, \cdots, x_n)$$

we want a method whereby the derivatives

$$p_i = \partial y / \partial x_i$$

are used as independent variables in a new function containing the same information as the original. To begin, consider a function of a single independent variable

$$y = y(x)$$

Geometrically, $p = dy/dx$ is a tangent (Figure C.7a). We might be tempted to simply eliminate $x$ and find $y = y(p)$, but this would lose some information since knowing $y$ as a function of the slope does not give us $y$ as a function of $x$ (Figure C.7b).

For example, let

$$y = x^3$$

then

$$\begin{aligned} p &= dy/dx \\ &= 3x^2 \end{aligned}$$

and

$$y = (p/3)^{1.5}$$

We now have $y$ as a function of $dy/dx$, but it is not what we want. However, if we knew the *intercept* as a function of the slope, we would have the

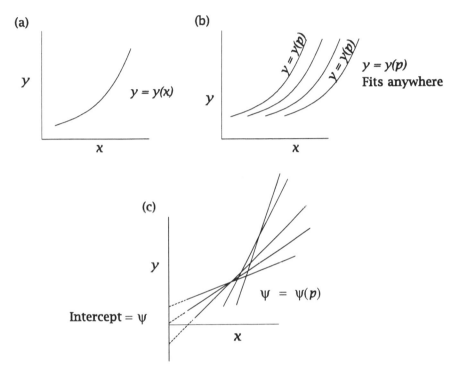

Figure C.7: Illustration of why one must know the slope $(p)$ as a function of the $y$-intercept $(\Psi)$ to have the same information as one has in the function $y = y(x)$. (a) The function $y = y(x)$. $y$ is known for any $x$. (b) $y$ is known as a function of the slope $(p = dy/dx)$ of $y = y(x)$. This does not fix the position of the curve with respect to the $x$-axis. (c) The slope $p$ as a function of the $y$-intercept of the slope $(\Psi)$. Defines an infinite set of tangents that outline the original function $y = y(x)$.

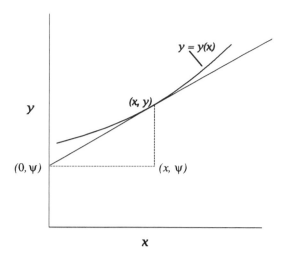

Figure C.8: The function $y = y(x)$ and the tangent to the function at an arbitrary point. $\Psi$ is the $y$-intercept of the tangent, and the slope $p = (y - \Psi)/(x - 0)$.

same information we started with, since the original curve $y = y(x)$ can be considered as being defined or outlined by an infinite number of tangents, each uniquely defined by a slope and intercept (Figure C.7c).

Thus, if $\Psi$ is the intercept, then

$$\Psi = \Psi(p)$$

is the relation we want.

Now since, as shown in Figure C.8,

$$p = \frac{y - \Psi}{x - 0}$$

then

$$\Psi = y - px$$

and, in case you didn't notice, the Legendre Transform has been found. It can be shown that in the general case

$$y = y(x_1, x_2, x_3, \cdots, x_n)$$

the Legendre Transform is

$$
\begin{aligned}
\Psi &= y - p_1 x_1 - p_2 x_2 - p_3 x_3 \cdots - p_n x_n \\
&= y - \sum_i p_i x_i
\end{aligned}
\tag{C.10}
$$

That is, to form the Legendre Transform of a function, subtract from the original function the products of each variable to be changed and the derivative of the function with respect to that variable. After that, one can proceed to tidy up by eliminating $y$ in the new function by differentiating. Thus in the case of $y = y(x)$,

$$\Psi = y - px$$

$$
\begin{aligned}
d\Psi &= dy - p\,dx - x\,dp \\
&= -x\,dp \quad (dy = p\,dx \text{ by definition})
\end{aligned}
$$

or

$$\frac{d\Psi}{dp} = -x$$

and in the general case

$$d\Psi = -\sum_i x_i dp_i$$

For example, if

$$y = x^3$$

then

$$p = 3x^2$$

and

$$
\begin{aligned}
\Psi &= y - x \cdot 3x^2 \\
&= x^3 - 3x^3 \\
&= -2x^3
\end{aligned}
$$

or since

$$x = (p/3)^{\frac{1}{2}}$$

$$\Psi = -2(p/3)^{\frac{3}{2}}$$

and

$$
\begin{aligned}
d\Psi/dp &= -(p/3)^{\frac{1}{2}} \\
&= -x
\end{aligned}
$$

Consider another example. If we have

$$x = y^3 - 3z^2$$

we know $x$ as a function of the two independent variables $y$ and $z$. If we need a function not of $y$ and $z$ but $y$ and $(\partial x/\partial z)$, we transform one variable, as above. If we need a function not of $y$ and $z$ or $y$ and $(\partial x/\partial z)$ but of $(\partial x/\partial y)$ and $(\partial x/\partial z)$, we transform both variables, and invent the new function $h$, such that

$$h = x - y(\partial x/\partial y)_z - z(\partial x/\partial z)_y$$

For thermodynamic purposes, that goes far enough, but we can demonstrate that $h$ is a function of $(\partial x/\partial y)$ and $(\partial x/\partial z)$. Thus, since

$$x = y^3 - 3z^2$$

Let

$$\begin{aligned} p_1 &= (\partial x/\partial y)_z \\ &= 3y^2 \end{aligned}$$

and

$$\begin{aligned} p_2 &= (\partial x/\partial z)_y \\ &= -6z \end{aligned}$$

then

$$\begin{aligned} h &= y^3 - 3z^2 - y(3y^2) - z(-6z) \\ &= -2y^3 + 3z^2 \\ &= -2(p_1/3)^{\frac{3}{2}} + 3(-p_2/6)^2 \\ &= (-2/\sqrt{27})p_1^{\frac{3}{2}} + p_2^2/12 \end{aligned}$$

Thus $h$ is a function of $p_1$ and $p_2$, and

$$(\partial h/\partial p_1)_{p_2} = -y$$

and

$$(\partial h/\partial p_2)_{p_1} = -z$$

## C.2  IDEAL SOLUTIONS

In Chapter 7 we mentioned that there were two kinds of ideal solutions, leading to two kinds of activity coefficients—Raoultian and Henryan. To understand the difference, you need to know Raoult's Law and Henry's Law.

### C.2.1  Henry's Law

Henry's Law in it's original form stated that the solubility of a gas in a liquid is proportional to the pressure on the gas. In Figure C.9 is shown an apparatus for controlling the pressure on a gas $i$ in contact with a liquid. As the pressure on the gas $P_i$ increases, more of it dissolves in the liquid, and so $X_i$ increases. When $X_i$ is sufficiently small, it is directly proportional to $P_i$, and the constant of proportionality is called the Henry's Law constant, $K_{H_i}$. As $X_i$ gets larger, there is inevitably some deviation from strict proportionality, as shown. Henry's Law has been generalized to refer not only to gas concentrations and pressures, but to a linear proportionality between any component activity and concentration, as discussed below.

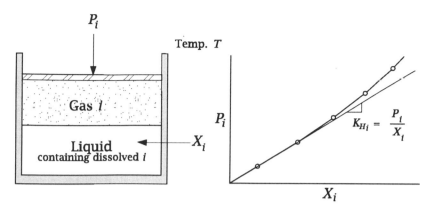

Figure C.9: Illustration of Henry's Law. As the pressure $P_i$ on the gas $i$ increases, more of it goes into solution in the liquid, increasing $X_i$.

## C.2.2 Raoult's Law

Raoult's Law originally concerned the composition of a vapor phase in equilibrium with a solution of two or more components. This sounds quite different from the Henry's Law situation, but the two are intimately related. Many combinations of components A and B (e.g., water and alcohol, or two organic liquids) were dissolved into one another in various proportions, and the composition and pressure of the coexisting vapor phase was measured (Figure C.10). The results of these measurements varied widely, but a very few systems showed a particularly simple relationship. When the two liquids A and B were very similar, the vapor pressure of their mixture was a simple function of the vapor pressures of the pure liquids,

$$P_{mixture} = X_A P_A^\circ + X_B P_B^\circ \tag{C.11}$$

and the partial pressures [equations (7.13)] of A and B in the vapor were found to be directly proportional to their concentration in the liquid (Figure C.11).

$$P_A = X_A^{liquid} P_A^\circ$$
$$P_B = X_B^{liquid} P_B^\circ$$

The only way that these simple relationships can hold is for the intermolecular forces between A–A, B–B, and A–B to be identical, so that a molecule of A behaves exactly the same way whether it is surrounded mostly by A or mostly by B. Solutions like this are called ideal solutions, and

$$P_i = X_i^{liquid} P_i^\circ$$

can be taken as a statement of Raoult's Law. In Figure C.11b we see that in an ideal solution of this type, $P/P^\circ$ for both components (i.e., $P_A/P_A^\circ$ and $P_B/P_B^\circ$) are

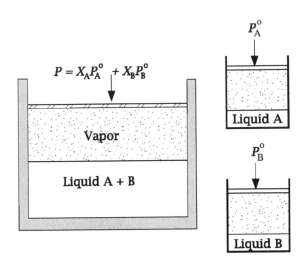

Figure C.10: The vapor pressure of a solution of A and B that obeys Raoult's Law.

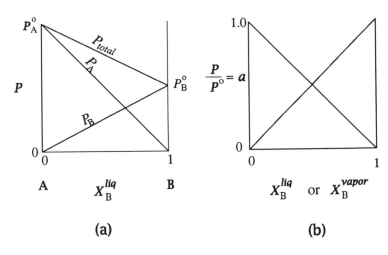

**(a)**                                              **(b)**

Figure C.11: (a) The vapor pressure ($P_{total}$) of a binary solution that obeys Raoult's Law is $P = P_A + P_B = X_A P_A^\circ + X_B P_B^\circ$. The partial pressure of each component is given by the diagonal lines, e.g., between 0 at $X_B = 0$ and $P_B^\circ$ at $X_B = 1$. (b) The partial pressure of each component divided by the vapor pressure of the pure component.

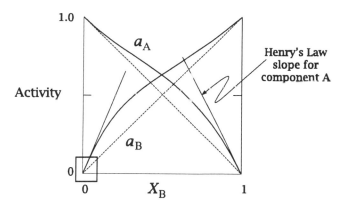

Figure C.12: Activities in a nonideal binary solution showing positive deviation from Raoultian behavior. The area of the box is expanded in Figure C.13.

represented by diagonal lines, meaning that activity equals mole fraction ($a_A = X_A$; $a_B = X_B$) in Raoultian systems. Raoult's Law has therefore been generalized to refer, not only to the partial pressures of gases, but to any solution in which component activities equal their mole fractions.

Solutions in which the solutes obey Henry's Law are also called ideal solutions, and so we must always be clear whether we are referring to Raoultian or Henryan ideality.

Figure C.11b is expanded in Figure C.12 to show not only Raoultian ideal behavior but the positive deviation from ideal behavior that a real solution might show. Negative deviation is, of course, also possible. The Raoultian activity coefficient, $y_R$, is a measure of this deviation. For example, at $X_B = 0.5$, both $a_A$ and $a_B$ are about 0.6, whereas ideal behavior would give $a = 0.5$ for both. Therefore, $y_{R_A} = y_{R_B} = 0.6/0.5 = 1.2$. In dilute solutions of A in B and of B in A, Henryan behavior is observed. That is, there is a region in the lower left and lower right corners of Figure C.12 where activity is directly proportional to concentration. This region of proportionality is indicated by the two tangents to the activity curves at $X_A = 0$ and $X_B = 0$. The area enclosed by the box in the $X_B = 0$ corner is expanded as Figure C.13.

In Figure C.13 the $x$-axis units are changed from mole fraction to molality. Shown are the lines for the activity of component B and the tangent to this line. Component B evidently obey's Henry's Law ($a \propto m$) up to about 0.4 $m$, beyond which it deviates in a negative sense. Positive deviation is also possible. The tangent indicates activities that would be observed if Henry's Law were obeyed at higher concentrations. At a concentration of 1 molal on this tangent is the ideal 1 molal standard state. Deviations from this tangent are measured by $y_{II}$. For example, the activity of B at 1 molal is 0.9, so $y_{H_B} = 0.9$. In contrast, $y_{R_B}$ would be greater than 1 at all concentrations.

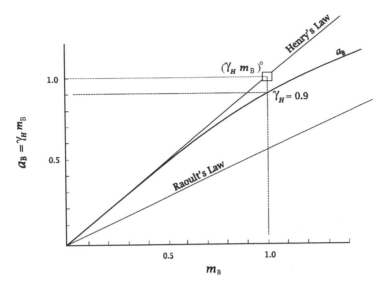

Figure C.13: Activities of component B. Henry's Law behavior is shown as the tangent to the activity curve. The ideal 1 molal standard state is at 1 molal on this tangent.

## C.3 ACTIVITY COEFFICIENTS

In Figures C.12 and C.13 you can see how the activity coefficients $y_R$ and $y_H$ can be determined graphically. All you need is some measurements of activity of some species in solution, plus a line indicating the activity that species would have if it followed Raoult's Law or Henry's Law. The activity coefficient is then a factor that accounts for the difference between the two. For a complete understanding, you need to be aware of the various methods by which activities are measured and the complications introduced for activities of ionic species, but what we have presented is sufficient for an introduction.

### C.3.1 Solute–Solute vs. Solute–Solvent Interaction

In Chapter 7 (§7.2.2) we said that Raoultian behavior in a system A–B implied that A–A, A–B, and B–B interactions (attraction, repulsion) at the molecular level were all the same. Henryan behavior, on the other hand, meant that A–B interaction remains the only significant factor, whatever the concentrations of A and B. Looking at Figures C.12 and C.13 we can see a little more clearly what this means. When $X_B$ is very small (left side of Figure C.12) every molecule of B is naturally surrounded by molecules of A, and so there is only A–A and A–B interaction, no B–B interaction. When B is added to this system, there must eventually be some B–B interaction because as $X_B$ increases, B molecules will get closer and closer together. A similar situation exists at the other side of the diagram—there is no A-A interaction when $X_A$ is very small, but it increases as $X_A$ increases. But pure Raoultian behavior is shown by the diagonal lines—there is no change in behavior of $a_A$ or $a_B$ from side to side. This could be true only if all three interactions are the same. True Raoultian behaviour, of course, does not exist, though some systems come quite close, when A and B are very similar.

Usually we refer to the component present in a small amount relative to other components a solute, and the dominant component the solvent. In system A-B, let's say that component B is the solute, and A the solvent. This means we are looking at the left side of Figure C.12. Interaction A-B is now solvent-solute behavior, and B-B is solute-solute behavior. The normal situation will be that there is A-B interaction that is quite different from A-A interaction (the normal solvent interactions, such as between $H_2O$ molecules in water). However, this A-B interaction will be uncomplicated by any B-B interaction at very low concentrations of B. Therefore, on adding B, we have only A-B interactions (plus a more or less constant or background A-A interaction) up until the concentration of B is such that B-B interactions become important. Because the interactions are constant, the line ($a_B$ vs. $m$ or $X_B$) is straight, but because A-B is different from A-A, its slope is different from the Raoult's Law slope. The slope is controlled by the nature of the A-B interactions. Eventually, as $m_B$ increases, B-B becomes significant, and the activity must deviate from a straight line. However, everywhere on this straight line, the "Henry's Law slope," the only interactions are solute-solvent and solvent-solvent, which is often referred to as "infinite dilution" behavior, as theoretically there must always be some solute-solute interaction if there is more than one molecule of solute. Extrapolating this line to a concentration of 1 molal ($m_B = 1$) gives the "ideal 1 molal standard state" (Figure C.13), which is a state (hypothetical, but useful) in which the solute has a concentration of 1 molal, but at the same time shows no solute-solute interactions—it behaves as if at "infinite dilution."

## C.3.2 Calculated Activity Coefficients

Activity coefficients ($\gamma_H$) for ions can be calculated for relatively low concentrations by variations of the Debye-Hückel equation. These all make use of the ionic strength $I$, defined as

$$I = \frac{1}{2} \sum_i m_i z_i^2 \tag{C.12}$$

where $m_i$ is the molality of ionic species $i$, and $z_i$ is its charge. For very low concentrations,

$$\log \gamma_{H_i} = -A z_i^2 \sqrt{I} \tag{C.13}$$

where $A$ is a constant (0.5092 at 25°C, 0.5998 at 100°C). This is called the Debye-Hückel Limiting Law, because it works in the limit of essentially zero concentration of $m_i$ (< 0.01 $m$ for univalent ions). It gives the Henry's Law slope. At higher concentrations, up to about $0.1\ m$, the Debye–Hückel expression is

$$\log y_{H_i} = \frac{-Az_i^2\sqrt{I}}{1 + B\mathring{a}\sqrt{I}} \tag{C.14}$$

where $B$ is another constant (0.3283 at $25°C$, 0.3422 at $100°C$), as long as $\mathring{a}$ is measured in angstroms (1 angstrom $= 10^{-8}$ cm); $\mathring{a}$ is theoretically a distance of closest approach between ions of opposite charge, but is in practice an adjustable parameter. Values of $\mathring{a}$ for various ions can be found in tables in books on physical chemistry.

At concentrations beyond about $0.1\ m$, a variety of methods have been used to calculate $y_H$. Most of them add a term or terms to the Debye–Hückel expression. A common one is

$$\log y_{H_i} = \frac{-Az_i^2\sqrt{I}}{1 + \sqrt{I}} + 0.2Az_i^2 I \tag{C.15}$$

Here the $B\mathring{a}$ term in the denominator has been changed to 1.0, partly because that is its approximate value and partly because it is theoretically undesirable to have a solute-specific term ($\mathring{a}$) in a general equation for $y_H$.

### Activity Coefficient Units

As mentioned in §7.4.2, neither activities nor activity coefficients have units—they are dimensionless. You may see in other books statements to the effect that activities are dimensionless, but that activity coefficients have units of inverse concentration. This is because they define activity as $a_i = m_i y_{H_i}$, rather than our more complete definition, $a_i = (m_i y_{H_i})/(m_i y_{H_i})°$. Believing activity coefficients to have units does no real harm because it does not affect any calculations, but it is strictly speaking incorrect.

## C.4  ACTIVITIES AND STANDARD STATES

### C.4.1  Derivation of $\mu_i - G_i° = RT \ln X_i$

In Chapter 7 we said that the relationship between the concentration of a component $i$ and its free energy is

$$\mu_i - G_i° = RT \ln X_i \tag{7.7}$$

and that a more general form of this equation is

$$\mu_i - G_i° = RT \ln a_i \tag{7.20}$$

where $a_i$ is the activity of component $i$. We also gave some rules for determining $a_i$ depending on whether $i$ is in a gaseous, liquid, solid, or dissolved form. In this section we must show why it is that an expression of the form $RT\ln(some\ function\ of\ concentration)$ gives us a $\Delta G$, and we must discuss the various forms of the activity and its corresponding standard states. The idea is not to be mathematically rigorous, but to show where the equations come from, and that they make sense.

We start with the effect of $P$ on $G$,

$$\partial G/\partial P = V \tag{4.10}$$

Integrating from $P_1$ to $P_2$, we have

$$G_{P_2} - G_{P_1} = \int_{P_1}^{P_2} V\,dP \tag{C.16}$$

or, for an ideal gas in which $V = RT/P$,

$$
\begin{aligned}
G_{P_2} - G_{P_1} &= \int_{P_1}^{P_2} (RT/P)\,dP \\
&= \int_{P_1}^{P_2} RT\,d\ln P \\
&= RT\ln(P_2/P_1)
\end{aligned}
\tag{C.17}
$$

If $P_1$ is 1 bar and this is designated a standard or reference state denoted by a superscript $^\circ$, then $P_2$ becomes simply $P$, and

$$
\begin{aligned}
G - G^\circ &= RT\ln(P/P^\circ) \tag{C.18} \\
&= RT\ln P \text{ since } P^\circ = 1 \tag{C.19}
\end{aligned}
$$

Thus for ideal gases $RT\ln P$ all by itself gives the value of $\int_{P=1}^{P} dG$, or in other words of $\int_{P=1}^{P} V\,dP$. Unfortunately, this doesn't work for real gases, although it's not a bad approximation at low pressures and high temperatures where real gases approach ideal behavior. However, the *form* of the relationship

$$\int_{P=1}^{P} V\,dP = \int_{P=1}^{P} dG = \int_{P=1}^{P} RT\,d\ln P$$

is sufficient to suggest that we could define a function such that the relationship *would* hold true for real gases. This function is the fugacity, $f$, where

$$V\,dP = dG = RT\,d\ln f \tag{C.20}$$

and

$$
\begin{aligned}
\int_{P_1}^{P_2} V\,dP &= \int_{P_1}^{P_2} dG \\
G_{P_2} - G_{P_1} &= RT\ln(f_{P_2}/f_{P_1})
\end{aligned}
\tag{C.21}
$$

To complete the definition, we stipulate that $f$ becomes equal to $P$ at low pressures,

$$\lim_{P \to 0}(f/P) = 1 \tag{C.22}$$

and the ratio $f/P$ is called the fugacity coefficient, $\gamma_f$.

Fugacity is thus a quantity that we use in place of pressure for gases when we want to calculate a free energy difference. Think of it as a pressure or as a partial pressure.

Equation (C.21) has a meaning more general than just giving the difference in $G$ on raising the pressure of a gas from $P_1$ to $P_2$. Changing the pressure from $P_1$ to $P_2$ is just one way of affecting the free energy of a component. Actually, the equation can be applied to *any* two states at the same $T$. If the first state is a reference state, denoted by superscript $°$, and the second state is unsuperscripted, we have

$$\mu_i - G_i^° = RT \ln(f_i/f_i^°) \tag{C.23}$$
$$= RT \ln(P_i \gamma_{f_i})/(P_i \gamma_{f_i})^° \tag{C.24}$$

where we change $G_i$ to $\mu_i$ because of the generalization from one pure component to one component in any situation, including solutions. We have added the subscript $i$ to emphasize that we are talking about a specific component, not some mixture of components (although that is not disallowed, it's not very useful).

One definition of activity is

$$a_i = f_i/f_i^° \tag{C.25}$$

and so we arrive at equation (7.20),

$$\mu_i - G_i^° = RT \ln a_i \tag{7.20}$$

Also, $f_i/f_i^°$ is of course equal to $P_i/P_i^°$ for an ideal gas, so if $P_i^°$ refers to pure gaseous $i$, it is $P_{total}$, and

$$\begin{aligned}
\frac{f_i}{f_1^°} &= \frac{P_i}{P_i^°} \\
&= \frac{P_i}{P_{total}} \\
&= \frac{X_i \times P_{total}}{P_{total}} \quad \text{by (7.13)} \tag{C.26} \\
&= X_i \tag{C.27}
\end{aligned}$$

So, for ideal gaseous solutions, and by extension, for any ideal solution,

$$\mu_i - G_i^° = RT \ln X_i \tag{7.7}$$

## C.4.2   A Different Route to $RT \ln X_i$

This derivation of the basic relationship between Gibbs energy and concentration (7.7) is straightforward but rather lengthy, and it's easy to lose the overall perspective: to see only trees and not the forest.

A shortcut through the forest is to simply ask what is the value of $(\partial \mu_i / \partial n_i)_{\hat{n}_i}$? That is, how does $\mu_i$ (the $G$ of $i$ in solution) vary with the amount of $i$ in solution in the ideal case, other things being constant? $n_i$ is the number of moles of $i$, and $\hat{n}_i$ means all components other than $i$ are held constant.

Well, some mathematical intuition tells us that it would be useful to expand $(\partial \mu_i / \partial n_i)_{\hat{n}_i}$, so

$$\left( \frac{\partial \mu_i}{\partial n_i} \right)_{\hat{n}_i} = \frac{\partial \mu_i}{\partial P_i} \frac{\partial P_i}{\partial n_i}$$

where $(\partial \mu_i / \partial P_i) = \overline{V}_i$, and $(\partial P_i / \partial n_i) = P_i / n_i$ (Henry's Law, Figure C.9[2]). In the ideal case, $\overline{V}_i = V_i = RT / P_i$, so, combining all this we get

$$\left( \frac{\partial \mu_i}{\partial n_i} \right)_{\hat{n}_i} = \frac{RT}{n_i} \tag{C.28}$$

for ideal solutions. Integrating this equation between two values of $n_i$, $n_i'$ and $n_i''$, we get

$$\mu_i'' - \mu_i' \; = \; RT \ln \frac{n_i''}{n_i'}$$

$$= \; RT \ln \frac{P_i''}{P_i'} \quad \text{by Henry's Law}$$

$$\mu'' - \mu^\circ \; = \; RT \ln \frac{P_i''}{P_i^\circ} \quad \text{if state ' is pure } i$$

$$\mu - \mu^\circ \; = \; RT \ln X_i \quad \text{(state '' no longer needs superscripts)}$$

or

$$\mu_i - G_i^\circ = RT \ln X_i \tag{7.7}$$

So $(\partial \mu_i / \partial T) = -\overline{S_i}$, $(\partial \mu_i / \partial P) = \overline{V}_i$, and $(\partial \mu / \partial n_i) = RT / n_i$ are therefore the three relationships we need to be able to evaluate (by integration) to know the Gibbs energy of any substance as a function of $T$, $P$, and composition. A complete understanding of these relationships is therefore quite fundamental.

## C.4.3   Standard States

Standard states are a kind of carefully defined reference state, made necessary, ultimately, by the fact that we have no absolute values for $U$, and hence none

---

[2]Strictly speaking, Figure C.9 illustrates that $P_i$ is proportional to $X_i$ at low concentrations, not to $n_i$. In very dilute solutions, where $n_i$ is small, $\hat{n}_i$ is virtually constant, and $P_i$ is proportional to both $n_i$ and $X_i$.

for $G$ or $H$. This means that for any substance $i$ we can only use $G_i$ and $\mu_i$ in difference terms, such as $\Delta_f G_i^\circ$, or equation (7.7). In $\Delta_f G_i^\circ$, the difference is between the $G$ of $i$ in its standard state at some $T$ and $P$, and the sum of the $G$'s of its constituent elements.[3] In $\mu_i - G_i^\circ = RT \ln a_i$ the difference is between $\mu$ of $i$ in some real state of interest, and $\mu$ (or $G$) of $i$ in some standard state at the same $T$ but not necessarily the same $P$. Evidently we must choose this standard state, both in order to tabulate values of $\Delta_f G^\circ$ and $\Delta_f H^\circ$ and to calculate activities.

Students usually suppose, before learning better, that the standard state for a compound X is simply pure X at 25°C, 1 bar. Unfortunately, the subject is considerably more complex in two ways. First, the $T$ and $P$ of the standard state is not restricted to 25°C and 1 bar, but can have any values; and second, the state must usually be specified more accurately than just "pure X." We have already mentioned that when X is a gas, the standard state is not just the pure gas at $T$ and $P$, but the pure *ideal* gas at $T$ and $P = 1$ bar, and when X is a solute, the standard state is a solution of 1 molal X acting ideally (obeying Henry's Law). It may seem natural to have different standard states for gases and solutes than for solids (after all, you cannot have a "pure solute" because the solvent is always there too), but why insist on ideal conditions, which mean that the standard states are quite hypothetical? Its really quite simple. Let's consider solids and liquids first, then gases, then solutes.

### Solids and Liquids

For solids and liquids, both of which act as solvents, mole fraction is a convenient concentration unit and (7.7) is a convenient relation between free energy and concentration:

$$\mu_i - G_i^\circ = RT \ln X_i \tag{7.7}$$

The standard state for $i$ is the state for which $G_i^\circ$ is the Gibbs energy, so when $i$ is in this state, $\mu_i = G_i^\circ$, and $X_i = a_i = 1$. In other words, the standard state is pure $i$ (the pure solid or liquid) at some chosen $T$ and $P$. No problem.

### Gases

For gases we could use mole fraction, but it has proved convenient to use fugacities, and (C.23) is the relation between fugacity and the Gibbs function:

$$\mu_i - G_i^\circ = RT \ln(f_i / f_i^\circ) \tag{C.23}$$

Here, the $G_i^\circ$ term in $\mu_i - G_i^\circ$ refers to the same state as the denominator in the activity term ($f_i^\circ$),

$$\underbrace{\mu_i - G_i^\circ = RT \ln \frac{\overbrace{f_i}}{f_i^\circ}}$$

---

[3] The elements may be at $T$, 1 bar, or they may be at 25°C, 1 bar, giving two varieties of $\Delta_f G^\circ$, both of which are in common use, but which are mutually incompatible. See Anderson and Crerar (1993), Chapter 7, for details.

and because it is convenient to have the denominator equal to 1.0 (1 bar) for any state of interest, the standard state must then be the state in which gaseous $i$ has a fugacity of 1.0 at any $T$. There is only one such state, hypothetical ideal gaseous $i$ at $T$ and $P = 1$ bar, and so this becomes the common standard state for gases.

**Aqueous Solutes**

For solutes, we start with (C.28),

$$\left(\frac{\partial \mu_i}{\partial n_i}\right)_{\hat{n}_t} = \frac{RT}{n_i} \tag{C.28}$$

Let's consider a binary solution having $n_1$ moles of solvent (water) and $n_2$ moles of solute. To get a kilogram of solvent, we let $n_1 = 1000/18.015 = 55.51$, and then $n_2$ becomes $m_2$, the molality of the solute. Now,

$$\left(\frac{\partial \mu_2}{\partial m_2}\right) = \frac{RT}{m_2} \tag{C.29}$$

and, dropping the subscript 2 and integrating between two molalities $m'$ and $m''$, we have

$$\mu'' - \mu' = RT \ln(m''/m')$$

for ideal solutions. For real solutions we introduce the Henryan activity coefficients, and

$$\mu'' - \mu' = RT \ln(y_H m)''/(y_H m)'$$

If we now designate the denominator as a standard state, in complete analogy with (C.23), we get

$$\mu - \mu^\circ = RT \ln(y_H m)/(y_H m)^\circ \tag{C.30}$$

Again, it is convenient to define the standard state such that the denominator is 1.0 at all $T$ and $P$, and so we insist not only that $(y_H m)^\circ = 1$, but that $y_H = 1$ and $m = 1$ simultaneously. The only state for which this is true at all $T$ and $P$ is of course the ideal (Henryan) 1 molal solution, and that is our chosen standard state for solutes.

So we have three variations of the activity,

$$a_i = \frac{f_i}{f_i^\circ}$$

$$a_i = \frac{(m_i y_{H_i})}{(m_i y_{H_i})^\circ}$$

$$a_i = X_i$$

each of which will give $\mu_i - G_i^\circ$. However, each uses a different standard state. What we are saying is that we can calculate a free energy term from concentrations, all right, but every time we change our method of measurement of concentration, we use a different reference state for free energy, because the $G_i^\circ$ term in $\mu_i - G_i^\circ$ refers to the same state as the denominator in the activity term $[f_i^\circ$ or $(m_i \gamma_{H_i})^\circ]$, or to $X_i = 1$.

Is this OK? Can we have so many different standard states for activities? How does it work in practice? And besides, at least two of these standard states are hypothetical—there is no such thing as an ideal gas or an ideal 1 molal solution. How can this be useful? You can use thermodynamics without really understanding these things—you just follow the rules we adopted. But let's take a look behind the scenes.

First, let's get rid of the seemingly sensible idea that a standard state should be a physically real state. The purpose of the standard state is (1) so that we can easily calculate activities, and (2) so that we can measure and tabulate the properties ($\Delta_f G^\circ$, $\Delta_f H^\circ$, etc.) of this state. We never have to actually have a system in this state, as long as we can do these two things. Calculating activities is simple using hypothetical states, because $f_i^\circ = 1$, $m_i^\circ = 1$, and $\gamma_{H_i}^\circ = 1$ at all $T$ for any $i$ by definition. And measuring and tabulating properties for these states is not difficult either. Basically it is done by measuring properties in real states, and extrapolating to ideal conditions. Things could be done differently, but there is no reason, theoretical or practical, for not using hypothetical standard states.

As for having a different standard state for each kind of concentration unit, this is simply the most convenient thing to do and creates no difficulties, because ultimately what we really want is the difference between $G_i$ or $H_i$ in the system and the combined $G$ or $H$ of the elements in $i$. The elements, of course, all have the same standard state—the most stable form of the element at 1 bar and either $T$ or 25°C (see footnote 3). In other words, it doesn't matter at all what standard states we choose, and so we might as well choose convenient ones.

### An Example of Changing Standard States

For example, let's say we have a real system consisting of a gas in contact with a liquid. Methane in the gas has a measured partial pressure of 0.01 bar, which we take to be its fugacity, and we want to calculate how much methane is dissolved in the water. In the gas phase, $f_{CH_4} = 0.01$ bar, and using a standard state of ideal methane gas at 1 bar ($f_{CH_4}^\circ = 1.0$ bar), we have an activity of methane of 0.01 ($a_{CH_4} = 0.01$). The relevant reaction is

$$CH_4(g) = CH_4(aq)$$

and getting numbers from the tables we have,

$$
\begin{aligned}
\Delta_r G^\circ &= \Delta_f G^\circ_{CH_4(aq)} - \Delta_f G^\circ_{CH_4(g)} \\
&= -34451 - (-50720) \\
&= 16269\,J
\end{aligned}
$$

This means that, *because the elements all cancel out* (§5.4),

$$G^\circ_{CH_4(aq)} - G^\circ_{CH_4(g)} = 16269\,J$$

where $G^\circ_{CH_4(aq)}$ is the absolute $G$ of methane in an ideal 1 molal solution, and $G^\circ_{CH_4(g)}$ is the absolute $G$ of methane as an ideal gas at 1 bar. Then

$$
\begin{aligned}
\log K &= -\Delta_r G^\circ / (2.30259\,RT) \\
&= -16269 / (2.30259 \times 8.31451 \times 298.15) \\
&= -2.850
\end{aligned}
$$

where

$$K = \frac{a_{CH_4(aq)}}{a_{CH_4(g)}}$$

Therefore if $a_{CH_4(g)} = 0.01$, then

$$10^{-2.850} = \frac{a_{CH_4(aq)}}{0.01}$$

and

$$
\begin{aligned}
a_{CH_4(aq)} &= 0.01 \times 10^{-2.850} \\
&= 10^{-4.850}
\end{aligned}
$$

So the calculated activity of dissolved methane is $10^{-4.85}$. With a standard state of ideal 1 molal methane, this means $m_{CH_4}\gamma_{CH_4} = 10^{-4.85}$, and on the reasonable assumption that $\gamma_{CH_4} = 1.0$, then $m_{CH_4} = 10^{-4.85}$. So in spite of the fact that two different standard states are used for the same component in the same reaction, we arrive at a useful answer. This is because the standard states used do not influence the final result in any way. They are simply useful fictions to hang tabulated data on.

To illustrate this, we can choose a completely different standard state for the gaseous methane and see what happens. Our new standard state is ideal gaseous methane at 978.4852 bar.[4] First, we need $\Delta_f G^\circ$ of methane in this new state. Because the gas is ideal, this is easy. From equation (C.17), and because for an ideal gas, $f = P$,

$$
\begin{aligned}
G_{978.4852\ bars} - G_{1\ bar} &= RT\ln(978.4852/1) \\
&= 17070.2\,J
\end{aligned}
$$

This is the change in $G$ as ideal methane is squeezed from 1 bar to 978.4852 bar. The difference in $G$ between ideal methane at 1 bar and the elements C and $2\,H_2$ at 1 bar is $-50720\,J$ (Figure C.14). Therefore, the difference in $G$ between methane ideal gas at 978.4852 bar and the elements at 1 bar is

---

[4]Standard states are always chosen with great care. For example, 978.4852 is my phone number.

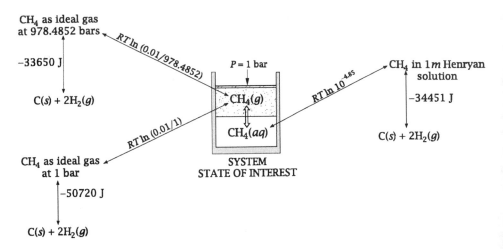

Figure C.14: The pressure on the gas phase is 1 bar, but the partial pressure of $CH_4$ in the gas is 0.01 bar. The free energy ($\mu$) of $CH_4$ is the same in each phase, and a variety of standard state data may be used. Standard states are convenient places to hold data, but do not affect calculations of model equilibria.

$-50720 + 17070 = -33650$ J. We can call this $\Delta_f G^\star$, our new standard state free energy. Our calculation is now

$$
\begin{aligned}
\Delta_r G^{\circ\star} &= \Delta_f G^\circ_{CH_4(aq)} - \Delta_f G^\star_{CH_4(g)} \\
&= -34451 - (-33650) \\
&= -801 \, J
\end{aligned}
$$

$$
G^\circ_{CH_4(aq)} - G^\star_{CH_4(g)} = -801 \, J
$$

Then

$$
\begin{aligned}
\log K &= -\Delta_r G^{\circ\star} / (2.30259 \, RT) \\
&= 801 / (2.30259 \times 8.31451 \times 298.15) \\
&= 0.140
\end{aligned}
$$

or

$$
\begin{aligned}
K &= \frac{a_{CH_4(aq)}}{a_{CH_4(g)}} \\
&= 10^{0.140} \\
&= 1.38
\end{aligned}
$$

Now $a_{CH_4(g)}$ is no longer 0.01. Because of the change of standard state,

$$
\begin{aligned}
a_{CH_4(g)} &= \frac{f_{CH_4(g)}}{f^\star_{CH_4(g)}} \\
&= \frac{0.01}{978.4852} \\
&= 10^{-4.99}
\end{aligned}
$$

and so, solving for $a_{CH_4(aq)}$ with our new $K$,

$$
\begin{aligned}
\frac{a_{CH_4(aq)}}{a_{CH_4(g)}} &= 1.38 \\
a_{CH_4(aq)} &= 1.38 \times a_{CH_4(g)} \\
&= 1.38 \times 10^{-4.99} \\
&= 10^{-4.85}
\end{aligned}
$$

as before. So using any arbitrary standard state makes no difference at all. If you follow the calculations closely, you will see that the properties of the standard state cannot affect the results. They are a convenient repository for tabulated data derived from experimental work. As shown in Figure C.14, all paths from the elements to the equilibrium state must give the same total change in $G$, because $G$ of all products and reactants is fixed in both states. The standard state is merely a repository somewhere along the elements→equilibrium state path.

### Units Again

A reminder—

- Activities and activity coefficients $(a, \gamma)$ have no units, but fugacity $(f)$ does.

- Activities have standard states, but fugacities do not.

## C.5  EFFECT OF TEMPERATURE ON EQUILIBRIUM CONSTANTS

In Chapter 8 we saw that $\ln K = -\Delta_r G^\circ / RT$, so that the effect of temperature on $K$ evidently depends on the effect of temperature on $\Delta_r G^\circ$. In Chapter 6 we saw that $\partial \Delta_r G^\circ / \partial T = -\Delta_r S^\circ$, and we treated $\Delta_r S^\circ$ as a constant, unaffected by $T$. This results in a very simple linear relationship between $\ln K$ and $1/T$,

$$
\ln K = \frac{-\Delta_r H^\circ_{298}}{RT} + \frac{\Delta_r S^\circ_{298}}{R} \tag{8.13}
$$

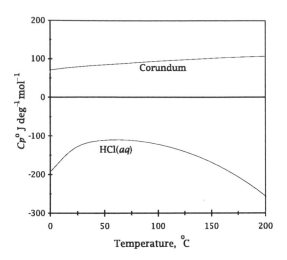

Figure C.15: The heat capacities of corundum ($Al_2O_3$) and HCl($aq$) as a function of temperature.

However, we also pointed out in §4.4.4 that

$$\frac{d\Delta H}{dT} = \Delta C_p \qquad (4.20)$$

or

$$\frac{d\Delta_r H°}{dT} = \Delta_r C_p° \qquad (C.31)$$

and

$$\frac{d\Delta_r S°}{dT} = \frac{\Delta_r C_p°}{T} \qquad (C.32)$$

In other words, what we have really assumed is that $\Delta_r C_p°$ is zero, or that the heat capacities of reactants and products are equal. However, if you look at measured heat capacities, they look something like those shown in Figure C.14. Pure solids, liquids, and gases generally have $C_p°$ values that increase with $T$ as illustrated by the mineral corundum. Over a temperature range of only 200°C, the variation of $C_p°$ is fairly linear; over a larger range of $T$, it shows a distinct curvature. If the particular values of $C_p°$ of products and reactants in the reaction of interest happen to be about the same and so cancel out, the $\Delta_r C_p° = 0$ approximation works well, but realistically this only happens in fairly simple reactions such as calcite⇌aragonite, or by chance.

For accurate work it is best to have, for each product and reactant, an expression that describes the variation of $C_p°$ with temperature. Several such expressions are in common use, the most common probably being the Maier-Kelley expression, which is

$$C_p° = a + bT - cT^{-2} \qquad (C.33)$$

Values of $a$, $b$, and $c$ are tabulated for most solids, liquids, and gases, so that $\Delta_r C_p^\circ$ is readily calculated for any reaction involving only pure phases, giving

$$\Delta_r C_p^\circ = \Delta_r a + \Delta_r b T - \Delta_r c T^{-2} \tag{C.34}$$

for a complete reaction. Inserting this expression into (C.30) and (C.31) gives

$$\int_{T_r}^{T} d\Delta_r H^\circ = \int_{T_r}^{T} \Delta_r C_p^\circ \, dT$$

$$\Delta_r H_T^\circ - \Delta_r H_{T_r}^\circ = \int_{T_r}^{T} (\Delta_r a + \Delta_r b T - \Delta_r c T^{-2}) dT$$

$$= \Delta_r a (T - T_r) + \frac{\Delta_r b}{2}(T^2 - T_r^2) + \Delta_r c \left(\frac{1}{T} - \frac{1}{T_r}\right) \tag{C.35}$$

and

$$\int_{T_r}^{T} d\Delta_r S^\circ = \int_{T_r}^{T} \frac{\Delta_r C_p^\circ}{T} dT$$

$$\Delta_r S_T^\circ - \Delta_r S_{T_r}^\circ = \int_{T_r}^{T} \left(\frac{\Delta_r a}{T} + \Delta_r b - \frac{\Delta_r c}{T^3}\right) dT$$

$$= \Delta_r a \ln\left(\frac{T}{T_r}\right) + \Delta_r b (T - T_r) + \frac{\Delta_r c}{2}\left(\frac{1}{T^2} - \frac{1}{T_r^2}\right) \tag{C.36}$$

Knowing both $\Delta_r H^\circ$ and $\Delta_r S^\circ$ as a function of $T$ means we also know $\Delta_r G^\circ$ and $\log K$ as a function of $T$. Simple, if somewhat tedious, manipulations thus give

$$\begin{aligned}
\Delta_r G_T^\circ = {} & \Delta_r G_{T_r}^\circ - \Delta_r S_{T_r}^\circ (T - T_r) \\
& + \Delta_r a \left[T - T_r - T \ln\left(\frac{T}{T_r}\right)\right] \\
& + \Delta_r \frac{b}{2}\left(2TT_r - T^2 - T_r^2\right) \\
& + \frac{\Delta_r c \, (T^2 + T_r^2 - 2TT_r)}{2TT_r^2}
\end{aligned} \tag{C.37}$$

and

$$\begin{aligned}
\ln K_T = {} & \ln K_{T_r} - \frac{\Delta_r H^\circ}{R}\left(\frac{1}{T} - \frac{1}{T_r}\right) + \frac{\Delta_r a}{R}\left(\ln\frac{T}{T_r} + \frac{T_r}{T} - 1\right) \\
& + \frac{\Delta_r b}{2R}\left(T + \frac{T_r^2}{T} - 2T_r\right) \\
& + \frac{\Delta_r c}{R}\frac{(-T^2 - T_r^2 + 2TT_r)}{2T^2 T_r^2}
\end{aligned} \tag{C.38}$$

These rather long equations are simple to use if the Maier-Kelley $a$, $b$, and $c$ values are available for all products and reactants, and give accurate results. However, if your reactions contain aqueous species, the Maier-Kelley formulation does not apply, and $a$, $b$, and $c$ values are not available because they don't work. The $C_p^\circ$ of HCl($aq$) shown in Figure C.15 is fairly typical of aqueous species, which are convex upward and have a maximum somewhere around 50 to 100°C.

### Isocoulombic Reactions

Obviously, quite a different function is required to fit this kind of curvature, but one fairly simple way around the problem is to write your reaction such that equal numbers of ions appear on each side of the reaction or, failing that, the same total charge on each side. Each ion will have some sort of convex-upward $C_p^\circ$ variation, and most of the variation will cancel out if the same number of ions occur on each side or if the same total charge is on each side. Such reactions are called "isocoulombic." It is understood that the isocoulombic reaction has no solids or gases. The transformation from the reaction you are interested in to an isocoulombic reaction is made by combining your reaction with another one, which is often the ionization of water reaction. For example,

$$H_3PO_4(aq) = H^+ + H_2PO_4^-$$

is a typical ionization reaction with all charges appearing on the right side. The equilibrium constant for this reaction shows a considerable curvature, plotted as $\log K$ vs. $1/T$. Add to this the reaction

$$OH^- + H^+ = H_2O(l)$$

and we have

$$H_3PO_4(aq) + OH^- = H_2PO_4^- + H_2O(l)$$

which has the same charge on both sides, and $\log K$ vs. $1/T$ is rather close to a straight line. Of course, to recover the desired ionization constant at some higher temperature, you need to know the ionization constant for water as a function of $T$ in order to be able to "uncombine" it at the higher $T$. Linear extrapolations of isocoulombic reactions can be surprisingly accurate in many cases (Gu et al., 1994).

## C.6   SOME CONVENTIONS REGARDING COMPONENTS

### C.6.1   Solutes Such As $SiO_2(aq)$ and $H_4SiO_4(aq)$

In Chapter 8, we wrote the dissolution of quartz as

$$SiO_2(s) + 2H_2O = H_4SiO_4(aq) \tag{C.39}$$

and we calculated the equilibrium concentration of $H_4SiO_4(aq)$ to be 5.2 ppm. Another way to write the same reaction is

$$SiO_2(s) = SiO_2(aq) \tag{C.40}$$

The only difference between these two ways of writing a reaction for the dissolution of quartz is that in (C.39) we have assumed that the dissolved silica is in the form of a molecule containing one $SiO_2$ attached to two $H_2Os$, whereas in (C.40) we have made no assumption as to the form of the dissolved silica. The important properties of this aqueous molecule are (1) that it has only one Si (i.e., it is *monomeric*, not *polymeric*), and (2) it is uncharged. We might as well then write a formula for this species that is as simple as possible, while observing these two facts, and $SiO_2(aq)$ does this. Therefore, $SiO_2(aq)$ does *not* refer to a species of dissolved silica which is not attached to any $H_2O$ or other molecules; it refers to the silica that exists as a monomeric uncharged species of whatever nature in solution. It might be attached to two $H_2Os$, or six $H_2Os$, or be a mixture of several such species; it doesn't matter. Actually, if you think about it, the same remarks could be made about $H_4SiO_4(aq)$. Whatever monomeric uncharged silica species exist in solution, we *call* them $H_4SiO_4(aq)$, which originates historically in the belief that Si in water must be tetrahedrally coordinated by oxygens, as it is in crystals. That may well be true, but there may be other oxygens in the form of $H_2Os$ also attracted to the Si. The exact nature of the complexes of Si and many other elements of interest is a continuing research topic. However, we know that one of these complexes is monomeric and uncharged, and what we call it is a matter of convention. The two common things we call it are $H_4SiO_4(aq)$ and $SiO_2(aq)$.

Because these two formulae refer to the same physical substance, their concentrations and activities are identical. But because they are related by the equation

$$SiO_2(aq) + 2H_2O(l) = H_4SiO_4(aq)$$

their standard state properties such as $\Delta_f G°$ and $\Delta_f H°$ must be different by exactly twice the corresponding property of $H_2O(l)$. Thus

$$\Delta_f G°_{H_4SiO_4(aq)} = \Delta_f G°_{SiO_2(aq)} + 2\Delta_f G°_{H_2O}(l)$$

Thus the Gibbs energy of formation of $H_4SiO_4$ is *defined* as the sum of the Gibbs energies of $SiO_2(aq)$ and (twice that of) $H_2O(l)$. In other words, the relationship between $H_4SiO_4(aq)$ and $SiO_2(aq)$ is strictly a formal one. They are derived from the same experimental data and will yield the same results in calculations.

The same relationship also holds for other species. For example, when $CO_2$ gas dissolves in water, it hydrolyzes (reacts with water) to a very small extent, forming some $H_2CO_3$ molecules in solution. It is rather difficult to determine the exact amount of $H_2CO_3$, and this problem is avoided by simply calling the total amount of carbon dioxide in solution either $CO_2(aq)$ or $H_2CO_3(aq)$, exactly as the dissolved silica is called $SiO_2(aq)$ or $H_4SiO_4(aq)$. Then for the same reason as before, we find that $\Delta_f G^\circ_{H_2CO_3(aq)} = \Delta_f G^\circ_{CO_2(aq)} + \Delta_f G^\circ_{H_2O(l)}$. Again, this relationship is strictly formal, although in this case it can be more confusing, because there is in fact some literature on the subject of how much dissolved $CO_2$ actually hydrolyzes to the species $H_2CO_3$ and how much remains as $CO_2$ molecules. In other words, $H_2CO_3$ is sometimes used as a species, and sometimes in the conventional sense we are discussing. The thermodynamic properties of $H_2CO_3$ in these two senses will of course be completely different. In this book we use the conventional sense for $H_2CO_3(aq)$.

Another example is the aluminum species $Al(OH)_3(aq)$. Again, this is a monomeric uncharged species of Al in solution. There is really no need to assume that it has three oxygens and three hydrogens attached to it. Whatever is attached to it, we can *call* it $AlO_{1.5}(aq)$, whose properties will differ from those of $Al(OH)_3(aq)$ by those of $1.5\,H_2O(l)$, because

$$AlO_{1.5}(aq) + 1.5\,H_2O(l) = Al(OH)_3(aq)$$

and similarly for other aqueous aluminum species.

## C.6.2   The Choice of the Mole of a Component

We have now seen that commonly we have more than one choice of formula to represent solute species and that these formulae commonly differ by some number of $H_2O$ molecules. Another way that the choice of formula can differ is that some choices can be multiples of other choices. This is most often seen in choosing solute species in solids, because there are no "real" species, just a crystal structure that is a solid solution. For example, the mineral olivine is a solid solution of two components forsterite ($Mg_2SiO_4$) and fayalite ($Fe_2SiO_4$). The solution is represented by $(Mg, Fe)_2SiO_4$, because the Mg and Fe atoms share the same positions in the crystal structure.

But what reason do we have to choose $Mg_2SiO_4$ and $Fe_2SiO_4$ as our components? The formula simply shows us the stoichiometry of the components—the ratios or relative amounts of the elements. Why not $MgSi_{0.5}O_2$, or $Mg_4Si_2O_8$? The same question could also arise in discussing aqueous species, except in that case we often have experimental evidence about the nature of the species in solution. That kind of evidence does not exist for the three-dimensional crystal structures of solid solutions—we are free to choose any component that is stoichiometrically correct. Does it make any difference? Yes.

Suppose our system consists of a crystal of pure forsterite, $Mg_2SiO_4$. The Gibbs energy of the system is a finite, unknown quantity, which depends on the mass of the crystal. A crystal with twice the mass has a G twice as large. But the molar $G$ does not vary with the size of the crystal. The molar $G$ is defined as $G = G/n$ (§2.4.1), where $n$ is the number of moles in the crystal. The point is, the number of moles of *what*? Obviously the number of moles of $Mg_2SiO_4$ in the crystal will be exactly half the number of moles of $MgSi_{0.5}O_2$ in the crystal, because $Mg_2SiO_4$ contains twice the number of atoms that $MgSi_{0.5}O_2$

does. Therefore, $G_{Mg_2SiO_4} = 2\,G_{MgSi_{0.5}O_2}$. Or, if you prefer, you can say that $G_{Mg_2SiO_4} = 2\,G_{MgSi_{0.5}O_2}$ simply because it contains twice the mass and, therefore, twice the energy of whatever kind.

This difference in the Gibbs energy of the mole is translated into a difference in activities. Because $G_{Mg_2SiO_4} = 2\,G_{MgSi_{0.5}O_2}$ and $G^{\circ}_{Mg_2SiO_4} = 2\,G^{\circ}_{MgSi_{0.5}O_2}$, then

$$G_{Mg_2SiO_4} - G^{\circ}_{Mg_2SiO_4} = 2\,(G_{MgSi_{0.5}O_2} - G^{\circ}_{MgSi_{0.5}O_2})$$

and

$$RT \ln a_{Mg_2SiO_4} = 2\,RT \ln a_{MgSi_{0.5}O_2}$$

and therefore

$$a_{Mg_2SiO_4} = a^2_{MgSi_{0.5}O_2}$$

The problem this poses can be seen in considering Raoult's Law, which we said was $a_i = X_i$. But if $a_i = a^2_{0.5i}$ we have a problem. Because $X_i$ is independent of how we write the formula for $i$, we see that $a_i$ and $a_{0.5i}$ cannot both be equal to $X_i$, even if Raoult's Law is followed exactly. This is a well known problem, and generally the formula for components is chosen such that the simple statement of Raoult's Law is followed as closely as possible. Again, this relationship between activities is entirely formal and tells us nothing about forsterite or olivine. However, it is important to remember that choosing a formula for your components has consequences for activities.

## C.7 THE PHASE RULE

The total differential of $G$ in terms of $T$ and $P$ is

$$dG = (\partial G/\partial T)_P dT + (\partial G/\partial P)_T dP \qquad (C.41)$$

from which we got

$$dG = -S\,dT + V\,dP \qquad (4.8)$$

in Chapter 4. This is one version of the Fundamental Equation. Equation (C.41) shows $G$ to be a function of only two variables, $T$ and $P$, and therefore it refers to a system of fixed composition. If we are dealing with a system of variable composition, we must explicitly include the compositional variables in our Fundamental Equation, in the same way as $T$ and $P$. Thus, if we add $n_1$ moles of component 1, $n_2$ moles of component 2, and so on, we have

$$dG = (\partial G/\partial T)_{P,n} dT + (\partial G/\partial P)_{T,n} dP$$
$$+ (\partial G/\partial n_1)_{T,P,\hat{n}_1} + (\partial G/\partial n_2)_{T,P,\hat{n}_2} + \cdots + (\partial G/\partial n_c)_{T,P,\hat{n}_c}$$

or

$$dG = -S\,dT + V\,dP + \sum_{i=1}^{c} (\partial G/\partial n_i)_{T,P,\hat{n}_i} dn_i$$
$$= -S\,dT + V\,dP + \sum_{i=1}^{c} \mu_i dn_i \qquad (C.42)$$

where $n$ means $n_1, n_2, \ldots n_c$ (all components), $n_i$ means any individual component $i$, and $\hat{n}_i$ means all components except $n_i$. Note that we are using absolute numbers of moles of components 1, 2, and so on, so we must use the total G, total S, etc., of the system, not the molar $G$, $S$, and so on. This equation still refers to a tangent plane, as discussed in §C.1.2, but in multidimensional space. Equation (C.42) is a "more complete" version of a Fundamental Equation, in the sense that it includes the effects of changing composition as well as $T$ and $P$ on G. It refers to any homogeneous (one phase) system. If there is more than one phase, we have an equation (C.42) for each phase, where $T$, $P$, and $\mu_i$ are the same in every phase at equilibrium.

Dividing through by $n_1 + n_2 + \cdots + n_c$, we get the intensive form

$$dG = -S\,dT + V\,dP + \sum_{i=1}^{c} \mu_i dX_i \tag{C.43}$$

where $X_i = n_i/(n_1 + n_2 + \cdots + n_c)$, the mole fraction of $i$. From (C.43), we can see the number of independent intensive variables in any homogeneous phase. There are $c$ mole fraction terms, therefore $c - 1$ independent compositional terms, as the final mole fraction is determined if all but one are known. In addition there are three other intensive terms, $G$, $T$, and $P$, for a total of $c - 1 + 3 = c + 2$ intensive variables. We don't count $S$ and $V$ because these are just $T$ and $P$ derivatives of $G$, hence not independent terms.

In a single homogeneous phase, these $c + 2$ variables are linked by one equation (C.42), so only $c + 2 - 1$ of them are independent. If there are $p$ phases, there are still only $c + 2$ intensive variables, because they have the same value in every phase, but now there is one equation (C.42) for each phase, and $c + 2 - p$ independent intensive variables. These independent intensive variables are called degrees of freedom, $f$, so

$$f = c - p + 2 \tag{C.44}$$

which is the Phase Rule.

By and large, natural systems obey the Phase Rule. You might reflect now and then on why natural systems should care about the results of this piece of mathematical reasoning.

## C.8   CONSTRAINTS

In real systems, constraints are physical devices or circumstances that control the state of the system, such as thermostats and pressure regulators. In model systems, constraints are state variables that are specified in order to define the system. The minimum number of constraints needed to define a thermodynamic system is two. This number arises from our decision to limit the number of ways of changing the energy of a system to two—heat and one form of work. For every additional type of work we use, we have an additional state variable, and an additional constraint on the system. In our usage, systems having more than the minimum two constraints are called metastable.

In real systems, we often have a constraint over which we have no control, such as a degree of disorder in crystals, or an activation energy barrier. In model systems, we always have complete control over our constraints. Imposing a third (or higher) constraint on a system entails doing work on the system, raising the appropriate thermodynamic potential. Releasing a constraint enables work to be done, as the system lowers its thermodynamic potential. The process of releasing a third constraint in a model system employs the progress variable, as discussed in §12.5 and §C.1.2.

The subject of constraints is one of the most fundamental and least discussed subjects in thermodynamics. Reiss (1965) is a valuable exception.

# C.9 COMPUTER PROGRAMS AND DATABASES

If you do any calculations at temperatures much different from 25°C, they can become very time-consuming. Numerous computer programs have been written to do these types of calculations, of course, some with extensive databases. One of the most general is SUPCRT92 (Johnson et al., 1992), which is freely available, runs on all common computer platforms, and has an extensive database of minerals, gases, aqueous species, and organic substances. It uses the Maier-Kelley formulation for minerals, and complex equations of state for water and aqueous species, both organic and inorganic. It is widely used over a large range of temperatures and pressures. An explanation of the equations used in SUPCRT92, plus additional discussion of other points in this Appendix, is given in Anderson and Crerar (1993).

A number of programs have been written to carry out the speciation of solutions discussed in §10.2.5. The one used in Chapter 10 is SOLMINEQ88 (Kharaka et al., 1988). A personal computer version for this program that makes it easier to use has been written by one of the authors, Dr. E.H. Perkins, Alberta Research Council, P.O. Box 8330, Postal Stn. F, Edmonton, Alberta, from whom it is available. The main frame version is available at media cost, and the PC version is $500 for industry, $100 for academic use.

Programs to carry out reaction path modeling in muticomponent systems such as discussed in §12.5 are also quite numerous. One of the more versatile of these is CHILLER, that allows calculation of the effects of fluid cooling, boiling, mixing and other processes in quite complex systems. It has a companion program, SOLVEQ, a speciation program, that is useful for preparing input for CHILLER. These are available from the author, Prof. Mark H. Reed, Department of Geological Sciences, University of Oregon, Eugene, Oregon 97403. With complete documentation for academic use, the cost is $200.

Speciation programs such as SOLMINEQ88 and SOLVEQ come with their own databases, and each database is different in various respects. As emphasized in Chapter 10, the user is responsible for the data she uses—you cannot always trust the database. In addition, the database may not always contain the data you need, particularly if you are interested in supercritical conditions. A very small and simple speciation program which gives the user complete control over the data used (you must enter it yourself) is program EQBRM, found in Appendix E of Anderson and Crerar (1993).

A number of other speciation and reaction path programs, their capabilities and references to relevant literature are discussed in Bassett and Melchior (1990).

# D

# ANSWERS TO PROBLEMS

## Chapter 2

1. $\Delta_r V° = -15.867 \, \text{cm}^3$.

2. $\Delta_r V° = -7.934 \, \text{cm}^3$. Note that you must pay attention to how the equation is written when you calculate reaction properties.

3. $\Delta_r V° = -475.348 \, \text{cm}^3$.

4. Because the reference or standard state of gases in the tables is the ideal gas. In other words, all the properties are those the gas would have if it were ideal. This is discussed further in Appendix C. From Appendix A, the molar volume of ideal gas at 273.15 K, 1 bar is $0.0224140 \, \text{m}^3\text{mol}^{-1}$. By the ideal gas law, the volume at 298.15 K is $0.0224140 \times 298.15/273.15 = .024465.55 \, \text{m}^3\text{mol}^{-1}$, or $24465.6 \, \text{cm}^3\text{mol}^{-1}$.

5. $\Delta_r V° = 17.3 + 0 - 17.3 = 0 \, \text{cm}^3$. The properties of individual ions such as $Cl^-$ cannot be measured. In order to have properties for them, the properties of the hydrogen ion $H^+$ are defined as zero, and so the properties of $Cl^-$ become the same as the properties of $HCl(aq)$.

6. $\Delta_r V° = -10.915 \, \text{cm}^3$. The properties of aqueous (dissolved) species in the tables are actually *partial molar* properties, discussed in Chapter 7. There you will learn how they can be negative.

## Chapter 3

1. $w = -644.4 \, \text{J}$.

2. $w = -1243.7 \, \text{J}$.

3.

$$
\begin{aligned}
w_{max} &= -\int_{V_1}^{V_2} P \, dV \\
&= -\int_{V_1}^{V_2} (RT/V) \, dV \\
&= -RT \ln(V_2/V_1) \\
&= -8.31451 \times 298.15 \times \ln(2000/1000) \\
&= -1718.29 \, J
\end{aligned}
$$

4. $+2200 \, J$.

5. The essential part of doing these work calculations is to know the gas volume as a function of pressure. This is known as the "equation of state" of the gas. These are sufficiently hard to use that they become a topic of study in themselves, and attention would be diverted from the simple idea of work being discussed. The *principles* of $w$ and $w_{max}$ are illustrated just as well with an ideal gas as with a real gas.

6. $w = +1.608 \, J$. Don't forget the conversion from 1 atm to 1 bar (Appendix A).

# Chapter 4

1. By equation (4.17), $\Delta_r U^\circ = -53480 - 1.608 = -53481.6 \, J$. Notice how small the $P\Delta V$ term is at low pressures.

2. $\Delta_r H^\circ = -89475 \, J$.

3. $w_{net} = \Delta_r G^\circ = -13,903 \, J$   reaction (2.2).
   $w_{net} = \Delta_r G^\circ = -267,550 \, J$   reaction (2.4).

4. $w = -9782.6 \, J$ at 1 bar, $-9912.2 \, J$ at 1 atm. Notice that this is considerably larger than $w$ for reaction (2.2) (1.608 J) because gases are among the products. It is still much smaller than the maximum useful work, $\Delta_r G^\circ = -267,550 \, J$, which is typical of chemical reactions.

5. $12.782/1500 = .00852$ days, or about 12 minutes. Unfortunately, humans cannot digest corundum.

# Chapter 5

1. $\Delta_f S^\circ = -757.012 \, J \, mol^{-1}$ of $CaAl_2Si_2O_8$. This gives a $\Delta_f G^\circ = -4002.2 \, kJ$ when combined with $\Delta_f H^\circ$.

2. $\Delta_r H^\circ = -4.9$ kJ. Exothermic. $\Delta_r G^\circ = -5.1$ kJ. Albite should form spontaneously. Nepheline and quartz together at 25°C will not react; it is a truly metastable assemblage. They will react at high temperatures.

4. $\Delta_r G^\circ = -2.613$ kJ. Nesquehonite more stable.

5. Diaspore>gibbsite>boehmite>corundum is the order of stability, with diaspore the most stable.

6. $\Delta_r G^\circ = -5.289$ kJ. Hematite more stable.

7. 1.99 kcal.

## Chapter 6

1. 15418 bar at 25°C. 16759 bar at 100°C.

2. 9691 bar.

3. 158.5°C.

5. 1465°C. $\alpha$-Quartz is metastable at this temperature. See Figure 11.8.

## Chapter 7

1. Brush up on your calculus, differentiate the expression with respect to $X_A$, put this expression equal to zero, and solve for $X_A$. You'll find it is 0.5. Do it.

2. $\Delta G_{mix} = -1668.4$ J.

3. The chemical potential of A is 7979.5 J less in the solution than it would be in an ideal 1 molal solution.

4. In an ideal gas solution, the volume % is proportional to the number of moles of gas, or to the mole fractions. Therefore the mole fraction of each gas is its volume % divided by 100. This also equals its fugacity and partial pressure in bar.

5. At 100 bar, $f_{CH_4} = 1.7 \times 10^{-4}$ bar.

6. $f_{CH_4} = 1.615 \times 10^{-4}$ bar.

# Chapter 8

1. $\log K = 143.29$. $\log f_{CH_4} = -145.4$ bar, equivalent to zero molecules per liter. Atmospheric $CH_4$ is continuously produced by life processes and is obviously not at equilibrium with atmospheric oxygen.

2. The idea was to see that $\Delta_r G^\circ$ (J) $= -168600 + 75.729\,T$ could be compared to $\Delta_r G^\circ = \Delta_r H^\circ - T\,\Delta_r S^\circ$, giving $\Delta_r H^\circ = -168.60$ kJ mol$^{-1}$, $\Delta_r S^\circ = -75.729$ J deg$^{-1}$ mol$^{-1}$. $f_{O_2,600C} = 10^{-12.26}$ bar.

3. $f_{H_2O,25C} = 10^{-1.499}$ bar, or 0.0317 bar.
   $f_{H_2O,100C} = 1.123$ bar.

4. $Al_2O_3 \cdot H_2O(s) = Al_2O_3(s) + H_2O(g)$. $\log K = -5.415$. $0.0317 > 10^{-5.4}$, and so diaspore is stable. Or use $\Delta_r G^\circ + RT \ln Q$. $30908 + 5708.04 \log(0.0317)$ is positive, and so diaspore is stable.

5. $m_{H_4SiO_4} = 10^{-3.120}$.

6. $\log m_{SiO_2(aq)} = -5.83$. at 25°C.

7. At a given $T$, $P$, only one concentration of aqueous silica can coexist at equilibrium with pure albite and pure nepheline. If the silica concentration is too high, nepheline will react to form albite until silica is reduced to the equilibrium level, and vice versa. Many assemblages buffer the concentration of some species in this manner.

8. (a) $\Delta_r G^\circ = -73070$ J. This shows that the reaction will indeed go in the direction of producing calcite in the cement blocks, at least if the $CO_2$ pressure is 1 bar, and the other phases are pure, and for $CO_2$ pressures down to that calculated in part (b).

   (b)
   $$
   \begin{aligned}
   \log K &= 73070/(2.30259 \times 8.31451 \times 298.15) \\
   &= 12.801 \\
   K &= 6.33 \times 10^{12} \\
   &= \frac{a_{H_2O}\, a_{CaCO_3}}{a_{Ca(OH)_2}\, a_{CO_2}} \\
   &= 1/f_{CO_2} \\
   f_{CO_2} &= 10^{-12.80} \text{ bar} \\
   &= 1.58 \times 10^{-13} \text{ bar}
   \end{aligned}
   $$

   This is the fugacity or partial pressure of $CO_2$ in equilibrium with both portlandite and calcite at 25°C. Any greater $CO_2$ pressure will cause formation of calcite from portlandite. Any lesser pressure will cause formation of portlandite from calcite plus water. However, this is a very low value of fugacity, and so virtually any $CO_2$ would react with portlandite.

(c) The reaction to write is

$$CaO(s) + H_2O(l) = Ca(OH)_2(s)$$

for which $\Delta_r G° = -57.331$ kJ. This means, with no ambiguity, that CaO and water are not stable together. And, as water is present, CaO cannot be. Or, in other words, if CaO happened to be there, it would react with water to form portlandite until it was gone. Note that CaO and $Ca(OH)_2$ could theoretically coexist in the absence of water. However, as shown in part (d), it would have to be an extremely dry environment.

(d) The reaction for this question is the same as the last one, *except* that the water must be gaseous, not liquid, because we want to calculate a water fugacity, not a water mole fraction. Only gas activities are in the form of fugacities or partial pressures. Thus the reaction is

$$CaO(s) + H_2O(g) = Ca(OH)_2(s)$$

$$
\begin{aligned}
\log K &= 65888/(2.30259 \times 8.31451 \times 298.15) \\
&= 11.543 \\
K &= 3.49 \times 10^{11} \\
&= \frac{a_{Ca(OH)_2}}{a_{CaO}a_{H_2O(g)}} \\
&= 1/f_{H_2O} \\
f_{H_2O} &= 10^{-11.54}\,\text{bar, or} \\
&= 2.86 \times 10^{-12}\,\text{bar}
\end{aligned}
$$

This is the fugacity or partial pressure of water vapor in equilibrium with CaO and portlandite. As in Question (b), it is a very low value. Most vacuum pumps cannot produce pressures this low. Therefore virtually any moisture in the air will cause lime to hydrate. Or you could use lime in a desiccator to produce extremely dry air.

(e) There are at least three ways to do this. They all give about the same answer.
$\Delta_r H° = -113151$ J.
$\Delta_r S° = -134.32$ J.

$$
\begin{aligned}
\log K_{423} &= \log K_{298} + \frac{-113151}{(2.30259 \times 8.31451)}\left(\frac{1}{423.15} - \frac{1}{298.15}\right) \\
&= 12.80 - 5.856 \\
&= 6.94 \\
K &= 10^{6.94} \\
&= 8.71 \times 10^6
\end{aligned}
$$

The reaction will be reversed when a temperature is reached at which $\Delta_r G^\circ = 0$, or $\log K_T = 0$. To find this $T$, write

$$0 = \log K_{298} + \frac{-113151}{(2.30259 \times 8.31451)} \left( \frac{1}{T} - \frac{1}{298.15} \right)$$
$$T = 841.6 \, \text{K}$$
$$= 568°\text{C}$$

Or

$$\Delta_r G_T^\circ = \Delta_r H_{298}^\circ - T \cdot \Delta_r S_{298}^\circ$$
$$= -113151 - 423.15 \times (-134.32)$$
$$= -56313.5 \, \text{J}$$

Then

$$\log K_{423} = \frac{-(-56313.5)}{2.30259 \times 8.31451 \times 423.15}$$
$$= 6.95$$

and

$$\Delta_r G_T^\circ - \Delta_r G_{298}^\circ = -\Delta_r S_{298}^\circ (T - 298.15)$$
$$0 - (-73070) = -(-134.32)(T - 298.15)$$
$$T = 842.2 \, \text{K}$$
$$= 569°\text{C}$$

Note that in doing this calculation we have ignored the fact that liquid water cannot exist at 1 bar at this high temperature. If we were really interested in this reaction at high $T$, we would switch to $H_2O(g)$, which *can* exist at 1 bar at high $T$. Also, we would have to check whether in fact portlandite was stable at this $T$. It is possible that the lime-portlandite reaction would reverse at high $T$, making CaO more stable than $Ca(OH)_2$, even in the presence of water at 1 bar.

This raises the general point that there are at least two distinct aspects to doing calculations of this type. One is learning the mechanics—getting numbers from tables and correctly calculating a number for a *given* reaction. The second, and more difficult, is finding or writing the reaction most appropriate for a given problem. It must not only be balanced, but must fairly closely match what would actually happen. In other words, your model must reflect reality. Thermodynamics allows you to calculate an infinite number of results which, though correct, mean virtually nothing, such as our answer to this question.

9. (a) $\log K_{150C} = 21.76$.

(b)

$$\Delta_r G_{150C} = \Delta_r G^\circ_{150C} + RT \ln Q$$
$$= -176243 + 2.303 \times 423.15 \times 8.315 \log \frac{1 \times 1}{10^{-3} \times 10^{-6}}$$
$$= -103332$$

Negative, so PbS precipitates. Or calculate the equilibrium $a_{Pb^{2+}}$ from $K$, and note that it is less than $10^{-6}$, and so PbS must precipitate.

(c) $a_{Pb^{2+}} = 10^{-9.31}$, PbS still precipitates.

(d) $\log K_{sp} = -7.74$. $a_{Pb^{2+}} \cdot a_{SO_4^{2-}} = 10^{-12.31}$, and so anglesite does not precipitate. $a_{Pb^{2+}}$ would have to be $> 10^{-4.74}$ to precipitate anglesite.

10. $2\,S\,(orthorhombic) = S_2(g)$.
$f_{S_2} = 10^{-12.15}$.

11. $Al(OH)_3\,(gibbsite) = Al^{3+} + 3\,(OH)^-$.
$\log K_{sp} = -34.75$.
$a_{Al^{3+}} = 10^{-10.75}$. (Don't forget that $a_{H^+} \cdot a_{OH^-} = 10^{-14}$.)

12. $\log K = a^4_{H^+}/a^2_{Zn^{2+}} = -17.84$. At $pH = 5.0$, $a_{Zn^{2+}}$ at equilibrium is 0.083. So if $a_{Zn^{2+}}$ is 0.173, reaction goes to the right, and willemite precipitates. Or

$$\Delta_r G = \Delta_r G^\circ + RT \ln Q$$
$$= 101858 + 5708.042 \log \frac{(10^{-5})^4}{(10^{-0.76})^2}$$
$$= -3626.6$$

Negative, reaction goes to the right, and willemite precipitates.

13. $\log f_{O_2} = -211.64$ for equilibrium between Ca and CaO, which is so low that metallic Ca never occurs naturally on Earth. Similar calculations and conclusions for BaS($s$) and other compounds.

14. The reaction is

$$Al_2Si_2O_5(OH)_4(s) + 5H_2O(l) = 2Al(OH)_3(s) + 2H_4SiO_4(aq)$$

The value of $\Delta_f G^\circ$ for $Al(OH)_3$ is one half the tabulated value for $Al_2O_3 \cdot 3\,H_2O$. $m_{H_4SiO_4} = 10^{-5.238}$.

15. 100 ppm $= 10^{-2.78}\,m$ $SiO_2$ or $H_4SiO_4$. At equilibrium, $a_{H_4SiO_4} = 10^{-5.238}$; therefore kaolinite is expected.

16. $CaMg(CO_3)_2(s) + 2\,SiO_2(s) = CaMgSi_2O_6(s) + 2\,CO_2(g)$.
$\Delta_r G^\circ = 55962$ J; reaction does not proceed. $f_{CO_2} = 10^{-4.90}$ at 1 bar, 25°C. $f_{CO_2} = 1$ bar at 194°C.

17. $m_{H_2S} = 0.0991$ at $25°C$, $0.020$ at $100°C$.

18. (a) $m_{SiO_2} = 10^{-3.029}$. (b) $\Delta_f G°_{H_4SiO_4(aq)} = -1302554\,J\,mol^{-1}$; $\Delta_f G°_{SiO_2(aq)} = -828246\,J\,mol^{-1}$.

19. (a) $H_2S(aq) = H^+ + HS^-$; $\log K_1 = -6.992$.
$HS^- = H^+ + S^{2-}$; $\log K_2 = -12.915$. Overall, $K = K_1 K_2 = 10^{-19.907}$.
(c) $PbS(s) = Pb^{2+} + S^{2-}$; $\log K_{sp} = -28.043$.
$m_{Pb^{2+}} = 10^{-16.136}$.

20. $CH_3OOH(aq) = CH_3OO^- + H^+$. $\log K_{25C} = -4.757$. $\log K_{100C} = -4.783$.

21. $(f_{NO_2}^2/f_{N_2O_4}) = 0.148$; $f_{NO_2} + f_{N_2O_4} = 1.0\,bar$; so $f_{NO_2} = X_{NO_2} = 0.318$; $f_{N_2O_4} = X_{N_2O_4} = 0.682$.

# Chapter 9

1.
$$UO_2^{2+}(aq) + 2e = UO_2(s)\quad \mathcal{E}° = 0.405\,V$$

$$UO_2CO_3°(aq) + 2H^+ + 2e = UO_2(s) + CO_2(g) + H_2O(l)\quad \mathcal{E}° = 0.497\,V$$

$$UO_2(CO_3)_2^{2-}(aq) + 4H^+ + 2e = UO_2(s) + 2CO_2(g) + 2H_2O(l)\quad \mathcal{E}° = 0.677\,V$$

$$UO_2CO_3°(aq) + 2H^+ = UO_2^{2+}(aq) + CO_2(g) + H_2O(l)\quad pH = 2.06.$$

$$UO_2(CO_3)_2^{2-}(aq) + 2H^+ = UO_2CO_3°(aq) + CO_2(g) + H_2O(l)\quad pH = 3.53.$$

Boundary 2: $Eh = 0.349 - 0.0592\,pH$.
Boundary 3: $Eh = 0.558 - 0.1184\,pH$.

2. (d) At this $Eh$ and sulfate concentration, the equilibrium $H_2S(aq)$ activity is enormous. However, it cannot exceed about 0.1 (the solubility of $H_2S$ in water), and so sulfate is being reduced to try to reach the equilibrium concentration.

3. $Eh = -0.141\,V$; $f_{O_2} = 10^{-68.6}\,bar$.

4. $f_{O_2} = 10^{-29.0}\,bar$.

5. I in $I^-$ is $-1$; I in $IO_3^-$ is $+5$.
$IO_3^- + 6H^+ + 6e = I^- + 3H_2O(l)$.
$\mathcal{E}° = 1.097\,V$; $Eh = 0.742\,V$.

6. $IO_3^- = I^- + 1.5\,O_2(g)$.
$f_{O_2} = 10^{-8.93}\,bar$; magnetite only stable if $f_{O_2} \leq 10^{-68.6}\,bar$, so expect iodide, $I^-$.

# Chapter 10

1. $CaCO_3(calcite) + H^+ = Ca^{2+} + HCO_3^-$.

$$\Delta_r G = \Delta_r G^\circ + RT \ln Q$$
$$= -11560 + 5708.042 \log \left( \frac{10^{-6} \times 0.09}{10^{-8}} \right)$$
$$= -6113.14 \text{ J}$$

Negative, therefore calcite will not precipitate. Or calculate $IAP = 10^{-9.38} < K_{sp}$.
$Ca^{2+} + Mg^{2+} + 2\,HCO_3^- = CaMg(CO_3)_2 + 2\,H^+$.
$\log K = -3.245$.
$a_{Mg^{2+}} = 10^{-4.66}$ at equilibrium, so $a_{Mg^{2+}} > 10^{-4.66}$ will precipitate dolomite. This is much less than the actual magnesium ion activity, and so the oceans are supersaturated with dolomite.

2. (a) $a_{Ca^{2+}} = 0.249$. (b) $a_{Ca^{2+}} = 0.378$. With both minerals in the same solution at 1 bar, aragonite will always require a greater $a_{Ca^{2+}}$ for equilibrium than will calcite, no matter what the conditions. Therefore, it should continue to dissolve while calcite precipitates until it is used up. This is consistent with the free energy relations, which say that aragonite should change to calcite spontaneously at 1 bar. The dissolution/precipitation simply provides a mechanism to do this.

3. (a) The reaction given suggests $\log(a_{Na^+}^2/a_{Ca^{2+}})$ vs. $\log a_{SiO_2(aq)}$. Slope $= -1$. (b). Only the ratio $(a_{Na^+}^2 a_{SiO_2(aq)}/a_{Ca^{2+}})$ is buffered. If quartz is present, then the ratio $(a_{Na^+}^2/a_{Ca^{2+}})$ is buffered.

4. (b) $\log(a_{Al^{3+}}/a_{H^+}^3) = 6.506$, $\log a_{SiO_2(aq)} = -5.518$.

5. (b) $\log a_{Al^{3+}} = -3\,pH + 7.23$
$\log a_{Al(OH)^{2+}} = -2\,pH + 2.253$
$\log a_{Al(OH)_2^+} = -pH + 2.903$
$\log a_{Al(OH)_3^\circ} = -9.53$
$\log a_{Al(OH)_4^-} = pH - 14.932$

6. This diagram represents equilibrium conditions among pure minerals. If the natural system (in this case granite and hot-spring water) is not near equilibrium, then the solution composition will not fall on the mineral field boundaries, and there may be too many minerals (the phase rule is not obeyed; see Chapter 11). In this case, with four components $(K_2O, Al_2O_3, SiO_2, H_2O)$ and two degrees of freedom ($T$ and $P$), there should be no more than four phases coexisting, that is, three minerals plus water. Therefore the four-mineral granite mentioned, plus water, cannot fit the

model. This is also shown by the diagram, in which not more than three minerals can coexist.

However, we know that if reactions occur, they will be in the direction toward making the diagram true. So, for example, if a solution having values of $\log(a_{K^+}/a_{H^+})$ and $\log a_{SiO_2}$, which fall in the kaolinite field, is in contact with microcline, we can predict that microcline will react to form kaolinite and that the solution composition will move towards the microcline–kaolinite boundary. Most groundwaters do in fact plot in the kaolinite field, which is why feldspars are unstable in the weathering environment.

(a)   i. If quartz *and the other minerals* do not equilibrate with the solution, we can say little, except that the solution composition will not be on the quartz saturation line, except by accident. At low temperatures ($<\sim$ 150°C) solutions are quite often supersaturated and sometimes undersaturated with respect to quartz. At higher temperatures, saturation with quartz is the rule, and silica content may even be used as a geothermometer. However, if we assume that quartz is the only recalcitrant mineral, then the solution composition should migrate toward the intersection of the microcline, muscovite, and kaolinite fields.

    ii. If quartz does equilibrate, then the solution composition lies on the quartz saturation line, but we can say little else. One of the other three minerals (either microcline or kaolinite) must disappear before equilibrium can be achieved. If microcline disappears, the solution could equilibrate at the muscovite- kaolinite-quartz intersection.

    iii. The answer depends on what stable or metastable assemblage you assume would exist. If, for example, you assume that Kspar changes metastably to kaolinite while at equilibrium with muscovite, $a_{K^+} \leq 10^{-2.45}$. Other "correct" answers are possible.

(b) You would not expect to find much quartz in bauxite. Any you did see should be in the process of dissolving. Kaolinite could be present at equilibrium with gibbsite, but it might be disappearing too, if meteoric water with essentially zero silica is flowing through the bauxite.

You also can say nothing about the mineral *proportions*, even at equilibrium. That is, it makes no sense to say "the granite would be mostly microcline, with some muscovite and kaolinite." If microcline is present in the system, its activity in the model is 1.0, whether it makes up 1% or 99% of the mineral mass present.

8. $K = 10^{12.241}$; $a_{K^+} = 4.17$; redox conditions have no effect; lack of equilibrium, possibly poor thermodynamic data.

# Chapter 11

1. $\Delta_f H_{\beta}^{\circ} = -18890 \, \text{cal mol}^{-1}$. There are still a lot of data in cal and kcal. Get used to it.

2. $V_{H_2O(s)}^{\circ} = 19.688 \, \text{cm}^3 \, \text{mol}^{-1}$.

3. (a) Point 2 has four coexisting phases; maximum is three for one component. Point 2 may actually be two points close together.
   Point 5 has a triple point with one angle $> 180°$, which is not possible. Each metastable extension must lie between two stable curves. You can prove this to yourself by drawing $G$-$T$ or $G$-$P$ sections.

   (d) $V_E^{\circ} = 14.01 \, \text{cm}^3 \, \text{mol}^{-1}$.

   (e) $S_C^{\circ} = 21.26 \, \text{J mol}^{-1}$.

   (f) $P_7 = 501.7 \, \text{bar}$.

   (g) $15.0 > V_B^{\circ} > 11.62 \, \text{cm}^3 \, \text{mol}^{-1}$.

   (h) Point 6 is a critical point. E is liquid. F is vapor or gas. Others are solids. Solid A floats. Solids B, C, D, G sink.

   (i) The slope $dP/dT$ is not constant because $\Delta S/\Delta V$ is not constant, and this, in turn, is because the properties of phase F (a gas) change more rapidly with $P$, and to a lesser extent $T$, than do the properties of solids and liquids.

4. PERFECT EQUILIBRIUM COOLING HISTORY OF COMPOSITION 6.

   - Homogeneous liquid of composition 75% B, 25% A cools to top of miscibility gap, and a second liquid ($L_1$, composition 86% B, 14% A) separates. On cooling, compositions of original liquid (now called $L_2$) and $L_1$ follow sides of miscibility gap. Given enough time, these two liquids would probably split into separate layers, like oil and water.

   - At $T_1$, $L_1$ is 91% B, 9% A; $L_2$ is 68% B, 32% A. Proportions are about 70% $L_2$ and 30% $L_1$.

   - On further cooling, liquid compositions follow miscibility gap to the eutectic temperature between $T_1$ and $T_2$.

   - At the eutectic $T$, crystals of pure B and liquid $L_2$ form from liquid $L_1$. The reaction can be written $L_1 \rightarrow L_2 + B$. During this reaction, the phase proportions change from 31% $L_1$, 69% $L_2$ to 23% crystals of B, 77% liquid $L_2$. The temperature stays constant until liquid $L_1$ is used up.

   - When the last drop of $L_1$ disappears, cooling resumes. Crystals of B form as the liquid becomes richer in A, following the liquidus. As there are no longer two liquids, we refer simply to the liquid, rather than to $L_2$.

- At $T_2$ liquid composition is 60% B, 40% A. Proportions are about 38% crystals of B, 62% liquid.

- At $T_3$, liquid composition is 55% B, 45% A. Proportions are about 44% crystals of B, 56% liquid.

- Cooling continues to the eutectic below $T_3$, where the liquid has composition 53% B, 47% A. At the eutectic, crystals of solid solution $\alpha_2$ (composition 48% B) and crystals of B form simultaneously while the temperature remains constant until the liquid is all gone (L $\rightarrow$ $\alpha_2$ + B). During this reaction, the phase proportions change from 53% liquid, 47% crystals of B to about 48% crystals of $\alpha_2$, 52% crystals of B.

- When the last drop of liquid disappears, cooling of the now completely solid mass of crystals continues. The crystals of $\alpha_2$ become steadily richer in A while remaining completely homogeneous. This requires that component B diffuse out of the crystals of $\alpha_2$ and form crystals of B.

- At $T_4$, $\alpha_2$ composition is 43% B, 57% A. Proportions are about 56% crystals of B, 44% crystals of $\alpha_2$.

- Cooling continues to the eutectoid below $T_4$, where $\alpha_2$ has composition 42% B, 58% A. At this temperature, crystals of solid solution $\beta_2$ (41% B, 59% A) and crystals of B form simultaneously ($\alpha_2$ $\rightarrow$ $\beta_2$ + B). During this reaction, the phase proportions change from 57% crystals of B, 43% solid $\alpha_2$, to 58% crystals of B, 42% solid $\beta_2$. (The terms *peritectoid* and *eutectoid* refer to the same geometrical relationships on diagrams as peritectic and eutectic, except that all phases are solids.)

- When the last crystal of $\alpha_2$ disappears, cooling of the crystals of $\beta_2$ and B continues, while $\beta_2$ exsolves B and becomes steadily richer in A.

PERFECT FRACTIONAL COOLING HISTORY OF COMPOSITION 2.

We will discuss fractional crystallization in terms of zoned crystals. As mentioned in §11.4.8, the same result can be obtained by physically separating the newly-formed crystals from the liquid.

- Starting homogeneous liquid composition is 80% A, 20% B. At $T_A$, first crystals of solid solution $\alpha_1$ appear, composition 98% A, 2% B. On cooling, liquid composition follows the liquidus, while the *surface* of existing crystals and any new crystals of $\alpha_1$ have compositions given by the solidus. No homogenization within the crystals takes place, so that crystals are *zoned*, with centers richer in A than the margins. The liquid is always completely homogeneous. Alternatively, we could say that crystals are separated from the liquid as soon as they form. The effect of separation or zoning is that once a crystal forms, it does not later react with the liquid.

- On continued cooling, the surface composition of crystals follows the solidus, liquid composition follows the liquidus.

- At $T_B$ a peritectic is reached. Liquid (composition 35% B, 65% A) co-exists with crystals of $\alpha_1$, which are zoned in composition, the outermost layer of all crystals being 10% B, 90% A. These react with the liquid to form a layer of solid solution $\beta_1$ (composition 12% B, 88% A) on the surface of all crystals, so that the liquid now "sees" only crystals of $\beta_1$. No plateau on the cooling curve is observed, that is, *there is no 3-phase equilibrium.* (In equilibrium cooling, the reaction at $T_B$ would be L + $\alpha_1 \rightarrow \beta_1$.)

- Cooling continues. Layers of solid solution $\beta_1$ of composition shown by the solidus continue to coat the crystals, while the liquid gets richer in B.

- Cooling continues to $T_C$ (another peritectic). Liquid (39% B) coexists with zoned crystals, some of which have centers of composition 2% B (only "some," because some crystals nucleate during the cooling history and therefore would have center compositions somewhere between 2% B and 13% B). The surfaces of all crystals are $\beta_1$ with composition 13% B, 87% A. At $T_C$ a layer of solid solution $\alpha_2$ is deposited on all crystals. The liquid is *not* used up (as it would be in equilibrium cooling), and no plateau on the cooling curve is observed. (In equilibrium cooling the reaction at $T_C$ would be L + $\beta_1 \rightarrow \alpha_2$.)

- Cooling continues. Layers of solid solution $\alpha_2$ of composition shown by the solidus continue to coat the crystals, while the liquid follows the liquidus, getting richer in B.

- At $T_D$ (eutectic), liquid reaches the eutectic composition 53% B, 47% A, which coexists with zoned crystals, some of which contain $\beta_1$ and $\alpha_1$ all the way back to 2% B. The surfaces of all crystals are $\alpha_2$ with composition 48% B, 52% A. Crystals of B and $\alpha_2$ (48% B) crystallize simultaneously until the liquid disappears (L $\rightarrow$ $\alpha_2$ + B). The temperature remains constant during the coexistence of these three phases. The crystals produced during crystallization at the eutectic temperature are not zoned (there is no difference in crystallization at a eutectic in fractional and equilibrium processes). The proportions of phases produced during crystallization at $T_D$ are 90% $\alpha_2$, 10% crystals of B. When the last drop of liquid disappears, the *average* composition of all solid phases (the composition you would get if you ground up all the solids together and analyzed the mixture) is the same as the bulk composition (composition 2). Therefore, a (curved) line showing the average solid composition would start at $T_A$, 2% B, and end at $T_D$, 20% B.

- After disappearance of the liquid, cooling of the crystals of B and the zoned crystals continues with no further changes in composition.

This is because the crystals can only change composition (stay on the solvus lines) if components A and B can diffuse through the crystal structures. However, this has been ruled out in the case of perfect fractional crystallization. If it was possible, the zoned crystals would have homogenized.

- Note that in fractional crystallization, some crystals of pure B appear, whereas in equilibrium crystallization no phase having more than 39% B would ever appear. This illustrates the importance of fractional crystallization in natural processes, that is, a crystallizing liquid can generate phases of much greater compositional variation during fractional crystallization than during equilibrium crystallization.

Note: These cooling histories are called "Perfect" Fractional Crystallization and "Perfect" Equilibrium Crystallization because they are the two hypothetical extremes of a complete spectrum of possible (or imperfect) cooling histories. In possible or realistic cooling histories, complete equilibrium is not attained, but neither are crystals separated from the melt immediately on forming, and some solid-state diffusion and crystal homogenization takes place.

### A Mass Balance

Consider composition 2 at $T_1$, at equilibrium (no fractional crystallization). We have

| Phase | Composition | Proportion |
|-------|-------------|------------|
| solid $\alpha_1$ | 4% B, 96% A | 32% |
| liquid | 28% B, 72% A | 68% |

Mass balance on component B:

$$(0.04 \times 32) + (0.28 \times 68) = 20.32$$

which is approximately the percentage of B in the bulk composition. Try the same thing for component A. The two answers, each slightly inexact because you cannot read the diagram with perfect accuracy, should add to exactly 100%.

5. (b) Eutectic at approximately 38°C, 35% B.

   (d) Liquid proportion: $(35/45) \times 1.5\,g = 1.17\,g$.
       Solid proportion: $(10/45) \times 1.5\,g = 0.33\,g$.

6. (c) $T_mA = 60°C$. $T_mB = 880°C$. Two polymorphs, transition $T$ 275°C.
       $S_\beta^\circ = 6.0\,J\,mol^{-1}\,K^{-1}$.

7. Any way of looking at it that obeys the Phase Rule is right. Choose the one that suits you. The phase transition loop becomes detached from the temperature axis for component A when solution 1 is liquid, solution 2 is vapor, and the pressure of the $T$-$X$ section is above the critical pressure of A, and similarly for B. Therefore, the loop may become detached from one or both axes.

8. There are reasons why ice-IX does not exist, but they are not within classical thermodynamics, the subject of this book. Thermodynamics does not so much "explain" energetic relationships as define parameters and methods of measurement and calculation that bring order to our knowledge of existing substances. To understand why this order exists, and not some other kind of order, you must go to statistical and quantum chemistry. If ice-IX did exist, its melting temperature would have to be $> 0°C$ or $32°F$, and $114.4°F$ is as good a number as any.

# Chapter 12

1. First order; $k = 0.000490\,s^{-1}$.

2. $f_{O_2} = 0.1996\,bar$; $f_{NO_2} = 0.7985\,bar$; $f_{N_2O_5} = 0.001855\,bar$.

3. $m_{O_2} = 0.00134$.

4. $K_{12.27} = 10^{213.97}$; $K_{12.28} = 10^{95.099}$; $K_{12.29} = 10^{7.77}$. No amount of aqueous molecular oxygen can equilibrate with pyrite under these conditions.

5. Rate $= 10^{-9.407}\,mol\,m^{-2}\,s^{-1}$.

6. $\Delta_r H° = -1402.66\,kJ$ under standard conditions.

8. $E_a = 53\,kJ$. $A = 0.85$. The activated state would appear to be the same for all silica polymorphs.

9.

| Level no. | Temperature $2T$ | | | Temperature $3T$ | | |
|---|---|---|---|---|---|---|
| | $\epsilon/2kT$ | $e^{-\epsilon/2kT}$ | $N_i$ for $N = 1000$ | $\epsilon/3kT$ | $e^{-\epsilon/3kT}$ | $N_i$ for $N = 1000$ |
| 0 | 0.0 | 1.0 | 395 | 0.0 | 1.0 | 291 |
| 1 | 0.5 | 0.6065 | 240 | 0.33 | 0.7165 | 208 |
| 2 | 1.0 | 0.3679 | 145 | 0.67 | 0.5134 | 149 |
| 3 | 1.5 | 0.2231 | 88 | 1.00 | 0.3679 | 107 |
| 4 | 2.0 | 0.1353 | 53 | 1.33 | 0.2636 | 77 |
| 5 | 2.5 | 0.0821 | 32 | 1.67 | 0.1889 | 55 |
| 6 | 3.0 | 0.0498 | 20 | 2.00 | 0.1353 | 39 |
| 7 | 3.5 | 0.0302 | 12 | 2.33 | 0.0970 | 28 |
| 8 | 4.0 | 0.0183 | 7 | 2.67 | 0.0695 | 20 |
| 9 | 4.5 | 0.0111 | 4 | 3.00 | 0.0498 | 14 |
| 10 | 5.0 | 0.0067 | 3 | 3.33 | 0.0357 | 10 |
| partition function | | 2.53111 | | | 3.43755 | |

# References

Anderson, G.M., and Crerar, D.A., 1993, Thermodynamics in Geochemistry—The Equilibrium Model: Oxford University Press, New York, 588 pp.

Bassett, R.L., and Melchior, D.C., 1990, Chemical modeling of aqueous systems: an overview. Chap. 1 *In:* Chemical Modeling of Aqueous Systems, D.C. Melchior and R.L. Bassett, eds. ACS Symposium Series 416. American Chemical Society, Washington, D.C., pp. 1-14.

Callen, H.B., 1960, Thermodynamics: New York, Wiley & Sons, 376 pp.

Drever, J.I, 1988, The Geochemistry of Natural Waters, 2nd ed.: Englewood Cliffs, N.J., Prentice-Hall, 437 pp.

Feynman, R.P., Leighton, R.B., and Sands, M., 1963, The Feynman Lectures on Physics, Vol. 1, pp. 4-2: New York, Addison Wesley Publishing Co.

Garrels, R.M., and Christ, C.L., 1965, Solutions, Minerals, and Equilibria: New York, Harper & Row, 450 pp.

Garrels, R.M., and Thompson, M.E., 1962, A chemical model for seawater at 25°C and one atmosphere total pressure. Amer. Jour. Sci., v. 260, pp. 57-66.

Gu, Y., Gammons, C.H., and Bloom, M.S., 1994, A one-term extrapolation method for estimating equilibrium constants of aqueous reactions at elevated temperatures. Geochim. et Cosmochim. Acta, v. 58, pp. 3545-3560.

Helgeson, H.C., 1979, Mass transfer among minerals and hydrothermal solutions. *In:* H.L. Barnes, ed., Geochemistry of Hydrothermal Ore Deposits, 2nd ed.: New York, Wiley-Interscience, pp. 568-610.

Helgeson, H.C., 1991, Organic/inorganic reactions in metamorphic processes. Can. Mineralogist, v. 29, pp. 707-739.

Johnson, J.W., Oelkers, E.H., and Helgeson, H.C., 1992, A software package for calculating the standard molal thermodynamic properties of minerals, gases, aqueous species as functions of temperature and pressure: Computers Geosc., v. 18, pp. 899-947.

Kharaka, Y.K., Gunter, W.D., Aggarwal, P.K., Perkins, E.H., and DeBraal, J.D., 1988, SOLMINEQ88: A computer program for geochemical modeling of water-rock interactions: U.S. Geol. Survey, Water Resource Investigations, Report 88-4227, Menlo Park, CA, 420 pp.

Kivelson, D., and Oppenheim, I., 1966, Work in irreversible expansions. Jour. Chem. Education, v. 43, pp. 233-235.

Lasaga, A.C., and Kirkpatrick, R.J., 1981, Kinetics of Geochemical Processes. Reviews in Mineralogy, v. 8, 398 pp. Mineralogical Society of America.

Merino, E., 1975, Diagenesis in Tertiary sandstones from Kettleman North Dome, California—II. Interstitial solutions: distribution of aqueous species at 100°C and chemical relation to the diagenetic mineralogy. Geochim. et Cosmochim. Acta, v. 39, pp. 1629-1645.

Morey, G.W., Fournier, R.O., and Rowe, J.J., 1962, The solubility of quartz in the temperature interval from 25° to 300°C. Geochim. et Cosmochim. Acta, v. 26, pp. 1029-1043

Nicholson, R.V., 1994, Iron sulfide oxidation mechanisms: laboratory studies. Chapter 6 In: J.L. Jambor and D.W. Blowes, eds. Short Course Handbook on Environmental Geochemistry of Sulfidic Mine-Wastes. Mineralogical Association of Canada, Short Course Handbook Vol. 22, pp. 163-183.

Reiss, H., 1965, Methods of Thermodynamics. New York, Blaisdell, 217 pp.

Renders, P.J.N., Gammons, C.H., and Barnes, H.L., 1995, Precipitation and dissolution rate constants for cristobalite from 150 to 300°C. Geochim. et Cosmochim. Acta, v. 59, pp. 77-85.

Robie, R.A., and Hemingway, B.S., 1972, Calorimeters for Heat of Solution and Low-Temperature Heat Capacity Measurements. U.S. Geological Survey Prof. Paper 755, 32 pp.

Singer, P.C., and Stumm, W., 1968, Acidic mine drainage: The rate-determining step. Science, v. 167, pp. 1121-1123.

Tuttle, O.F., and Bowen, N.L., 1958, Origin of granite in the light of experimental studies in the system $NaAlSi_3O_8-KAlSi_3O_8-SiO_2-H_2O$. Geol. Soc. Amer., Memoir 74, 153 pp.

Williamson, M.A., and Rimstidt, J.D., 1994, The kinetics and electrochemical rate-determining step of aqueous pyrite oxidation. Geochim. et Cosmochim. Acta, v. 58, pp. 5443-5454.

# Index

.